T0249634

Please note that the previous printing included a CD-ROM.

The material is now only available on the companion website:
http://www.elsevierdirect.com/companion.jsp?ISBN=9780750678100

Analog and Digital Circuits for Electronic Control System Applications

Using the TI MSP430 Microcontroller

by

Jerry Luecke

AMSTERDAM • BOSTON • HEIDELBERG • LONDON
NEW YORK • OXFORD • PARIS • SAN DIEGO
SAN FRANCISCO • SINGAPORE • SYDNEY • TOKYO

Newnes is an imprint of Elsevier

Newnes is an imprint of Elsevier
200 Wheeler Road, Burlington, MA 01803, USA
Linacre House, Jordan Hill, Oxford OX2 8DP, UK

 Recognizing the importance of preserving what has been written, Elsevier prints its books on acid-free paper whenever possible.

Library of Congress Cataloging-in-Publication Data

Luecke, Gerald.
   Analog and digital circuits for electronic control system applications : using the TI MSP430 microcontroller / by Gerald Luecke.
      p. cm.
   ISBN: 978-0-7506-7810-0
   1. Electronic circuit design. 2. Electronic control. 3. Programmable controllers. I. Title.

TK7867.L84 2004
629.8'9--dc22   2004054669

British Library Cataloguing-in-Publication Data
A catalogue record for this book is available from the British Library.

For information on all Newnes publications
visit our Web site at www.books.elsevier.com

Transferred to Digital Printing in 2012

The book is dedicated to my wife Velma and our grandchildren:

From the Luecke side:
Cameron, Graham, Andy, Alex, Alyssa,
Brent, Jacob, Harper, Arielle, Emery.

From the Hubbard side:
Jared, Garrett, Matthew, Ashton, Audrey.

Contents

Foreword

February 2004

The concept of a programmable system-on-chip (SoC) started in 1972 with the advent of the unassuming 4-bit TMS1000 microcomputer—the perfect fit for applications such as calculators and microwave ovens that required a device with everything needed to embed electronic intelligence. Microcomputers changed the way engineers approached equipment design; for the first time they could reuse proven electronics hardware, needing only to create software specific to the application. The result of microcomputer-based designs has been a reduction in both system cost and time-to-market.

More than thirty years later many things have changed, but many things remain the same. The term microcomputer has been replaced with microcontroller unit (MCU)—a name more descriptive of a typical application. Today's MCU, just like yesterday's microcomputer, remains the heart and soul of many systems. But over time the MCU has placed more emphasis on providing a higher level of integration and control processing and less on sheer computing power. The race for embedded computing power has been won by the dedicated digital signal processor (DSP), a widely used invention of the '80s that now dominates high-volume, computing-intensive embedded applications such as the cellular telephone. But the design engineer's most used tool, when it comes to implementing cost effective system integration, remains the MCU. The MCU allows just the right amount of intelligent control for a wide variety of applications.

Today there are hundreds of MCUs readily available, from low-end 4-bit devices like those found in a simple wristwatch, to high-end 64-bit devices. But the workhorses of the industry are still the versatile 8/16-bit architectures. Choices are available with 8 to 100+ pins and program memory ranging from <1 KB to >64 KB. The MCU's adoption of mixed-signal peripherals is an area that has greatly expanded, recently enabling many new SoC solutions. It is common today to find MCUs with 12-bit analog-to-digital and digital-to-analog converters combined with amplifiers and power management, all on the same chip in the same device. This class of device offers a complete signal-chain on a chip for applications ranging from energy meters to personal medical devices.

Modern MCUs combine mixed-signal integration with instantly programmable Flash memory and embedded emulation. In the hands of a savvy engineer, a unique MCU solution can be developed in just days or weeks compared to what used to take months or years. You can find MCUs everywhere you look from the watch on your wrist to the cooking appliances in your home to the car you drive. An estimated 20 million MCUs ship every day, with growth forecast for at least a decade to come. The march of increasing silicon integration will continue offering an even greater variety of available solutions—but it is the engineer's creativity that will continue to set apart particular system solutions.

Mark E. Buccini
Director of Marketing
MSP430
Texas Instruments Incorporated

Preface

Analog system designers many times in the past avoided the use of electronics for their system functions because electronic circuits could not provide the dynamic range of the signal without severe nonlinearity, or because the circuits drifted or became unstable with temperature, or because the computations using analog signals were quite inaccurate. As a result, the design shifted to other disciplines, for example, mechanical.

Today, young engineers requested by their superiors to design an analog control system, have an entirely new technique available to them to help them design the system and overcome the "old" problems. The design technique is this: sense the analog signals and convert them to electrical signals; condition the signals so they are in a range of inputs to assure accurate processing; convert the analog signals to digital; make the necessary computations using the very high-speed IC digital processors available with their high accuracy; convert the digital signals back to analog signals; and output the analog signals to perform the task at hand.

Analog and Digital Circuits for Control System Applications: Using the TI MSP430 Microcontroller explains the functions that are in the signal chain, and explains how to design electronic circuits to perform the functions. Included in this book is a chapter on the different types of sensors and their outputs. There is a chapter on the different techniques of conditioning the sensor signals, especially amplifiers and op amps. There are techniques and circuits for analog-to-digital and digital-to-analog conversions, and an explanation of what a digital processor is and how it works. There is a chapter on data transmissions and one on power control.

And to solidify the learning and applications, there is a chapter that explains assembly-language programming, and also a chapter where the reader actually builds a working project. These two chapters required choosing a digital processor. The TI MSP430 microcontroller was chosen because of its design, and because it is readily available, it is well supported with design and applications documentation, and it has relatively inexpensive evaluation tools.

The goal of the book is to provide understanding and learning of the new design technique available to analog system designers and the tools available to provide system solutions.

Acknowledgments

Mark Buccini, Product Line Marketing Manager for the MSP430 in the Semiconductor Group for Texas Instruments Incorporated and his staff deserve much credit for the project in Chapter 10, and for the thoroughness and accuracy of the MSP430 information. Special thanks go to Neal Frager, an applications expert, for writing the program for the Chapter 10 project, for designing the PCB breadboard, arranging meetings and for researching many inquiries as the book developed. Others that deserve mention for their assistance: Cornelia Huellstrunk, Byron Alsberg who helped develop the initial schematic, Dale Wellborn, Dan Harmon, Rajen Shah, Zack Albus, Modupe Ajibola, Mike Mitchell for his excellent reviews, and Neal Brenner and for helping clean up the last details. A hearty "Thank You" to all!

What's on the companion website?

■ A user's guide to the MSP430x1xx family of microcontrollers.

■ Layout wiring of PCB interconnection layers.

Signal Paths from Analog to Digital

Introduction

Designers of analog electronic control systems have continually faced the following obstacles in arriving at a satisfactory design:

1. Instability and drift due to temperature variations.
2. Dynamic range of signals and nonlinearity when pressing the limits of the range.
3. Inaccuracies of computation when using analog quantities.
4. Adequate signal frequency range.

Today's designers, however, have a significant alternative offered to them by the advances in integrated circuit technology, especially low-power analog and digital circuits. The alternative new design technique for analog systems is to sense the analog signal, convert it to digital signals, use the speed and accuracy of digital circuits to do the computations, and convert the resultant digital output back to analog signals.

The new design technique requires that the electronic system designer interface between two distinct design worlds. First, between analog and digital systems, and second, between the external human world and the internal electronics world. Various functions are required to make the interface. First, from the human world to the electronics world and back again and, in a similar fashion, from the analog systems to digital systems and back again. *Analog and Digital Circuits for Control System Applications* identifies the electronic functions needed, and describes how electronic circuits are designed and applied to implement the functions, and gives examples of the use of the functions in systems.

A Refresher

Since the book deals with the electronic functions and circuits that interface or couple analog-to-digital circuits and systems, or vice versa, a short review is provided so it is clearly understood what analog means and what digital means.

Analog

Analog quantities vary continuously, and analog systems represent the analog information using electrical signals that vary smoothly and continuously over a range. A good example of an analog system is the recording thermometer shown in *Figure 1-1*. The actual equipment is shown in *Figure 1-1a*. An ink pen records the

a. Recording thermometer
Photo courtesy of Taylor Precision Products

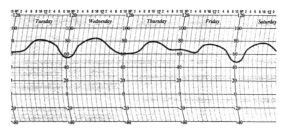

b. Plot of daily temperature variations
Courtesy of Master Publishing, Inc.

Figure 1-1: A recording thermometer is an example of an analog system

temperature in degrees Fahrenheit (°F) and plots it continuously against time on a special graph paper attached to a drum as the drum rotates. The record of the temperature changes is shown in *Figure 1-1b*. Note that the temperature changes smoothly and continuously. There are no abrupt steps or breaks in the data.

Another example is the automobile fuel gauge system shown in *Figure 1-2*. The electrical circuit consists of a potentiometer, basically a resistor connected across a car battery from the positive terminal to the negative terminal, which is grounded. The resistor has a variable tap that is rotated by a float riding on the surface of the liquid inside the gas tank.

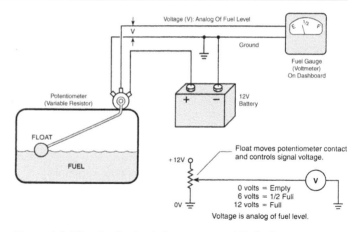

Figure 1-2: The simple circuit for an automobile fuel gauge demonstrates how an electrical quantity, a voltage, is an analog of the fuel level. *Courtesy of Master Publishing, Inc.*

A voltmeter reads the voltage from the variable tap to the negative side of the battery (ground). The voltmeter indicates the information about the amount of fuel in the gas tank. It represents the fuel level in the tank. The greater the fuel level in the tank the greater the voltage reading on the voltmeter. The voltage is said to be an analog of the fuel level. An analog of the fuel level is said to be a copy of the fuel level in another form—it is *analogous* to the original fuel level. The voltage (fuel level) changes smoothly and continuously so the system is an analog system, but is also an analog system because the system output voltage is a copy of the actual output parameter (fuel level) in another form.

Digital

Digital quantities vary in discrete levels. In most cases, the discrete levels are just two values—ON and OFF. Digital systems carry information using combinations of ON-OFF electrical signals that are usually in the form of codes that represent the information. The telegraph system is an example of a digital system.

The system shown in *Figure 1-3* is a simplified version of the original telegraph system, but it will demonstrate the principle and help to define a digital system. The electrical circuit (*Figure 1-3a*) is a battery with a switch in the line at one end and a light bulb at the other. The person

a. Electrical circuit

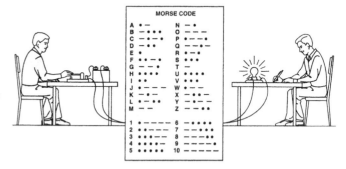

b. International Morse code

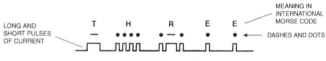

c. Digital information

Figure 1-3: The telegraph is a digital system that sends information as patterns of switched signals

at the switch position is remotely located from the person at the light bulb. The information is transmitted from the person at the switch position to the person at the light bulb by coding the information to be sent using the International Morse telegraph code.

Morse code uses short pulses (dots) and long pulses (dashes) of current to form the code for letters or numbers as shown in *Figure 1-3b*. As shown in *Figure 1-3c*, combining the codes of dots and dashes for the letters and numbers into words sends the information. The sender keeps the same shorter time interval between letters but a longer time interval between words. This allows the receiver to identify that the code sent is a character in a word or the end of a word itself. The T is one dash (one long current pulse). The H is four short dots (four short current pulses). The R is a dot-dash-dot. And the two Es are a dot each. The two states are ON and OFF—current or no current. The person at the light bulb position identifies the code by watching the glow of the light bulb. In the original telegraph, this person listened to a buzzer or "sounder" to identify the code.

Coded patterns of changes from one state to another as time passes carry the information. At any instant of time the signal is either one of two levels. The variations in the signal are always between set discrete levels, but, in addition, a very important component of digital systems is the timing of signals. In many cases, digital signals, either at discrete levels, or changing between discrete levels, must occur precisely at the proper time or the digital system will not work. Timing is maintained in digital systems by circuits called system clocks. This is what identifies a digital signal and the information being processed in a digital system.

Binary

The two levels—ON and OFF—are most commonly identified as 1(one) and zero (0) in modern binary digital systems, and the 1 and 0 are called **bi**nary dig**its** or **bits** for short. Since the system is binary (two levels), the maximum code combinations 2^n depends on the number of bits, n, used to represent the information. For example, if numbers were the only quantities represented, then the codes would look like *Figure 1-4*, when using a 4-bit code to represent 16 quantities. To represent larger quantities more bits are added. For example, a 16-bit code can represent 65,536 quantities. The first bit at the right edge of the code is called the *least significant bit* (LSB). The left-most bit is called the *most significant bit* (MSB).

Decimal (XX_{10})	Binary (XXXX_2)
	Most significant bit (MSB)
0	0000 ← Least significant bit (LSB)
1	0001
2	0010
3	0011
4	0100
5	0101
6	0110
7	0111
8	1000
9	1001
10	1010
11	1011
12	1100
13	1101
14	1110
15	1111

Figure 1-4: 4-bit codes to represent 16 quantities

Binary Numerical Quantities

Our normal numbering system is a decimal system. *Figure 1-5* is a summary showing the characteristics of a decimal and a binary numbering system. Note that each system in *Figure 1-5* has specific digit positions with specific assigned values to each position. Only eight digits are shown for each system in *Figure 1-5*. Note that in each system, the LSB is either 10^0 in the decimal system or 2^0 in the binary system. Each of these has a value of one since any number to the zero power is equal to one. The following examples will help to solidify the characteristics of the two systems and the conversion between them.

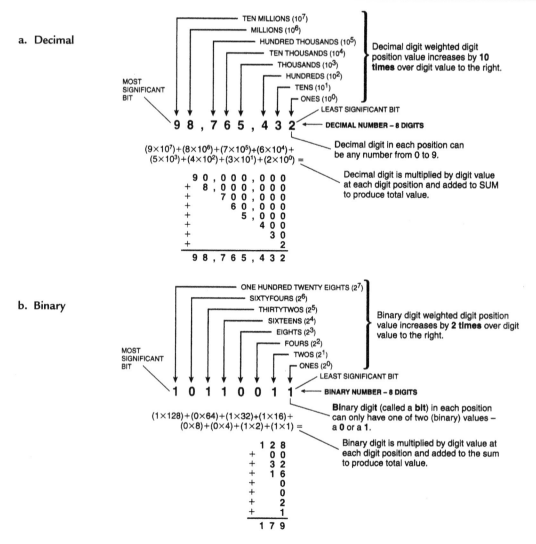

Figure 1-5: Decimal and binary numbering systems
Courtesy of Master Publishing, Inc.

Example 1. **Identifying the Weighted Digit Positions of a Decimal Number**

Separate out the weighted digit positions of 6524.

Solution:

$6524 = 6 \times 10^3 + 5 \times 10^2 + 2 \times 10^1 + 4 \times 10^0$

$6524 = 6 \times 1000 + 5 \times 100 + 2 \times 10 + 4 \times 1$

$6524 = 6000 + 500 + 20 + 4$

Can be identified as 6524_{10} since decimal is a base 10 system. Normally 10 is omitted since it is understood.

Example 2. Converting a Decimal Number to a Binary Number

Convert 103 to a binary number.

Solution:

$103_{10}/2$ = 51 with a remainder of **1**
51/2 = 25 with a remainder of **1**
25/2 = 12 with a remainder of **1**
12/2 = 6 with a remainder of **0**
6/2 = 3 with a remainder of **0**
3/2 = 1 with a remainder of **1**
1/2 = 0 with a remainder of **1** (MSB)
103_{10} = **1100111**

Example 3. Determining the Decimal Value of a Binary Number

What decimal value is the binary number 1010111?

Solution:

Solve this the same as *Example 1*, but use the binary digit weighted position values.
Since this is a 7-bit number:
And since the MSB is a 1, then MSB = $\mathbf{1} \times 2^6$ = 64
and   (next digit) $\mathbf{0} \times 2^5$ = 0
and   (next digit) $\mathbf{1} \times 2^4$ = 16
and   (next digit) $\mathbf{0} \times 2^3$ = 0
and   (next digit) $\mathbf{1} \times 2^2$ = 4
and   (next digit) $\mathbf{1} \times 2^1$ = 2
and   (next digit, LSB) $\mathbf{1} \times 2^0$ = 1
$$\overline{87}$$

Binary Alphanumeric Quantities

If alphanumeric characters are to be represented, then *Figure 1-6*, the ASCII table defines the codes that are used. For example, it is a 7-bit code, and capital M is represented by 1001101. Bit #1 is the LSB and bit #7 is the MSB. As shown, upper and lower case alphabet, numbers, symbols, and communication codes are represented.

Accuracy vs. Speed— Analog and Digital

Quantities in nature and in the human world are typically analog. The temperature, pressure, humidity and wind velocity in our

Bit Position

1	2	3	4		5	6	7	0	1	0	1	1	0	0	1
								0	0	1	1	1	1	0	0
								1	1	1	1	0	0	0	0
0	0	0	0					@	P	`	p	0	sp	NUL	DLE
1	0	0	0					A	Q	a	q	1	!	SOH	DC1
0	1	0	0					B	R	b	r	2	"	STX	DC2
1	1	0	0					C	S	c	s	3	#	ETX	DC3
0	0	1	0					D	T	d	t	4	$	EOT	DC4
1	0	1	0					E	U	e	u	5	%	ENQ	NAK
0	1	1	0					F	V	f	v	6	&	ACK	SYN
1	1	1	0					G	W	g	w	7	'	BEL	ETB
0	0	0	1					H	X	h	x	8	(	BS	CAN
1	0	0	1					I	Y	i	y	9	)	HT	EM
0	1	0	1					J	Z	j	z	:	*	LF	SUB
1	1	0	1					K	[	k	{	;	+	VT	ESC
0	0	1	1					L	\	l	|	<	,	FF	FS
1	0	1	1					M	]	m	}	=	-	CR	GS
0	1	1	1					N	^	n	~	>	.	SO	RS
1	1	1	1					O	_	o	DEL	?	/	SI	US

Figure 1-6: American Standard Code for Information Interchange—ASCII code

environment all change smoothly and continuously, and in many cases, slowly. Instruments that measure analog quantities usually have slow response and less than high accuracy. To maintain an accuracy of 0.1% or 1 part in 1000 is difficult with an analog instrument.

Digital quantities, on the other hand, can be maintained at very high accuracy and measured and manipulated at very high speed. The accuracy of the digital signal is in direct relationship to the number of bits used to represent the digital quantity. For example, using 10 bits, an accuracy of 1 part in 1024 is assured. Using 12 bits gives four times the accuracy (1 part in 4096), and using 16 bits gives an accuracy of 0.0015%, or 1 part in 65,536. And this accuracy can be maintained as digital quantities are manipulated and processed very rapidly, millions of times faster than analog signals.

The advent of the integrated circuit has propelled the use of digital systems and digital processing. The small space required to handle a large number of bits at high speed and high accuracy, at a reasonable price, promotes their use for high-speed calculations.

As a result, if analog quantities are required to be processed and manipulated, the new design technique is to first convert the analog quantities to digital quantities, process them in digital form, reconvert the result to analog signals and output them to their destination to accomplish a required task. The complete procedure is indicated in *Figure 1-7*, and the need for analog circuits, digital circuits and the conversion circuits between them is immediately apparent.

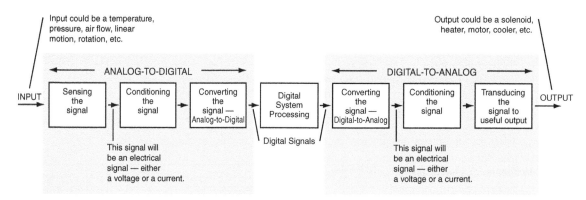

Figure 1-7: A typical system describing the functions in the analog-to-digital and digital-to-analog chain

Interface Electronics

The system shown in *Figure 1-7* shows the major functions needed to couple analog signals to digital systems that perform calculations, manipulate, and process the digital signals and then return the signals to analog form. This chapter deals with the analog-to-digital portion of *Figure 1-7*, and *Chapter 2* will deal with the digital-to-analog portion.

The Basic Functions for Analog-to-Digital Conversion

Sensing the Input Signal

Figure 1-8 separates out the analog-to-digital portion of the *Figure 1-7* chain to expand the basic functions in the chain. Most of nature's inputs such as temperature, pressure, humidity, wind velocity, speed, flow rate, linear motion or position are not in a form to input them directly to electronic systems. They must be changed to an electrical quantity—a voltage or a current—in order to interface to electronic circuits.

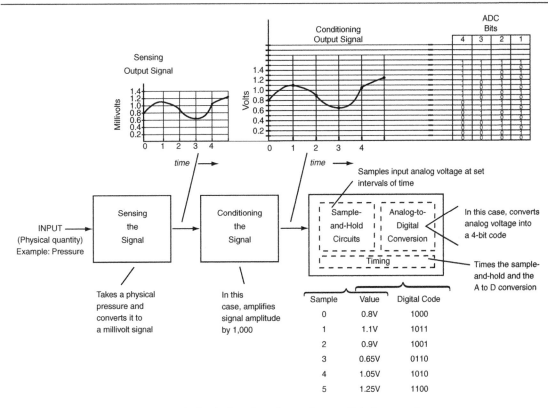

Figure 1-8: The basic functions for analog-to-digital conversion

The basic function of the first block is called sensing. The components that sense physical quantities and output electrical signals are called sensors.

The sensor illustrated in *Figure 1-8* measures pressure. The output is in millivolts and is an analog of the pressure sensed. An example output plotted against time is shown.

Conditioning the Signal

Conditioning the signal means that some characteristic of the signal is being changed. In *Figure 1-8*, the block is an amplifier that increases the amplitude of the signal by 1,000 times so that the output signal is now in volts rather than millivolts. The amplification is linear and the output is an exact reproduction of the input, just changed in amplitude. Other signal conditioning circuits may reduce the signal level, or do a frequency selection (filtering), or perform an impedance conversion. Amplification is a very common signal conditioning function. Some electronic circuits handle only small-signal signals, while others are classified as power amplifiers to supply the energy for outputs that require lots of joules (watts are joules/second).

Analog-to-Digital Conversion

In the basic analog-to-digital conversion function, as shown in *Figure 1-7*, the analog signal must be changed to a digital code so it can be recognized by a digital system that processes the information. Since the analog signal is changing continuously, a basic subfunction is required. It is called a *sample-and-hold* function. Timing circuits (clocks) set the sample interval and the function takes a sample of the input signal and holds on to it. The sample-and-hold value is fed to the analog-to-digital converter that generates a digital code whose value is equivalent to the sample-and-hold value. This is illustrated in *Figure 1-8* as the conditioned output signal is sampled at intervals 0, 1, 2, 3, and 4 and converted to the 4-bit codes shown. Because the analog signal changes continually, there maybe an error between the true input voltage and the voltage recorded at the next sample.

Example 4. A to D Conversion

For the analog signal shown in the plot of voltage against time and the 4-bit codes given for the indicated analog voltages, identify the analog voltage values at the sample points and the resultant digital codes and fill in the following table.

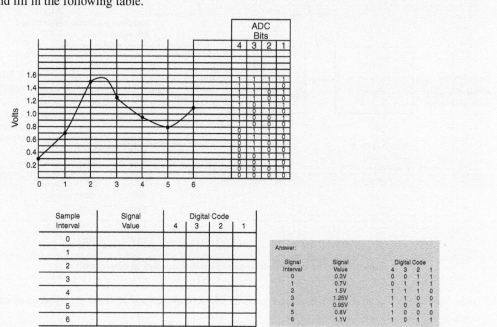

Sample Interval	Signal Value	Digital Code			
		4	3	2	1
0					
1					
2					
3					
4					
5					
6					

Answer:

Signal Interval	Signal Value	Digital Code			
		4	3	2	1
0	0.3V	0	0	1	1
1	0.7V	0	1	1	1
2	1.5V	1	1	1	0
3	1.25V	1	1	0	0
4	0.95V	1	0	0	1
5	0.8V	1	0	0	0
6	1.1V	1	0	1	1

Obviously, one would like to increase the sampling rate to reduce this error. However, depending on the code conversion time, if the sample rate gets to large, there is not enough time for the conversion to be completed and the conversion function fails. Thus, there is a compromise in the analog-to-digital converter between the speed of the conversion process and the sampling rate. Output signal accuracy also plays a part. If the output requires more bits to be able to represent the magnitude and the accuracy required, then higher-speed conversion circuits and more of them are going to be required. Thus, design time, cost, and all the design guidelines enter in. *Chapter 5* is a complete chapter on the conversion techniques to explore this function in detail. As shown in *Figure 1-8*, the bits of the digital code are presented all at the same time (in parallel) at each sample point. Other converters may present the codes in a serial string. It depends on the conversion design and the application.

Summary

This chapter reviewed analog and digital signals and systems, digital codes, the decimal and binary number systems, and the basic functions required to convert analog signals to digital signals. The next chapter will complete the look at the basic functions required to convert digital signals to analog signals. It will be important to have these basic functions in mind as the electronic circuits that perform these functions are discussed in the upcoming chapters.

Chapter 1 Quiz

1. A new design technique available to analog system designers is:
   a. Sense the analog, compute using analog, output analog.
   b. Sense the analog, convert to digital, compute digitally, convert to analog, output analog.
   c. Sense the analog, convert to digital, compute digitally, output digitally.
   d. Sense digitally, compute digitally, output digitally.
2. Analog quantities:
   a. vary smoothly, then change abruptly to new values.
   b. consist of codes of high-level and low-level signals.
   c. vary smoothly continuously.
   d. have periods of high-level and low-level signals, then change to continuous signals.
3. Digital signals:
   a. vary smoothly, then change abruptly to new values.
   b. consist of codes of high-level and low-level signals.
   c. vary smoothly continuously.
   d. have periods of high-level and low-level signals, then change to continuous signals.
4. Electronic system designers must interface between:
   a. the human world and the electronic world.
   b. the wholesale world and the retail world.
   c. the private business world and the government business world.
   d. the analog world and the digital world.
   e. a and d above.
   f. none of the above.
5. In analog electronic systems, analog quantities are:
   a. not analogous to the original quantity.
   b. are not a copy of the original quantity in another form.
   c. are output in digital form.
   d. are a copy of the analog physical quantity in another form.
6. Binary digital systems:
   a. have two discrete levels—1 or 0, high level or low level.
   b. have three or more discrete levels.
   c. have a level that varies continuously with time.
   d. have binary digits, or bits for short.
   e. none of the above.
   f. d and a above.
7. Decimal numbering systems have:
   a. weighted digit positions that vary randomly.
   b. weighted digit positions varying by powers of 10.
   c. weighted digit positions varying by powers of 2.
   d. weighted digit positions that remain constant at one value.
8. Decimal numbering systems have:
   a. weighted digit positions that vary randomly.
   b. weighted digit positions varying by powers of 10.
   c. weighted digit positions varying by powers of 2.
   d. weighted digit positions that remain constant at one value.

9. Physical quantities in the human world are typically:
   a. digital and analog.
   b. analog and digital.
   c. digital.
   d. analog.
10. Digital systems represent quantities:
   a. using combinations of binary digits in codes.
   b. using more bits in its binary codes as the quantity value increases.
   c. using more bits in its binary code as more accuracy is required.
   d. using binary codes with just two levels – 1 or 0, high level or low level.
   e. none of the above.
   f. all of the above.
11. Analog quantities:
   a. usually have slow response and less than high accuracy.
   b. can be maintained at very high accuracy at very high computing speeds.
   c. are impossible to compute.
   d. either have slow response or very high accuracy.
12. Digital quantities:
   a. usually have slow response and less than high accuracy.
   b. can be maintained at very high accuracy at very high computing speeds.
   c. are impossible to compute.
   d. either have slow response or very high accuracy.
13. The basic functions for A-to-D (analog-to-digital) conversions are:
   a. Sense, compute digitally, convert to analog.
   b. compute as analog, sense, convert to digital.
   c. convert to digital, sense, condition to analog.
   d. sense, condition, convert to digital.
14. Sensing:
   a. computes analog quantities in nature.
   b. separates out analog quantities into different categories.
   c. changes quantities in nature to electrical signals.
   d. detects analog quantities by their magnitude.
15. Conditioning signals:
   a. means that the signals are being exercised.
   b. means that some characteristic of the signal is being changed.
   c. means that the input signal may be increased or decreased in amplitude, filtered or its impedance changed.
   d. means that nothing is done to the input signal.
   e. b and c above.
   f. a and d above.

Answers: 1.b, 2.c, 3.b, 4.e, 5.d, 6.f, 7.b, 8.c, 9.d, 10.f, 11.a, 12.b, 13.d, 14.c, 15.e.

Signal Paths from Digital to Analog

Introduction

Refer back to *Figure 1-7*. In *Chapter 1*, the basic functions used for the analog-to-digital portion of *Figure 1-7* were discussed. In this chapter, the basic functions of the digital-to-analog portion will be discussed.

The Digital-to-Analog Portion

The digital-to-analog portion is separated out from *Figure 1-7* in *Figure 2-1*. After the digital processing system completes its manipulation of the signal, the output digital codes are coupled to a digital-to-analog converter that changes the digital codes back to an equivalent analog signal. From the output of the digital-to-analog converter, the analog signal is coupled to a signal conditioner that changes the characteristics of the signal. Just as in *Chapter 1*, as the application demands, the amplitude of the signal may be increased with amplification, or decreased with attenuation. Or maybe the power level of the signal is changed, or there may be an impedance transformation to fit the transducer to which the output signal couples.

The output of the system is to some real-world quantity external to the electronic system. As shown in *Figure 2-1*, the output might be a meter, a gauge, a motor, a lever arm to produce motion, a heater, or other similar output.

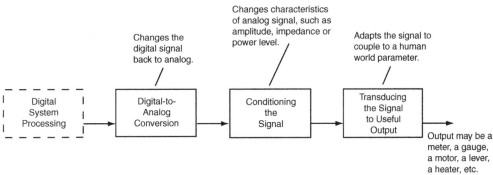

Figure 2-1: Digital-to-analog portion of the signal chain

Digital-to-Analog Conversion

Figure 2-2 illustrates the basic digital-to-analog function. The digital processing system outputs digital information in the form of digital codes, and as shown, the digital codes are usually presented to the input of the digital-to-analog converter in one of two ways.

Parallel Transfer of Data

The first way—parallel bit transfer—means that all bits of the digital code are outputted at the same time. In *Figure 2-2*, a 4-bit code is used as an example. The 4-bit codes are coupled out in sequence as they are processed by the digital processor. They arrive at a preset data interval. In *Figure 2-2*, the 4-bit code 1000 is outputted first, followed by 1011, 1001, 0110, 1010, and 1100, respectively. The digital-to-analog converter

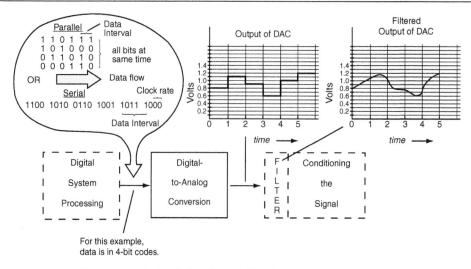

Figure 2-2: The basic function of digital-to-analog conversion

accepts all bits at the same time. It must have four input lines, the same number of input lines as the 4-bit code. In most modern day digital-to-analog converters the 4-bit codes of *Figure 2-2* are really 8-bit, or most likely 16-bit codes.

Example 1. Parallel Output

Refer to *Figure 2-2*. If the output of the digital-to-analog converter were an 8-bit code, what would the parallel bit codes be that are coupled out in sequence. Use the same value of analog signal.

Solution:

The analog values and the 4-bit codes are listed first. Since an 8-bit code can represent 256 segments, its codes for the same analog value are shown with the maximum analog signal of 1.5V equal to 255. Notice that the 8-bit code is two groups of 4-bit codes, which are also expressed in hexadecimal form.

Analog value	4-bit code	Hex	8-bit code		Hex
0	0000	0	0000	0000	00
0.1	0001	1	0001	0001	11
0.2	0010	2	0010	0010	22
0.3	0011	3	0011	0011	33
0.4	0100	4	0100	0100	44
0.5	0101	5	0101	0101	55
0.6	0110	6	0110	0110	66
0.7	0111	7	0111	0111	77
0.8	1000	8	1000	1000	88
0.9	1001	9	1001	1001	99
1.0	1010	A	1010	1010	AA
1.1	1011	B	1011	1011	BB
1.2	1100	C	1100	1100	CC
1.3	1101	D	1101	1101	DD
1.4	1110	E	1110	1110	EE
1.5	1111	F	1111	1111	FF

Serial Transfer of Data

The second way is serial transfer of data. As shown in *Figure 2-2*, the 4-bit codes are outputted one bit at a time, each following the other in sequence, and each group of four bits following each other in sequence. A clock rate determines the rate at which the bits are transferred. The digital-to-analog converter accepts the bits in sequence and reassembles them into the respective bit groups and then acts on them.

> ### Example 2. Bit Rate
>
> Refer to *Figure 2-2*. If the clock that outputs the bits in a serial output is 1 MHz, what are the serial bit transfer rate and the parallel bit transfer rate for a 4-bit and an 8-bit code?
>
> **Solution:**
>
> | Clock (Hz) | Serial | | Parallel | |
> | --- | --- | --- | --- | --- |
> | | 4-bit | 8-bit | 4-bit | 8-bit |
> | 1 MHz | 1 MHz | 1 MHz | 4 MHz | 8 MHz |

The Conversion

The digital codes received by the digital-to-analog converter are equivalent to a particular analog value. As shown in *Figure 2-2*, the input code is converted to and outputted as the equivalent analog value and held as this value until the next code equivalent value is outputted. Thus, as shown, the output of the digital-to-analog converter is a stair-step output that stays constant at a particular level until the next input digital code is received. The output resembles an analog signal but further processing is required in order to arrive at the final analog signal.

Filtering

A basic function required after the digital-to-analog conversion is filtering, or in more general terms, smoothing. As shown in *Figure 2-2*, such filtering produces an analog signal more equivalent to an analog signal that changes smoothly and continuously. The filter physically may be in the digital-to-analog converter or in the signal conditioner that follows it as shown in *Figure 2-2*. It was placed in the signal conditioner in *Figure 2-2* because it really is a signal conditioning function.

Conditioning the Signal

The function of conditioning the signal for the digital-to-analog portion can be the same as for the analog-to-digital portion. A most common function is amplification of the signal, but in like fashion, there is often the need to attenuate the signal; that is, to reduce the amplitude instead of increasing the amplitude. That is the function chosen for *Figure 2-3*. The output signal is attenuated to one-half the value of the input. No other characteristics of the signal are changed. The shape of the amplitude variations of the waveform with time are not changed, so the signal appears the same except its amplitude values are reduced.

Transducing the Signal

The output of the analog systems discussed is a human world parameter external to the electronic system. As mentioned previously several times, it may be a temperature, or a pressure, or a measure of humidity, or a linear motion, or a rotation. Thus, the electronic output of the signal conditioning function, in many cases, must be changed in form. It may be a voltage or a current out of the electronic system and must be changed to another form of energy.

A device to change or convert energy from one form to another is called a transducer. In *Figure 2-4*, the transducer is a meter that shows the amplitude of the output voltage on a voltage scale. The voltage output from the electronic system is converted to the rotation of a needle in front of a scale marked on the material

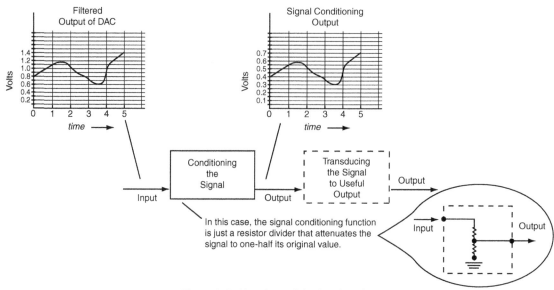

Figure 2-3: Signal conditioning function

behind the needle. The scale is calibrated so particular needle deflections represent specific voltage values. Thus, any deflection of the needle as a result of the electronic circuit output can be read as a particular voltage value at any instant of time. The electronic system output has been converted to a meter reading, and the meter reading can be calibrated into the type of parameter the system is measuring. It could be a fluid level, a rate of flow, a pressure, and so forth.

Similar changes in energy form occur in other types of transducers. The voltage or current output from the electronic system gets converted to all forms of human world parameters just by the choice of the transducer.

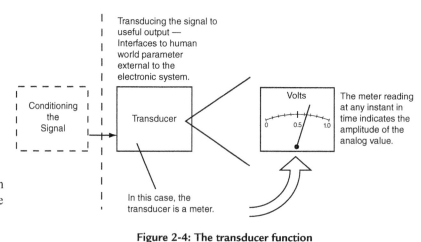

Figure 2-4: The transducer function

Examples of Transducers

Figure 2-5 shows examples of various types of transducers. *Figure 2-5a* is a picture of a speaker enclosure. Inside is what is called a driver. It is a common transducer that takes electrical audio signals and converts them into sound waves. The driver is placed inside a box to make it into a very good sounding speaker enclosure. Many times the driver only handles the low and mid-frequency audio signals, so another driver for the high frequencies, called a tweeter, is inserted into the speaker enclosure to allow the speaker to reproduce a broader range of audio frequencies.

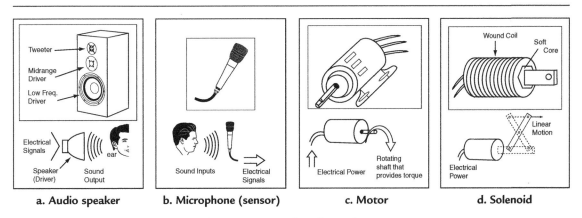

| a. Audio speaker | b. Microphone (sensor) | c. Motor | d. Solenoid |

Figure 2-5: Examples of transducers

There is a counterpart transducer to the speaker—a microphone—that is used as an input device for sensing the signal. It is shown in *Figure 2-5b*. The microphone converts sound signals into electrical signals so they may be inputted into an electronic system.

Figure 2-5c shows a motor. Normally a motor is not classified as a transducer, but it is. A motor takes electrical energy and converts it into rotational torque. Motors are used everywhere, from running machinery, to trimming grass, to providing transportation.

Figure 2-5d shows a solenoid. A solenoid is a transducer that converts electrical energy into linear motion. It consists of a coil of wire with a soft iron core inside of it. When current is passed through the coil, a magnetic field is produced that pulls on the soft iron core and draws it inside the core. The movement of the core can be used to move a lever arm, to close a door, to operate a shutter, and so forth.

There are many more examples of transducers that convert electrical energy into a pressure, a valve for controlling fluid flow, a temperature gauge, and so forth. As various applications are described in subsequent chapters many will use various types of transducers.

Summary

The discussion in this chapter covered the functions necessary to convert a digital signal into an analog output and then into a human world parameter. It completed the signal chain from an input analog signal, to a digital conversion, to computation in digital form, to a conversion back to an output analog signal, to an output human world parameter. The next chapter will examine the sensing function in detail.

Chapter 2 Quiz

1. A digital-to-analog converter:
   a. outputs a digital signal in serial form.
   b. outputs an analog signal in stair-step form.
   c. outputs a smooth and continuous analog signal.
   d. outputs one digital code after another.
2. The output of the digital-to-analog chain is:
   a. a serial digital code string.
   b. a parallel digital code stream.
   c. a real-world quantity.
   d. always a meter reading.
3. An input to a digital-to-analog converter may be:
   a. a parallel transfer of digital codes.
   b. an analog signal of suitable amplitude.
   c. an analog signal of discrete values.
   d. a serial transfer of digital codes.
   e. a and d above.
   f. b and c above.
4. In a parallel transfer of bits:
   a. all bits of a digital code are transferred at the same time.
   b. all bits of a digital code are transferred in a sequential string.
   c. all bits are filtered into an analog signal.
   d. all bits are signal conditioned one at a time.
5. In a serial transfer of bits:
   a. all bits of a digital code are transferred at the same time.
   b. all bits of a digital code are transferred in a sequential string.
   c. all bits are filtered into an analog signal.
   d. all bits are signal conditioned one at a time.
6. The output of the digital-to-analog converter is:
   a. a stair-step output that varies until the next input digital code is received.
   b. a stair-step output that changes between 1 and 0 until the next digital code is received.
   c. a stair-step output that stays constant at a particular level until the next digital code is received.
   d. a stair-step output that changes from maximum to minimum until the next digital code is received.
7. The digital-to-analog output must be filtered to:
   a. clarify the digital steps in the output.
   b. keep the stair-step digital output.
   c. make the analog output change smoothly and continuously.
   d. make the analog output more like a digital output.
8. A transducer is:
   a. a device to change or convert energy from one form to another.
   b. a device that maintains the analog output in digital steps.
   c. a device that converts analog signals to digital signals.
   d. a device that converts digital signals to analog signals.

9. A motor is:
   a. a transducer that changes digital signals into analog signals.
   b. a transducer that changes analog signals into digital signals.
   c. a transducer that raises the analog voltage output to a higher voltage.
   d. a transducer that changes electrical energy into rotational torque.
10. A meter is:
   a. a transducer that converts the analog output to the rotation of a needle in front of a scale.
   b. a transducer that changes analog signals into digital signals.
   c. a transducer that raises the analog voltage output to a higher voltage.
   d. a transducer that changes digital signals into analog signals.

Answers: 1.b, 2.c, 3.e, 4.a, 5.b, 6.c, 7.c, 8.a, 9.d, 10.a.

Sensors

Introduction

In *Chapter 1*, *Figure 1-8* shows the basic functions needed when going from an analog quantity to a digital output. The first of these is sensing the analog quantity. The device used in the function to sense the input quantity and convert it to an electrical signal is called a *sensor*—the main subject of this chapter.

A sensor is a device that detects and converts a natural physical quantity into outputs that humans can interpret. Examples of outputs are meter readings, light outputs, linear motions and temperature variations. *Chapter 1* indicated that a majority of these physical quantities are analog quantities; i.e., they vary smoothly and continuously. Sensors, in their simplest form, are devices that contain only a single element that does the necessary transformation. Although today, more and more complicated sensors are being manufactured; they cover more than the basic function, containing sensing, signal conditioning and converting all in one package.

In this chapter, in order to clearly communicate the sensing function, the majority of sensors will be single element sensors that output electrical signals—voltage, current or resistance. But also, closely coupled to sensors with electrical outputs, sensors are included that use magnetic fields for their operation.

Temperature Sensors

Oral Temperature

Everyone, sometime or another, has had the need to find out their body temperature or the body temperature of a member of their family. An oral thermometer like the one shown in *Figure 3-1* was probably used. Liquid mercury inside of a glass tube expands and pushes up the scale on the tube as temperature increases. The scale is calibrated in degrees (°F—Fahrenheit in this case) of body temperature; therefore, the oral thermometer converts the physical quantity of temperature into a scale value that humans can read. The oral thermometer is a temperature sensor with a mechanical scale readout.

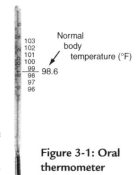

Figure 3-1: Oral thermometer

Indoor/Outdoor Thermometer

Another temperature sensor is shown in *Figure 3-2*. It is a bimetal strip thermometer. Two dissimilar metals are bonded together in a strip that is formed into a spring. The metals expand differently with temperature; therefore, a force is exerted between them that expands the spring and rotates the needle as the temperature increases. The thermometer scale is calibrated to known temperatures—boiling water and freezing water. These points establish a scale and the device is made into a commercial thermometer with Fahrenheit (°F) and/or Celsius (Centigrade—°C) scales. The one shown in *Figure 3-2* is for °F. The outdoor thermometer is another type of temperature sensor that converts the physical quantity of temperature into a meter reading easy for humans to see and interpret.

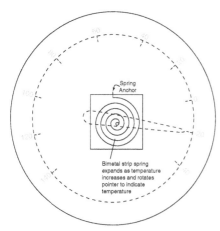

Figure 3-2: Rear-view of bimetal strip thermometer

Thermocouples

A thermocouple is another common temperature sensor. A place to find one is in a natural gas furnace in a home similar to that shown in *Figure 3-3*. It controls the pilot light for the burners in the furnace. The thermocouple is a closed tube system that contains a gas. The gas expands as it is heated and expands a diaphragm at the end of the tube that is in the gas control module.

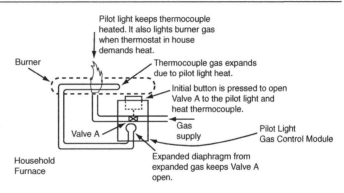

Figure 3-3: A residential furnace pilot light control

The system works as follows: A button on the pilot light gas control module is pressed to open valve A to initially allow gas to flow to light the pilot light. The expanded diaphragm of the thermocouple system controls valve A; therefore, the button for the pilot light must be held until the thermocouple is heated by the pilot light so that the gas expands and expands the diaphragm. The expanded diaphragm holds valve A open; therefore, the pilot light button can be released because the pilot light heating the thermocouple keeps the gas expanded. Since the pilot light is burning, any demand for heat from the thermostat will light the burners and the house is heated until the demand by the thermostat is met.

A thermocouple that puts out an electrical signal as temperature varies is shown in *Figure 3-4*. It is constructed by joining two dissimilar metals. When the junction of the two metals is heated, it generates a voltage, and the result is a temperature

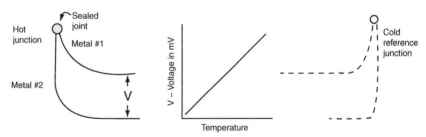

Figure 3-4: A bimetal thermocouple

sensor that generates millivolts of electrical signal directly. The total circuit really includes a cold-junction reference, but the application uses the earth connection of the package as the cold reference junction.

There may be a need to amplify the output signal from the sensor, as shown in *Figure 3-5*, because the output voltage amplitude must be increased to a useful level. This is the subject of *Chapter 4*, signal conditioning.

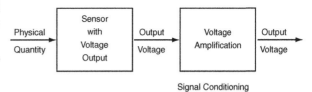

Figure 3-5: A sensor output signal may have to be increased to a useful level by amplification

Silicon-Junction Diode

Another sensor that produces a voltage directly as temperature varies is a silicon-junction diode. The characteristic curves for its forward and reverse voltage with current are shown in *Figure 3-6*. The forward current versus forward voltage for positive voltages increases little until the forward voltage reaches +0.7V, then it increases rapidly. Here the forward resistance is very small—in the order of 50 Ω to 80 Ω.

The reverse current for negative reverse voltage is 1,000 times and more smaller than the forward current. It stays relatively flat with reverse voltage until the magnitude reaches the reverse breakdown voltage. When the junction is reversed biased below the breakdown voltage, the reverse resistance is very large—in the order of megohms. The forward voltage and reverse breakdown voltage decrease as temperature is increased; thus, the diode junction has a negative temperature coefficient. The forward voltage has a much smaller voltage variation with temperature than does the reverse breakdown voltage. The reverse current below the breakdown region can also be used for a temperature sensor. A rule of thumb for the reverse current is that it doubles for every 10°C rise in temperature. The reverse conditions are used for temperature sensors, but the most common is to use the forward voltage change.

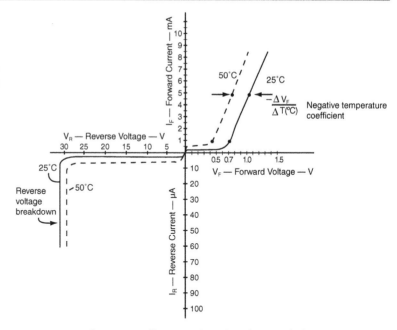

Figure 3-6: Silicon P-N junction characteristics

Example 1. Temperature Coefficient

Using *Figure 3-6*, calculate the temperature coefficient of the forward voltage of the diode and show that it is negative. The forward current $I_F = 5$ mA.

Solution:

T °C	V_F
25	1.02V
50	0.70V
$\Delta = 25$	−0.32V

Temp Coefficient $= \Delta V_F / \Delta T = -0.32V/25°C = -0.0128V/°C$

Thermistor

A thermistor is a resistor whose value varies with temperature. *Figure 3-7a* shows the characteristics of a thermistor readily available at RadioShack. Two circuits for the use of thermistors are shown in *Figure 3-7*. *Figure 3-7b* uses the thermistor in a voltage divider to produce a varying voltage output. *Figure 3-7c* uses a transistor to amplify the current change provided by the thermistor as

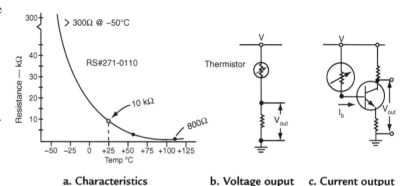

a. Characteristics   b. Voltage ouput   c. Current output

Figure 3-7: Thermistor temperature sensor

temperature changes. In some micromachined thermistors, the resistance at 25°C is of the order of 10 kΩ. One of the disadvantages of using a thermistor is that its characteristics with temperature are not linear. As a result, in order to produce linear outputs, the nonlinearity must be compensated for.

Angular and Linear Position

Position Sensor—Fuel Level

In *Chapter 1*, *Figure 1-2*, an automobile fuel gauge was used to demonstrate an analog quantity. That same example will be used, as shown in *Figure 3-8a*, to demonstrate the sensing function. The complete sensor consists of a float that rides on the surface of fuel in a fuel tank, a lever arm connected to the float at one end, and, at the other end, connected to the shaft of a potentiometer (variable resistor). As the fuel level changes, the float moves and rotates the variable contact on the potentiometer. The schematic of *Figure 3-8b* shows that the potentiometer is connected across the automobile battery from +12V to ground. The variable contact on the potentiometer moves in a proportional manner. When the contact is at the end of the potentiometer that is connected to ground, the output voltage will be zero volts from the variable contact to ground. At the other end, the one connected to +12V, there will be +12V from the variable contact to ground. For any position of the variable contact in between the end points, the voltage from the variable contact to ground will be proportional to the amount of the shaft rotation.

Calibrating it as shown in *Figure 3-8c* completes the liquid-level sensor. At a full tank, the float, lever arm and potentiometer shaft rotation are designed so that the variable contact is at the +12V end of the potentiometer. When the tank is empty, the same combination of elements results in the variable contact at the ground level (0V). Other positions of the float result in proportional output voltages between the variable contact and ground. As *Figure 3-8c* shows, a three-quarter full tank gives an output of 9V, a half-full tank will give an output of 6V, and a one-quarter full tank will give an output of 3V. Thus, adding a voltmeter to measure the voltage from the variable contact to ground, marked in liquid level, completes the automotive fuel gauge. Sensors that convert a physical quantity into an electrical voltage output are very common. The output voltage can be anywhere from microvolts to tens of volts.

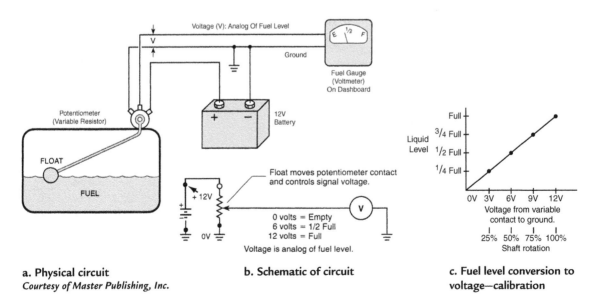

a. Physical circuit
Courtesy of Master Publishing, Inc.

b. Schematic of circuit

c. Fuel level conversion to voltage—calibration

Figure 3-8: Position sensor—fuel level gauge

Hall Effect—Position Sensor

The Hall effect is shown in *Figure 3-9a*. E.H. Hall discovered it. If there is current in a conductor and a magnetic field is applied perpendicular to the direction of the current, a voltage will be generated in the conductor that has a direction perpendicular to both the direction of the current and the direction of the magnetic field. This property is very useful in making sensors, especially when a semiconductor chip is used for the conductor. Not only can the semiconductor be used to generate the Hall voltage, but additional circuitry can be built into the semiconductor to process the Hall voltage. As a result, not only are there linear sensors that generate an output voltage that is proportional to the magnitude of the magnetic flux applied, but, because circuitry can be added to the chip, there are sensors that have switched logic-level outputs, or latched outputs, or outputs whose level depends on the difference between two applied magnetic fields.

Hall Effect—Switch

Figure 3-9b shows a Hall-effect switch and its output when used as a sensor. When the magnetic flux exceeds β_{ON} in maxwells, the output transistor of the switch is ON, and when the field is less than β_{OFF}, the output transistor is OFF. There is a hysteresis curve as shown. When the output transistor is OFF, the magnetic field must be greater than zero by β_{ON} before the transistor is ON, but will stay ON until the magnetic field is less than zero by B_{OFF}. The zero magnetic field point can be "biased" up to a particular value by applying a steady field to make $\beta_O = B_{STEADY\text{-}STATE}$.

Hall Effect—Linear Position

A linear Hall-effect sensor is shown in *Figure 3-9c*. Its output voltage varies linearly as the magnetic field varies. When the field is zero, there is a quiescent voltage = V_{OQ}. If the field is $+\beta$ (north to south), the voltage V_O increases from V_{OQ}; if the field is $-\beta$ (south to north), the voltage V_O decreases from V_{OQ}. The supply voltage is typically 3.8V to 24V for Hall-effect devices.

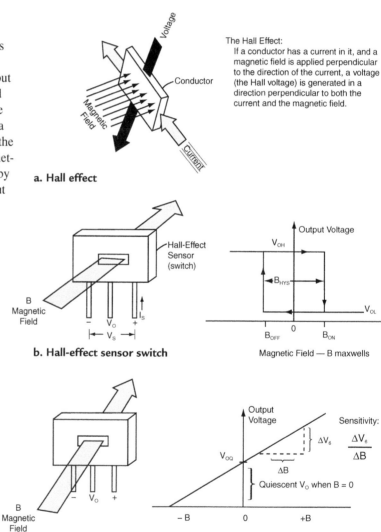

The Hall Effect:
If a conductor has a current in it, and a magnetic field is applied perpendicular to the direction of the current, a voltage (the Hall voltage) is generated in a direction perpendicular to both the current and the magnetic field.

a. Hall effect

b. Hall-effect sensor switch

c. Linear Hall-effect sensor

Figure 3-9: Hall-effect sensors

Hall Effect—Brake Pedal Position

A brake pedal position sensor is shown in *Figure 3-10a*. A Hall-effect switching sensor is used. Stepping on the brake moves a magnet away from the Hall-effect sensor and its output switches to a low voltage level turning on the brake light. When the brake is released, the magnetic field is again strong enough to switch the output V_O to a high level, turning off the brake light.

Hall Effect—Linear Position Sensor

In *Figure 3-10b*, as the magnet is moved over the sensor the magnetic field produces an output V_O that is proportional to the strength of the field. The linear output voltage can be converted to a meter reading that indicates the linear position of the assembly that moves the magnet. Amplifying V_O can increase the sensitivity of the measurement.

Hall Effect—Angular Position Sensor

A round magnet, half North pole and half South pole, is rotated in front of a linear Hall-effect sensor as shown in *Figure 3-10c*. As the magnet turns the magnetic field varies and produces an output V_O that is proportional to the angular rotation. V_O can be converted to a meter reading calibrated in degrees of rotation.

Hall Effect—Current Sensor

Current in a wire produces a magnetic field around the wire as shown in *Figure 3-10d*. If the wire is passed through a soft-iron yoke, the soft iron collects the magnetic field and directs it to a linear Hall-effect sensor. The magnetic field varies as the amplitude of the current varies, which produces a corresponding proportional V_O from the linear sensor, and, thus, a sensor that detects the amplitude of the current. An alternating current is shown in *Figure 3-10d*; therefore, the voltage V_O will be an alternating voltage. V_O is detected *in Figure 3-10d* using an oscilloscope.

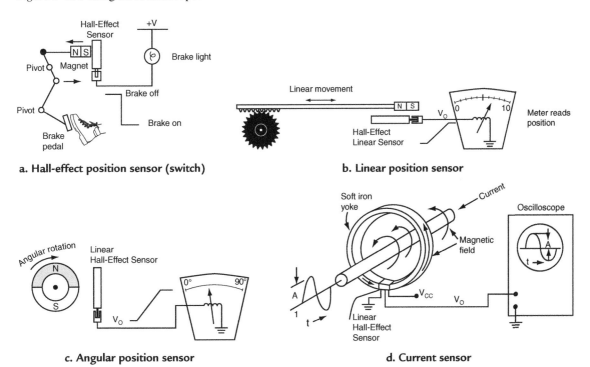

a. Hall-effect position sensor (switch)

b. Linear position sensor

c. Angular position sensor

d. Current sensor

Figure 3-10: Hall-effect sensor applications

Rotation

Variable Reluctance Sensor

Figure 3-11a shows the physical setup of an electromagnetic sensor that produces a continuous series of voltage pulses as a result of time-varying changes of magnetic flux. The magnetic flux path in *Figure 3-11a*, called the reluctance path, is through the iron core of the wound coil, through the cog on the rotating wheel and back to the coil. When the cog on the wheel is aligned with the iron core, the concentration of flux is the greatest. As the cog moves toward or away from the core of the coil, the concentration of flux is much less. Anytime magnetic flux changes and cuts across wires, it generates a voltage in the wires. The voltage produces a current in the circuit attached to the wires. As a result of the rotation of the wheel and the cog past the coil, a series of voltage pulses, as shown in *Figure 3-11b*, is generated. The time, t, between the pulses varies as the speed of the cogged wheel varies. Counting the pulses over a set period of time, say, a second, the speed (velocity) of the cogged wheel can be calculated. The variations of the speed can be calculated for acceleration, and of course, the presence of pulses means the wheel is in motion. The disadvantage of such a sensor is that there is no signal at zero speed, and the air gap between the mechanical moving part and the coil core must be small, usually equal to or less than 2–3 centimeters.

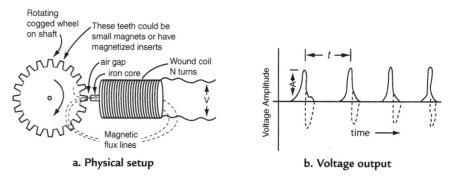

a. Physical setup   b. Voltage output

Figure 3-11: Variable reluctance rotation sensor

Example 2. RPM

A variable reluctance sensor outputs 120 pulses in a time period of 3 seconds. What is the rpm (revolutions per minute)?

Solution:

rps (revolutions per sec) = 120/3 = 40 × 60 sec = 2400 rpm

Magnetoresistor Sensor

A magnetoresistor sensor changes its resistance proportional to the magnetic flux density to which it is exposed. It is made of a nickel-iron (Permalloy) which is deposited as a thin film onto a semiconductor surface. It requires special fabrication of conducting strips on high-carrier mobility semiconductors such as Indium-Antimonide or Indium Arsenide. The basic principle is shown *in Figure 3-12a*. The thin film is deposited in a strong magnetic field that orients the magnitization M in a direction parallel to the length of the resistor. A current is then made to pass through the thin film at an angle θ to the M direction. If the angle is zero, the thin film will have the highest resistance. At an angle θ, it will have a lower resistance. When an external magnetic field is applied perpendicular to M, then θ changes and the resistance changes. This is the basic principle that produces a resistance change when a magnetic field is applied and allows the use of the thin film device as a sensor.

Figure 3-12b shows the change in resistance as the angle θ of the current in relationship to M varies. One of the advantages of using magnetoresistor is that other semiconductor circuits can be fabricated on and in the same semiconductor substrate. The resistor element is usually placed in a Wheatstone bridge circuit in order to make a more sensitive measurement.

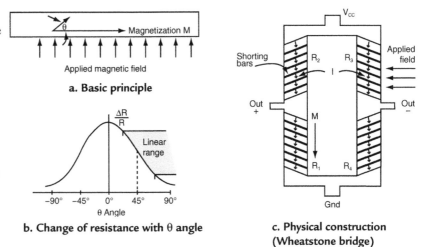

a. Basic principle

b. Change of resistance with θ angle

c. Physical construction (Wheatstone bridge)

Such a physical layout is shown in *Figure 3-12c*.

Figure 3-12: Magnetoresistor sensor

There are shorting bars deposited over the film to direct the bias current at an angle equal to 45°. This is to put the quiescent point in *the* center of the linear region of operation of the response curve of *Figure 3-12b*. When V_{CC} = 5V, the bridge sensitivity can be as much as 15 mV per Oersted of an applied field.

Pressure

Piezoresistive Diaphragm

The physical construction of a pressure sensor is shown in *Figure 3-13a*. A fluid or gas under pressure is contained within a tube the end of which is covered with a thin, flexible diaphragm. As the pressure increases the diaphragm deflects. The deflection of the diaphragm can be calibrated to the pressure applied to complete the pressure sensor characteristics.

Modern day semiconductor technology has been applied to the design and manufacturing of pressure sensors. A descriptive diagram is shown in *Figure 3-13b*. The thin diaphragm is micromachined from a silicon substrate on which a high-resistivity epitaxial layer has been deposited. The position of the diaphragm and its thickness on and in the substrate is defined using typical semiconductor techniques—form a silicon dioxide on the surface, coat it with photoresist, expose the photoresist with ultraviolet light through a mask to define the diaphragm area, and etch away the oxide and silicon to the correct depth for the thin diaphragm. The assembly is then packaged to allow pressure to deflect the diaphragm.

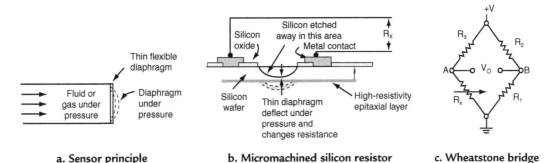

a. Sensor principle

b. Micromachined silicon resistor

c. Wheatstone bridge

Figure 3-13: Micromachined pressure sensor

Using integrated circuit metallization techniques, the thin diaphragm, which changes resistance as it deflects, is connected into a Wheatstone bridge circuit as shown in *Figure 3-13c*. This provides a very sensitive, temperature compensated, measuring circuit. R_X in the circuit is the thin diaphragm resistance exposed to pressure. R_1, R_2, and R_3 are similarly micromachined resistors but they are not exposed to pressure. As temperature changes all the resistors change in like fashion because they are located very close together on the small semiconductor surface and have the same temperature coefficient. As a result, the sensor is temperature compensated. And since the resistors are very close together on the substrate, and are machined at the same time, they are very uniform in value.

The Wheatstone Bridge

How does the Wheatstone bridge of *Figure 3-13c* work? The sensing voltage, V_O, is measured across the bridge from point A to point B. $V_O = 0$ when the bridge is balanced and is at its most sensitive measuring point. The circuit is analyzed as follows:

The voltage from point A to ground is:

$$V_A = R_X/(R_X + R_3) \times V$$

The voltage from point B to ground is:

$$V_B = R_1/(R_1 + R_2) \times V$$

When the bridge is balanced, $V_A = V_B$ and

$$R_X/(R_X + R_3) \times V = R_1/(R_1 + R_2) \times V$$

Cancelling V on both sides of the equation,

$$R_X/(R_X + R_3) = R_1/(R_1 + R_2)$$

and transposing,

$$R_X(R_1 + R_2) = R_1(R_X + R_3)$$

or

$$R_X R_2 = R_1 R_3 \text{ because } R_1 R_X \text{ cancels on each side of the equation.}$$

Therefore,

$$R_X = R_3 \times R_1/R_2$$

At balance, the unknown resistance is equal to R_3 times the ratio of R_1 to R_2.

As R_X changes, the bridge will become unbalanced and a voltage, V_O, other than zero results. The voltage, V_O, is calibrated to the pressure to complete the sensor characteristics. Pressures from 0–500 psi (pounds per square inch) can be measured with such a pressure sensor. If R_1, R_2, and R_3 all equal 10 kΩ, when R_X varies from 10 kΩ to 20 kΩ, the output voltage will be approximately from 10 mV to 20 mV per 1 kΩ of resistance change. One of the advantages of the silicon substrate sensors is that other integrated circuits can be in and on the silicon to provide signal conditioning to the voltage output, V_O.

Example 3. Wheatstone Bridge Characteristic Curve

In the Wheatstone bridge of *Figure 3-13c*, R1 = R2 = R3 = 10 kΩ and R_X varies with pressure from 10 kΩ to 15 kΩ. Plot the change in voltage, V_O, against the change in resistance, R_X. Make V = 1V.

Solution:

$$V_O = V_A - V_B = V(R_X/(R_X + R_3) - R_1/(R_1 + R_2)) = V(R_X/(R_X + 10 \text{ k}\Omega) - 0.5)$$

R_X	$R_X + 10\ k\Omega$	$R_X/(R_X + 10\ k\Omega)$	V_O
10k	20k	0.5	0
11k	21k	0.524	0.024
12k	22k	0.545	0.045
13k	23k	0.5652	0.065
14k	24k	0.5833	0.0833
15k	25k	0.600	0.100

Plot these numbers on an X and Y axis with X = V and Y = Ω and the characteristic curve of output voltage against resistance is obtained. The output voltage can then be calibrated to pressure to give a characteristic curve of voltage against pressure.

Capacitive Touch Diaphragm

The capacitive touch diaphragm sensor has the same micromachined structure as that shown in *Figure 3-13b*. However, its sensor principle, shown in *Figure 3-14a*, is different. The thin micromachined diaphragm is deflected as previously, but now the deflected diaphragm is designed to touch against a dielectric layer attached to a metal electrode. It forms a capacitor and as pressure increases, the capacitance between the diaphragm and the metal electrode, separated by the dielectric, increases linearly with pressure. The characteristic curve is shown in *Figure 3-14b*.

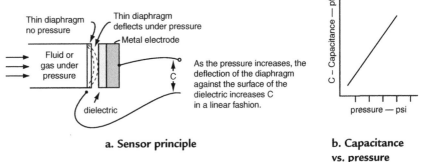

a. Sensor principle

b. Capacitance vs. pressure

Figure 3-14: Capacitive touch pressure sensor

Both of the micromachined sensors fabricated from silicon have –40° to +135° operation. For very extreme operating conditions of aircraft and automotive applications, there is a capacitive sensor with a ceramic diaphragm that deflects into a cavity. Its capacitance again increases with pressure.

Light Sensors

Light Basics (Review)

A brief review is presented of the principles of light and detection by photodiodes and phototransistors. For a more thorough review refer to *Basic Electronics*[1], *Chapter 11*. *Figure 3-15* shows how a reverse-biased photodiode has its reverse leakage increased by light shining on it. Photons,

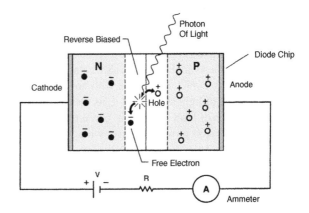

Figure 3-15: A reversed-biased photodiode light sensor
Courtesy of Master Publishing, Inc.

---

[1] *Basic Electronics*, G. McWhorter, A.J. Evans, © 1994, Master Publishing, Inc.

which are particles of light that are high-frequency electromagnetic waves, are absorbed in the reverse-biased diode depletion layer. They produce free electrons and holes that increase the reverse current. The more photons, the higher the intensity of light, the more energy is absorbed, and the larger the reverse current. Thus, the photodiode is a light sensor with a variable current output.

The Electromagnetic Spectrum

The electromagnetic spectrum is divided into radio waves and light waves by frequency. Light waves are further divided by into infrared, visible, ultraviolet and X-rays. The spectrum is either expressed in frequency or wavelength. Wavelength is the distance that an electromagnetic wave travels through space in one cycle of its frequency. Since distance is velocity multiplied by time, wavelength can be expressed as the velocity of electromagnetic waves multiplied by the time of one cycle of frequency f. Since the accepted speed of light is 186,000 miles per second or 300,000,000 meters per second, this is:

$$\lambda(\text{in meters}) = 300,000,000 \text{ meters/sec} \times 1/f(\text{in seconds})$$

or, $\lambda(\text{in meters}) = 300/f(\text{in MHz})$

If visible light (white light) is passed through a prism, as shown *in Figure 3-16*, the visible light separates into its color components. The frequency of visible light is from 400 million megahertz to 750 million megahertz. The wavelength is from 750 nanometers (10^{-9}) to 400 nanometers. Light sensors extend into the infrared frequency range below visible light and into the ultraviolet light frequency range above visible light. Cadmium sulfide sensors are most sensitive in the green light region of visible light, while solar cells and phototransistor sensors are most sensitive in the infrared region.

Photoresistor Sensor

A sensor that changes resistance as light is shined on it is made from Cadmium Sulfide (CdS), a semiconductor that is light sensitive. The characteristics of one available at RadioShack are

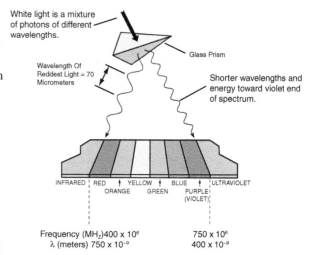

Figure 3-16: Visible light—its frequency and wavelength
Courtesy of Master Publishing, Inc.

shown in *Figure 3-17a*. In the dark with no light shining on it its resistance is greater than 0.5 MΩ. With one footcandle of light shining on it, its resistance is 1700 Ω, and the resistance is reduced to 100 Ω when 100 footcandles of light shine on it. Circuit applications are shown *in Figure 3-17b*. It can be used to change resistance values, to provide a sensor with a voltage output, or as a sensor supplying current to a load.

Example 4. Photoresistor Application

Use the circuit of *Figure 3-17b* that provides a voltage output. The resistor R1 = 200 Ω. What is the voltage out, V_O, when the supply voltage is 10V and the light shining on the sensor is 15 Ftc and 100 Ftc?

Solution:

Ftc	R_L	R_1	$R_1 + R_L$	V_O
15	800 Ω	200 Ω	1000 Ω	200/1000 × 10 = 2V
100	100 Ω	200 Ω	300 Ω	200/300 × 10 = 6.67V

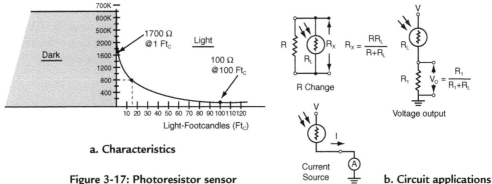

a. Characteristics

Figure 3-17: Photoresistor sensor

b. Circuit applications

Solar Cell

The solar cell is again a semiconductor PN junction that is light sensitive. It is made up of an N-type substrate, as shown in *Figure 3-18a*, with a very thin P region over the top surface. Most of the thin P surface is covered with narrow strips of metal that form the anode of the PN diode. A whole network of the narrow strips are interconnected on a silicon wafer to provide increased current output at the PN-junction voltage. The back of the silicon wafer is coated with metal to form the cathode of the diode. Light shining on the surface of the solar cell generates a maximum voltage of about 0.55V. Under load, the average voltage output is approximately 0.5V. A common characteristic curve of voltage plotted against current is shown in *Figure 3-18b*.

Solar cells can be applied in circuits, as shown in *Figure 3-18c*, by paralleling the cells for increased current output, or by connecting the cells in series for increased voltage output. Individual 2 × 4 cm solar cells are available at RadioShack that provide 300 mA at 0.55V, or there are enclosed modules that provide up to 6V at 50 mA.

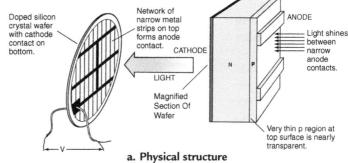

a. Physical structure
Courtesy of Master Publishing, Inc.

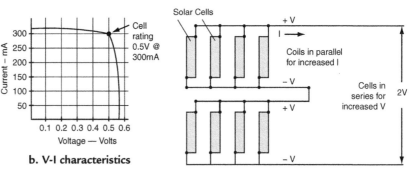

b. V-I characteristics

c. Circuit applications

Figure 3-18: Solar cell light sensors

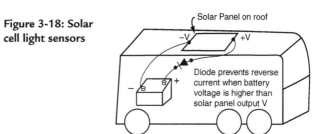

d. Trickle charge for RV coach batteries

A very common application for RV motorhomes is shown in *Figure 3-18d*. A solar panel is mounted on the roof of a motorhome and connected as shown to trickle charge the coach batteries when the RV is parked and under light load. Sunlight generates the voltage to supply the trickle current, which helps keep the batteries from discharging. Many units are available with power ratings from 2 to 50 watts.

Phototransistors

Figure 3-19 allows a quick review of the operation of bipolar transistors, both NPN and PNP. Recall that for an NPN grounded emitter stage shown in *Figure 3-19a* the emitter is tied to ground, and for active operation, the base voltage is at +0.7V above ground and forward-biases the base-emitter junction. The collector voltage is at a positive voltage above ground (+5V) so the collector-base junction is reverse biased. When there is a

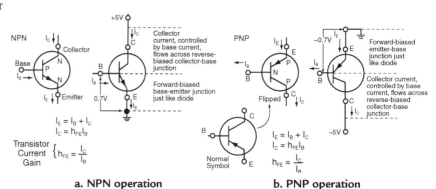

a. NPN operation   **b. PNP operation**

Figure 3-19: Bipolar transistor operation

current into the base, I_B, across the forward-biased base-emitter junction, a higher collector current, I_C, flows across the reverse-biased collector-base junction. There is a current gain through the transistor equal to the collector current divided by the base current, I_C/I_B. As shown in *Figure 3-19*, the current gain is h_{FE}. Everything is the same for the operation of the PNP transistor except the voltages are all negative with the emitter tied to ground. The h_{FE} is the same parameter as for the NPN.

A phototransistor, a transistor designed to be activated by light, has the same basic operation as the NPN and PNP transistor described except it has no base connection. Its wide base junction is left exposed to light. Phototransistors are most sensitive to infrared light. The symbols and voltages are shown in *Figure 3-20a*. Light rays that impact the base-emitter junction effectively produce base current that activates the phototransistor. Through transistor action a larger collector current is produced. As shown by the characteristic curves of *Figure 3-20b*, more light intensity produces more collector current.

A phototransistor can be coupled to the base of a driver transistor, as shown in *Figure 3-20c*, in order to make a linear

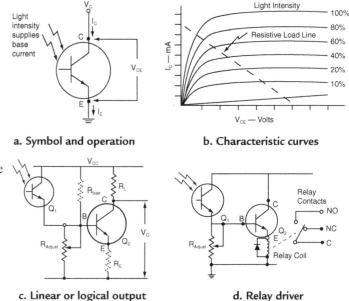

a. Symbol and operation   **b. Characteristic curves**

c. Linear or logical output   **d. Relay driver**

Figure 3-20: Phototransistor light sensor

driver or a logic-level driver. If a logic-level driver for ON-OFF applications is needed, R_{BIAS} and R_E are eliminated and the R_{ADJUST} used to set the desired sensitivity. R_{BIAS} and R_E set the operating point for Q_2 to obtain linear operation of the driver. *Figure 3-20d* shows a phototransistor sensing the presence of light to make a logic-level driver for a relay. The presence of light closes the normally-open contact to the center terminal to activate a connected circuit.

LED Light Source

Even though a light-emitting diode (LED) is not a sensor, it is a very important light source for light sensors. An LED is a forward-biased semiconductor diode as shown in *Figure 3-21a*. LEDs are made from special semiconductor materials other than silicon, but still have the same type of junction characteristics. When a rated amount of current is passed through the

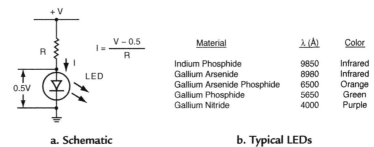

a. Schematic   b. Typical LEDs

Figure 3-21: LED light sources

Material	λ (Å)	Color
Indium Phosphide	9850	Infrared
Gallium Arsenide	8980	Infrared
Gallium Arsenide Phosphide	6500	Orange
Gallium Phosphide	5650	Green
Gallium Nitride	4000	Purple

forward-biased diode it emits light. The amount of current, I, through the diode can be adjusted by choosing the value of R when a given voltage, V, is used. The forward-biased voltage across the diode is approximately 0.5V, positive (+) on the anode and minus (−) on the cathode.

Various LEDs, the materials used to make them, and the color of their light output are shown in *Figure 3-21b*. The wavelength in this case is given in Angstroms (Å), where an Angstrom is 10^{-10} meters. When LEDs are used for light sources for phototransistors, the LED wavelength should be matched to the phototransistor. For example, *Figure 3-22* shows the relative output from a phototransistor using an LED as a source. An infrared LED with a wavelength of 898 nanometers (8980Å) provides almost three times as much output from the phototransistor as an LED with an orange light output of 650 nanometers (6500Å).

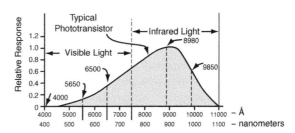

Figure 3-22: LEDs as light sources for a phototransistor sensor

---

Example 5. Wavelengths of LEDs

What is the wavelength in meters of the LEDs whose wavelength is given in Angstroms (Å)?

Solution: (a millionth of a meter (micron) equals 10,000Å)

Å	microns (divide by 10^4)	meters (divide by 10^6)	nanometers(10^{-9})
9850	0.9850	0.9850×10^{-6}	985
8980	0.8980	0.8980×10^{-6}	898
6500	0.6500	0.6500×10^{-6}	650
5650	0.5650	0.5650×10^{-6}	565
4000	0.4000	0.4000×10^{-6}	400

Other Sensors

The expansion of the types of sensors into modern day applications is almost mind boggling. The advent of micromachining using semiconductors and semiconductor fabrication techniques, and the ability to provide a sensor and its associated circuitry to signal condition the signal all in one package has expanded the types and variety of types of sensors. For example, in automotive and aircraft, there are sensors for mass air flow, exhaust gas and its properties, engine knock, linear acceleration, just to name a few. In fact, in the modern automobile there are over 100 sensors per car[2]. Such application explosions testify to the importance of the sensor in electronic circuitry.

Summary

Sensors of all types have been described in this chapter. The devices that convert an analog physical quantity into forms of energy that humans can understand and interpret. The sensors in this chapter had outputs of electrical signals—voltage, current, resistance, capacitance. Because the output signals are going to be used in other electronic circuitry to provide signals that can be converted to digital signals, changes must be made to the sensor output signals to adapt them to further use. That is the subject of the next chapter—signal conditioning.

Chapter 3 Quiz

1. A sensor:
    a. senses an output quantity and inputs an electrical signal.
    b. senses an output electrical signal and inputs a physical quantity.
    c. senses an input quantity and outputs an electrical signal.
    d. senses an output physical quantity and outputs a physical quantity.
2. Magnetic fields:
    a. are not important to the operation of sensors.
    b. play an important part in the operation of many sensors.
    c. are harmful to the operation of sensors.
    d. are generated by all sensors.
3. A thermocouple:
    a. senses temperature.
    b. senses voltage.
    c. senses current.
    d. senses impedance.
4. Silicon P-N junctions:
    a. use the reverse voltage variations to sense current.
    b. use both forward junction current variations for sensing voltage.
    c. don't use the junction voltage variations to sense temperature.
    d. use the forward voltage variations to sense temperature.
5. A thermistor:
    a. is a sensor that varies temperature as voltage is applied.
    b. is a sensor whose resistance varies with temperature.

---

[2] *Sensorsysteme fur das Auto,* Klaus-Dieter Linsmeier, ©1999, verlage moderne industrie.

    c.   is a sensor that has a linear variation with temperature.

    d.   is a sensor that varies temperature as current is applied.

6. In a Hall-effect sensor:
   a. a voltage is generated that is in the same direction as a current and a magnetic field.
   b. a voltage is generated that has no relationship to the direction of an applied current or magnetic field.
   c. a voltage is generated perpendicular to the direction of a current and perpendicular to the direction of a magnetic field.
   d. needs only a magnetic field for its operation.

7. Semiconductors are particularly useful for Hall-effect sensors because:
   a. other circuits useful for processing the sensor signal can be built into the semiconductor.
   b. they are isolated from the sensor.
   c. they can be manufactured in one step.
   d. there is no other way to make the sensor.

8. Hall-effect sensors can be used to sense:
   a. linear position.
   b. angular position.
   c. current.
   d. all of above.

9. A variable reluctance sensor:
   a. has zero output when the magnetic field is not changing.
   b. depends on time varying changes of a magnetic field.
   c. has high output when the magnetic field is not changing.
   d. doesn't need a magnetic field.
   e. a and b above.
   f. c and d above.

10. A magnetoresistor sensor:
    a. changes its resistance proportional to the magnetic field flux density to which it is exposed.
    b. changes its voltage output as a result of a magnetic field.
    c. changes its current output as a result of a magnetic field.
    d. doesn't require a magnetic field.

11. Micromachined sensors:
    a. are processed with micro machines.
    b. are machined using computer-controlled machines.
    c. are processed using semiconductor manufacturing techniques.
    d. don't need accurate machining techniques.

12. Micromachined sensors:
    a. measure pressure by applying a magnetic field.
    b. measure pressure by changing resistance.
    c. measure pressure by removing a magnetic field.
    d. measure pressure by changing capacitance.
    e. a and c above.
    f. b and d above.

13. The photo diode:
    a. is not sensitive to any light.

  b. is a light sensor whose output does not vary with light intensity.

  c. is a light sensor with a variable current output.

  d. is a light sensor with a variable voltage output.

14. The light spectrum:

  a. is below 400 megahertz.

  b. extends from infrared on the low end to ultraviolet on the high end.

  c. is above 1,000 million megahertz.

  d. is variable, not constant.

15. Wavelength:

  a. is the distance that an electromagnetic wave travels through space in one cycle of its frequency.

  b. is not a distance but a speed.

  c. is not a speed but a velocity.

  d. is a measure of time.

16. A sensor that changes resistance when light is illuminates it is:

  a. photo current sensor.

  b. photo voltage sensor.

  c. photo impedance sensor.

  d. photo resistor sensor.

17. Solar cells:

  a. are semiconductor PN junctions that are sensitive to light.

  b. can be connected in parallel to increase current output.

  c. can be connected in series to increase voltage output.

  d. a above.

  e. all of above.

18. A phototransistor, a light sensitive transistor:

  a. has normal base, collector, emitter connections.

  b. has the base and collector connected together.

  c. has no base connection.

  d. has the base and emitter connected together.

19. Phototransistors:

  a. are most sensitive to infrared light.

  b. are most sensitive to ultraviolet light.

  c. are most sensitive to 100 MHz light.

  d. are most sensitive to 10 MHz light.

20. LEDs (light emitting diodes) when used as light sources:

  a. may be used as random light sources for phototransistors.

  b. should be matched to their phototransistor sensor.

  c. are not important to phototransistor sensor applications.

  d. are not reliable light sources.

Answers: 1.c, 2.b, 3.a, 4.d, 5.b, 6.c, 7.a, 8.d, 9.e, 10.a, 11.c, 12.f, 13.c, 14.b, 15.a, 16.d, 17.e, 18.c, 19.a, 20.b.

Signal Conditioning

Introduction

Signal conditioning, as the name implies, means modifying the signal, changing its characteristics, adjusting it to the needs of the application. This may mean an increase or decrease in the magnitude of the voltage signal, or an increase (or decrease) in the magnitude of the current

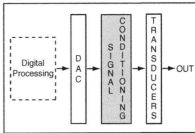

a. Sensor to digital   **b. Digital to transducer**

Figure 4-1: Signal conditioning function

signal, or a change in the ability of the signal to provide power. As shown in *Figure 4-1*, the signal conditioning function fits in two places in the chain from analog input to analog output. First, it is in the chain from sensor to the analog-to-digital conversion, second, it is in the chain from the digital-to-analog conversion to transducer. One of the most important electronic circuits to satisfy the signal conditioning function is the amplifier.

Amplification

An electronic circuit called an amplifier is used when a voltage or current signal needs to be increased in amplitude. An amplifier can be a single circuit with a single active device (transistor), or it can be a combination of circuits with many active devices. Recall that in *Chapter 3, Figure 3-19a*, the bipolar NPN transistor operation was dis-

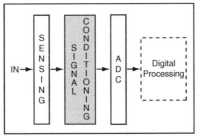

a. Schematic symbols of MOSFETs

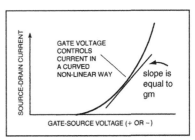

b. Characteristic curve of field-effect transistor

Figure 4-2: MOS (metal-oxide semiconductor) field-effect transistor
Courtesy of Master Publishing, Inc.

cussed, and in *Figure 3-19b*, the operation of a bipolar PNP transistor was discussed. There are also various types of field-effect transistors as shown in *Figure 4-2a*. These are called MOSFETs (metal-oxide semiconductor field-effect transistors). There are N-channel and P-channel devices that operate in the depletion mode or the enhancement mode. The characteristic curve of a field-effect transistor (FET) is shown in *Figure 4-2b*. A voltage from gate to source controls current from source to drain. Recall that for the bipolar transistors of *Figure 3-19*, *a current* into the base-emitter junction controls the collector-to-emitter *current*, while for field-effect transistors, a *voltage* from gate to source controls the *current* from drain to source. To understand the operation of an amplifier and how it might be used, an amplifier will be designed, and its characteristics examined, using a single NPN bipolar transistor in a common-emitter circuit. Common-emitter means the input signal is applied between base and emitter, and the output signal is taken between collector and emitter. The emitter is a common point between the two.

Bipolar NPN Amplifier

The design begins by choosing a device and looking at its characteristics shown in *Figure 4-3a*. The amplifier is going to be a "small-signal" linear amplifier. "Small-signal" means that the operating point will be set so that amplified output signals will be exactly the same as the input, with minimal distortion, but increased in amplitude. "Small-signal" means that the input signal amplitude only deviates the signal a small portion away from the steady-state operating point. As a result the amplification properties do not lose their linear qualities. The operating point—the no-signal steady-state operating

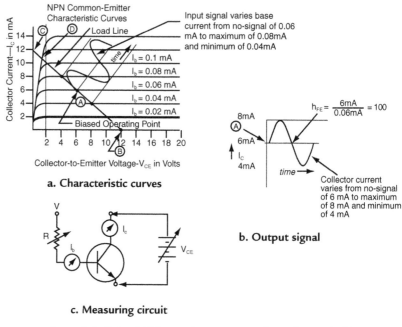

a. Characteristic curves

b. Output signal

c. Measuring circuit

Figure 4-3: Point "A" is steady-state operating point set by bias
Courtesy of Master Publishing, Inc.

point around which the small-signal ac signals vary—is chosen by some simple guidelines[1].

1. The operating point should be within the linear portion of the characteristic curves.

2. V_{CE} should be approximately $0.5V_{CC}$.

3. Emitter-to-ground voltage should be 10% to 15% of V_{CC}.

4. Base-to-ground voltage will be approximately 0.7V greater than the emitter-to-ground voltage.

The amplifier is going to be used in automotive applications so the supply voltage, V_{CC}, will be equal to +12V. The operating point is going to be set at point A, the biased operating point shown on the characteristic curves of *Figure 4-3a*. When there is no signal, point A says that the collector current will be 6 mA and the V_{CE} (voltage from collector to emitter) will be 6V.

Characteristic Curves

Look at the characteristic curves of *Figure 4-3a*. What do they mean? They were taken using the measuring circuit of *Figure 4-3c*. In this circuit, the base current, I_B, can be set to different values. The voltage V_{CE} can be varied, and the collector current, I_C, can be measured. I_B is set to 0.02 mA (20 microamperes) and V_{CE} is

---

[1] *Basic Communications Electronics*, J.W. Hudson, G. Luecke, ©1999, Master Publishing, Inc., Lincolnwood, IL.

varied from 0 to 20V. The heavy-line characteristic curve marked $I_B = 0.02$ mA is traced as I_C is measured during the variation of V_{CE}. As I_B is increased in 0.02mA steps and V_{CE} is varied, the other characteristic curves will be plotted.

The transistor will operate across these characteristic curves as driven by a signal that varies its base current and as regulated by the value of V_{CE}. At the operating point chosen (A), $I_C = 6$ mA and $I_B = 0.06$ mA, thus, as shown in *Figure 4-3b*, the steady-state common-emitter current gain (h_{FE}) equals 100 at this point. Small I_B changes produce I_C changes that are 100 times greater. The current gain, h_{FE}, is called a large-signal or DC current gain. There is an AC current gain, h_{fe}, that is used for small-signal AC circuit analysis. It may vary from the h_{FE} value because the variation of I_B is in very small increments.

Biasing

The operating point A is set at its operating characteristics by biasing the circuit. There are a number of biasing circuits: fixed-current I_B bias, voltage-divider bias, collector-feedback bias. This design will use voltage-divider bias. The circuit looks like the one in *Figure 4-4*. The resistor R_E is placed in the circuit to provide negative feedback. This feedback and its effect on circuit performance will be discussed when op amps and oscillators are discussed. For now, the presence of R_E in the circuit gives the circuit more stability against changes in temperature or parameter values.

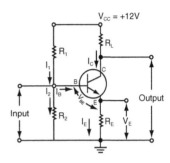

Figure 4-4: NPN common-emitter small signal amplifier

The Actual Design

Here are the design parameters that have been set:

$V_{CC} = 12$V
$I_C = 6$ mA
h_{FE} = The NPN transistor chosen has a specified minimum of 50; its actual hFE is 100.
$V_{BE} = 0.7$V for a silicon transistor
$V_E = 1.0$V (about 10% of VCC)
$V_{CE} = 6$V

1. The first step is to solve for R_L:

Since $V_{CE} = 6$V,

$$V_{CC} - I_C R_L = V_{CE}$$
$$12 - (6 \text{ mA} \times R_L) = 6\text{V}$$

therefore,

$$12 - 6 = 6 \text{ mA} \times R_L$$

and $\quad R_L = 6\text{V}/6 \text{ mA} = 6\text{V}/6 \times 10^{-3}$
$\quad\quad R_L = 1 \times 10^3 = 1000\ \Omega = 1\text{k}\ \Omega$

2. The next step is to solve for R_E:

Since in any transistor with reasonable gain, I_B is a small fraction of I_C, and since $I_E = I_C + I_B$, the approximation that I_E is equal to I_C is reasonably accurate.

Therefore,

$$V_E = R_E I_C$$

Therefore,

$$R_E = V_E/I_C$$

or $\quad R_E = 1\text{V}/6 \text{ mA} = 1\text{V}/6 \times 10^{-3} = 0.166 \times 10^3 = 166\ \Omega$

Resistors are manufactured in standard values of 150 Ω or 180 Ω. For this design $R_E = 150 \Omega$, and the actual $V_E = 0.9V$ that is: $(6 \times 10^{-3} \times 0.15 \times 10^{+3} = 0.9V)$.

3. The next step is to solve for R_2:

One of the rules for voltage-divider bias is that the current, I_2, through the divider should be at least 10 times the maximum base current. The maximum base current, I_{Bmax} is:

$$I_{Bmax} = I_C/h_{FE(min)} = 6 \text{ mA}/50 = 0.12 \text{ mA}$$

Thus, $I_2 = 1.2$ mA and the value or R_2 can be calculated since:

$$V_{R2} = V_{BE} + V_E = 0.7V + 0.9V = 1.6V$$

and

$$R_2 = V_{R2}/I_2 = 1.6V/1.2\text{mA} = 1.33 \times 10^3 = 1330 \ \Omega$$

A standard value is 1300 Ω, so $R_2 = 1.3$ kΩ

4. The next step is to solve for R_1:

R_1 in the voltage divider bias circuit will have the following current:

$$I_1 = I_2 + I_{Bmax} = 1.2 \text{ mA} + 0.12 \text{ mA} = 1.32 \text{ mA}$$

Since $V_{R2} = 1.6V$, the voltage across R_1 is:

$$V_{R1} = 12V - V_{R2} = 12 - 1.6 = 10.4V$$

Therefore,

$$R_1 = V_{R1}/I_1 = 10.4V/1.32 \text{ mA} = 7.88 \times 10^3 = 7880 \ \Omega$$

8,200 Ω is a standard value, so $R_1 = 8.2$ kΩ

The designed amplifier circuit using a 2N2222A transistor is shown in *Figure 4-5*. Its operating points are on the "load line" shown in *Figure 4-3a*.

5. The next step is to calculate the voltage gain.

The voltage gain[2] of a the common-emitter amplifier circuit shown in *Figure 4-5* is:

$$A_V = (R_L \times I_E)/0.026$$

where: R_L = total load resistance (R_L in parallel with any load across R_L)

I_E = DC emitter current in mA.

$I_E = I_C$ (approximately)

Substituting in the equation:

$$A_V = (1 \times 10^3 \times 6 \times 10^{-3})/2.6 \times 10^{-2} = (6 \times 10^2)/2.6 = 2.3 \times 10^2 = 230$$

The voltage gain in dB is:

$$DB = 20\log_{10}A_V = 20\log_{10}230 = 20 \times 2.76 = 47.2 \text{ dB}$$

The voltage gain is 230 or 47.2 dB.

---

[2] Ibid.

Amplifier Frequency Response

Not only must an amplifier amplify the voltage or current changes of an input signal, but it must be able to accurately reproduce these signals as the signal frequency changes. The capability of an amplifier to handle the signal over different frequencies is called its frequency response. An example of the frequency response of a common-emitter amplifier similar to the one in *Figure 4-5* is shown in *Figure 4-6*. It is a graph of an amplifier's gain, A_V, plotted against frequency with the input signal amplitude held constant as the signal frequency is varied. With the input signal amplitude constant, the output signal should remain constant if the gain, A_V, remains constant. The gain does remain constant in the midband as

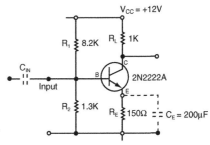

Figure 4-5: 2N2222A common-emitter small-signal amplifier

shown in *Figure 4-6*. However, for frequencies greater than f_H, the so called high-frequency corner frequency, the gain reduces as the signal frequency increases. This is due to circuit and device capacitance that is in parallel with R_L. A_V is reduced by 3 dB from its mid-band value at frequency f_H. The 3 dB point is also the frequency at which A_V has reduced to 0.707 of its midband value. For the amplifier in *Figure 4-6*, f_H is about 5 to 7 MHz.

When the amplifier is a DC amplifier, the midband value of A_V will extend down to zero frequency; however, if the amplifier only amplifies AC signals, A_V will reduce as the signal frequency is lowered below f_L, the low-frequency corner. Like f_H, f_L is the frequency where A_V is −3 dB (or 0.707) of its midband value. This reduction in gain is due to the coupling capacitors and capacitors across the emitter resistors used in AC amplifiers. For *Figure 4-6*, f_L is about 30 to 40 Hz.

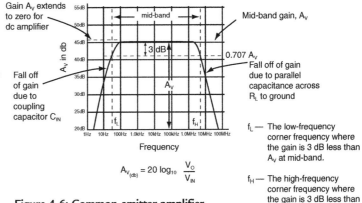

$$A_{V(db)} = 20 \log_{10} \frac{V_O}{V_{IN}}$$

Figure 4-6: Common-emitter amplifier frequency response

f_L — The low-frequency corner frequency where the gain is 3 dB less than A_V at mid-band.

f_H — The high-frequency corner frequency where the gain is 3 dB less than A_V at mid-band.

Example 1. Corner Frequencies

Show that the −3 dB point on a frequency response curve with a mid-band gain of 40 dB is the same point where the gain is 0.707 of the mid-band gain.

Solution:

Since the mid-band gain in dB is $A_V = 20\log 10\, V_O/V_{IN}$, then

$$40 = 20\log_{10}V_O/V_{IN}$$
$$2 = \log_{10}V_O/V_{IN}$$
$$10^2 = V_O/V_{IN}$$
$$100 = V_O/V_{IN}, \text{ the mid-band gain equals 100}$$

If the −3 dB point is $0.707A_V$, then the point is at $A_V = 70.7$

Therefore, $A_V = 20\log_{10}70.7$
$$A_V = 20 \times 1.85$$
$$A_V = 37 \text{ dB}$$

37 dB is −3 dB down from the mid-band gain of 40 dB.

For amplifier applications, it is very important to know the frequency range of the input signals that are to be handled and to examine the mid-band frequency range from f_L to f_H so that the proper amplifier can be used for an application.

Coupling

DC Coupling

When one circuit, such as the one in *Figure 4-5*, does not provide enough gain, circuits can be cascaded—coupled together—to provide more gain. The means of coupling are shown in *Figure 4-7*. *Figure 4-7a* shows a DC amplifier using two amplifier stages. The overall gain is equal to the first stage gain times the second stage gain. That is one advantage of expressing the amplifier gain in dB. The dB values of gain of each stage can be added to get the total gain in dB. The effect on frequency response is also shown. With DC coupling the amplifier has constant gain down to zero frequency. Special care must be taken in the design because the DC voltages couple from stage to stage so the proper operating voltages on the base, collector and emitter must be incorporated in the design.

AC Coupling

Figure 4-7b is AC coupling. A capacitor, C_C is used to couple the signal from the first stage to the second stage. There also is a capacitor, C_E, that is used to bypass the emitter resistor of second stage. The coupling capacitor prevents the DC voltages of stage 1 to couple through to stage 2, as they do in the DC case of *Figure 4-7a*. AC coupling allows the use of identical stages, makes the design easier, and even provides higher AC gain because the effect of the negative feedback of R_E is eliminated at frequencies above f_L. Below f_L, using R_E and C_C coupling causes a reduction in gain.

Figure 4-7c shows how inductance can be used to reduce the high-frequency response because of the increase in inductive reactance, and yet still maintain response at the low end down to zero frequency. *Figure 4-7d* shows transformer coupling. Transformer coupling provides DC isolation, but needs the AC time varying signal for its operation. The frequency response is like a capacitor-coupled amplifier.

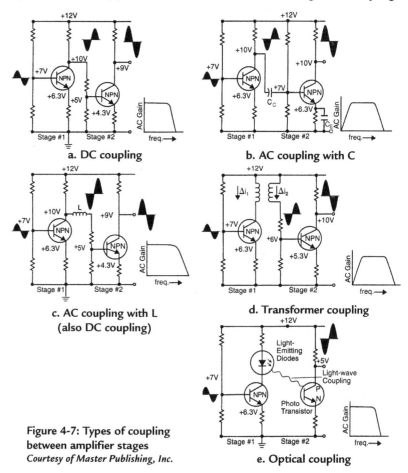

Figure 4-7: Types of coupling between amplifier stages
Courtesy of Master Publishing, Inc.

a. DC coupling

b. AC coupling with C

c. AC coupling with L (also DC coupling)

d. Transformer coupling

e. Optical coupling

Coupling Using Light

Figure 4-7e shows coupling using light. There is complete isolation between stage 1 and stage 2 using light coupling. The case shown uses a transistor to modulate the current through the LED, whose light emission is detected by a photo transistor. The light media can very easily be a fiber-optic cable.

When choosing an amplifier for an application, examine the transistor characteristic curves, know the type frequency response required and whether operation is required down to zero frequency, and determine if more than small-signal operation is needed.

Example 2. Cascaded Gain

Show than an amplifier with three stages each with a gain of 20dB per stage has an overall gain of 1000.

Solution:

A gain of 20 dB is

$$20 \text{ dB} = 20\log_{10}A_V$$
$$1 = \log_{10}A_V$$
$$A_V = 10$$

Three stages have overall gain of $10 \times 10 \times 10 = 1000$. It is interesting to note that the dB addition of $20 + 20 + 20 = 60$ dB for the overall gain. Therefore,

$$60 = 20\log_{10}A_V$$
$$3 = \log_{10}A_V$$
$$10^3 = A_V$$
$$1000 = A_V$$

Small-Signal vs. Large Signal

Return to *Figure 4-3a*, the common-emitter characteristic curves for the NPN transistor. The straight line plotted on the characteristic curves is a "load line" for the 1 kΩ load resistor used in the design of the amplifier stage of *Figure 4-5*. It represents the variation in collector voltage that occurs when collector current changes due to variations in base current. The operating point A is on the load line. If base current increases, the voltage drop across the 1 kΩ load resistor increases and the collector voltage is decreased. If the collector voltage is reduced to zero, the transistor would be shorted, and the operating point would be at point C. If the collector current is zero, the transistor is cutoff, and the operating point is at B. The operating point A is the no-signal steady-state operating point. When an input signal is applied that varies the base current a small-signal increment of 0.010 mA each side of the operating point A, the changes in the collector current will be 100 times (minimum of 50) greater or 1 mA. The collector voltage will swing 1V each side of the operating point A.

As shown in *Figure 4-3a*, an input signal is applied that varies I_B from 0.04 mA to 0.08 mA. The collector current varies from 4 mA to 8 mA as a result, and the collector voltage varies from 4V to 8V. There is a much larger output voltage swing, but the signal is still linear and not distorted. As the operation of the amplifier changes to a "large-signal" mode and more input base current change is supplied, the operating point on the load line runs into the nonlinear portion of the characteristic curves, point D, and distortion of the output waveform occurs. The distortion is shown in *Figure 4-8*. It is very important to the application, if linear operation is what is required, that the input signal does not drive the output of the amplifier into the distortion region.

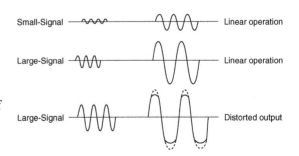

Figure 4-8: Linear small-signal and large-signal, and large-signal distorted operation

The signal range from small-signal until distortion at the output occurs is called the *dynamic range* of the amplifier. Unless the circuit is designed to operate outside the linear region, make certain amplifiers have enough dynamic range for the application.

Classes of Amplifiers

Figure 4-9 is the same sort of plot of characteristic curves as *Figure 4-3a*, but it defines the classes of amplifier circuits that can be designed. The operating points on the load line and the operating waveforms are shown. The small-signal amplifier of *Figure 4-5* at point A is called a Class A amplifier because the operation is totally linear—exact reproduction of the input at the output.

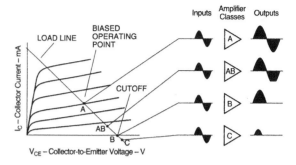

Figure 4-9: Bias points of various classes of transistorized amplifiers *Courtesy of Master Publishing, Inc.*

A Class B amplifier operates at point B on the load line. It is linear when it operates, but it operates only for 180° of the input cycle. This is a very important class for power amplifiers—amplifiers that must supply large amounts of current and have, at the same time, significant voltage swings.

A Class AB amplifier, as shown in *Figure 4-9*, has an operating point on the load line that is between Class A and Class B. It is used to eliminate crossover distortion in linear power amplifiers and in tuned amplifiers for communications circuits.

A Class C amplifier operates on the load line at a point where the transistor is cutoff and must be driven into conduction by the input signal. As shown in *Figure 4-9*, collector current flows for only a small portion of an input cycle. Class C amplifiers are used extensively in resonant tuned circuit amplifiers to provide outputs over a narrow band of frequencies, usually radio frequencies and above.

Field-Effect Transistor Amplifiers

Amplifiers are also designed using field-effect transistors. The symbols for MOS transistors were shown in *Figure 4-2*. There are also JFETs (junction field-effect transistors) that are made from semiconductor junctions rather than a layer of metal over oxide over silicon. They come in P-channel or N-channel devices operating in the depletion or enhancement mode. Depletion mode transistors have current from drain to source without any gate-to-source voltage; while enhancement mode transistors do not have any drain-to-source current unless a gate-to-source voltage is applied. Depletion mode JFETS are the most common type used for individual transistor amplifier stages. Most MOS transistors are enhancement mode devices.

JFET Characteristic Curves

Figure 4-10 shows characteristic curves of a depletion mode N-channel. The changes in drain-to-source current, I_D, are plotted against drain-to-source voltage, V_{DS}, as the gate-to-source voltage, V_{GS}, is varied. The curves are developed the same way as the bipolar transistor curves of *Figure 4-3*; however, note that for these curves, that a change in *voltage* from gate to source causes the change in *current* from drain to source.

To design an amplifier, a load line is plotted on these characteristic curves just as for the bipolar transistor. Before the amplifier is designed several important points are noted. There is a gate-to-source voltage that is called the "pinch-off" voltage. It is the gate-to-source voltage that starts conduction of drain-to-source current for enhancement mode transistors, and it is the gate-to-source voltage that causes zero drain-to-source current in depletion mode devices. Anytime the drain-to-source voltage is above the gate-to-source voltage by the pinch-off voltage, the transistor is operating in the pinch-off mode. The pinch-off mode

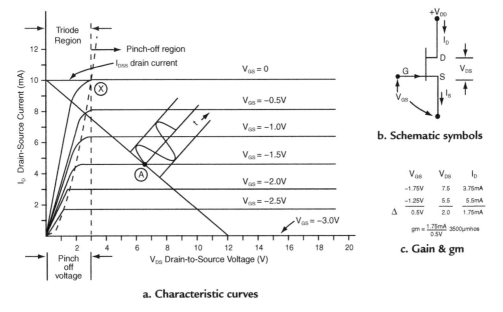

a. Characteristic curves

Figure 4-10: An NPN JFET transistor (depleting mode)

means the channel within the field-effect transistor is pinched off and the drain-to-source current, I_{DS}, remains essentially constant for further large variations of drain-to-source voltage, V_{DS}. For depletion mode JFETs, when V_{GS} equals zero, the device will be operating at point X, and the drain current will be I_{DSS}. The pinch-off voltage and I_{DSS} are important parameters specified by FET manufacturers.

A N-Channel JFET Amplifier Design

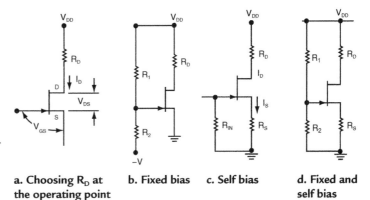

a. Choosing R_D at the operating point   b. Fixed bias   c. Self bias   d. Fixed and self bias

Figure 4-11: An NPN JFET amplifier design

The JFET amplifier design is again for an automotive small-signal amplifier. The device is an N-channel transistor operating in the depletion mode and the supply voltage is +12V. The design begins by plotting a load line on the characteristic curves of *Figure 4-10*. The point A is chosen as the operating point because it is in a nice linear region. At point A, $I_D = 4.6$ mA, $V_{GS} = -1.5$V and $V_{DS} = 6.5$V. The load line follows the equation:

$$I_D = V_{DD}/R_D - (1/R_D) \times V_{DS}$$

derived from $V_{DS} = V_{DD} - I_D R_D$ as follows:

$$I_D R_D = V_{DD} - V_{DS}$$

$$I_D = V_{DD}/R_D - V_{DS}/R_D = V_{DD}/R_D - (1/R_D) \times V_{DS}$$

Substituting in the values for I_D and V_{DS}, $R_D = 1,196$ ohms.

The slope of the load line is $1/R_D$, and the circuit corresponds to *Figure 4-11a*. R_D is chosen as 1.2 kΩ and the load line plotted through point A. Easy end points for the load line are determined when $I_D = 0$ then $V_{DS} = V_{DD}$ or +12V, and when $V_{DS} = 0$, $I_D = V_{DD}/R_D$ or 10 mA.

Biasing the Circuit

Figure 4-11 shows various ways of biasing the JFET at operating point A. A combination of fixed and self-bias shown in *Figure 4-11d* is chosen for the design. The self-bias requires that the value of R_S be calculated. It should be noted that with self-bias, the V_{DS} will be reduced by the amount of the voltage developed across R_S. Looking at the characteristic curves of *Figure 4-10*, I_D will not vary significantly if V_{DS} is reduced by several volts. The voltage across R_S is chosen to be +2V. With $I_D = 4.6$ mA, the value of R_S can be calculated as:

$$R_S = 2V/4.6 \text{ mA} = 0.435 \times 10^3 = 435 \ \Omega$$

A standard value of 430 Ω will be used.

Because the V_{GS} voltage must be –1.5V and the source is at the voltage across R_S, which is +2V, the gate voltage must be +0.5V. This voltage is provided by the resistor divider of R_1 and R_2. The input impedance of the JFET is very high so it will not load the resistor divider; therefore, the values of R_1 and R_2 can be quite high to reduce the power dissipation. The voltage across $R_2 = +0.5V$ and can be calculated as:

$$0.5V = R_2/(R_1 + R_2) \times V_{DD} = R_2/(R_1 + R_2) \times 12V$$
$$12R_2 = 0.5R_1 + 0.5R_2$$
$$11.5R_2 = 0.5R_1$$

Transposing,

$$R_1 = 11.5R_2/0.5$$
$$R_1 = 23 \ R_2$$

The value of R_1 is 23 times the value of R_2. R_2 is chosen as a standard value of 47 kΩ, and as a result, R_1 equals 1.1 MΩ. The completed design is shown in *Figure 4-12*. As noted in *Figure 4-10c*, for a 0.5V change in V_{GS}, there is a 2.0V change in V_{DS} and a 1.75 mA change in I_{DS} current. The amplifier has a voltage gain of 4.

Example 3. Calculating Ratio of R2 : R1

In *Figure 4-11b*, if $V_{DD} = +10V$ and $–V = –10V$, what ratio of $R_2 : R_1$ should be used to obtain a $V_{GS} = –1.5V$?

Solution:
Since $V_{DD} = +10V$ and $–V = –10V$ and $V_{GS} = –1.5V$, then the voltage across R_2 is:
$$8.5 = R_2/(R_1 + R_2) \times 20V$$
$$8.5R_1 + 8.5R_2 = 20R_2$$
$$8.5R_2 = 11.5R_1$$
$$R_2/R_1 = 8.5/11.5 = 0.74$$

Gm—Transconductance

There is a parameter, gm, for field-effect transistors called transconductance. It is defined as the change in drain-to-source *current* in amperes per *volt* of change in the gate-to-source voltage. It is a change in current, ΔI, over a change in voltage, ΔV; or the inverse of resistance. Thus, transconductance has the units of mhos, rather than ohms. For the amplifier of *Figure 4-12*, as shown in *Figure 4-10c*, there is a 1.75 mA change in drain current for a 0.5V change in gate-to-source voltage. This is a 3.5 mA per volt change or 3500 micromhos for gm.

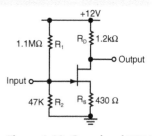

Figure 4-12: Completed NPN JFET small-signal amplifier

The voltage gain of an amplifier can be expressed as:

$$A_V = -gmR_L$$

Accordingly, the gain of the amplifier of *Figure 4-12* is:

$$A_V = 3500 \times 10^{-6} \times 1.2 \times 10^3 = 4200 \times 10^{-3} = 4.2$$

This matches the value computed from *Figure 4-10*. Examining the equation $A_V = -gmR_L$ one can see that the gain can be increased by increasing R_L. To do this one would need to increase V_{DD}. Of course, if gm is higher the gain is higher. Thus, devices are judged for amplifiers by their gm. *Figure 4-2b* shows an easy way to evaluate gm, it is the slope of a line tangent to I_{DS} vs V_{GS} curve.

An NPN MOSFET Amplifier

The same characteristic curves shown in *Figure 4-10* apply to an enhancement mode N-channel MOSFET with a couple of exceptions. The gate-to-source voltages are positive and equal to and greater than the threshold voltage, V_t. V_t is required to start the channel current from drain to source in the enchancement mode. The characteristic curve that plots on the V_{DS} axis has $V_{GS} = V_t$ and any additional curves have $V_{GS} = V_t + V$. The characteristic curves represent the full V_{GS} voltage, but only the V value above V_t is contributing to enhancing the channel current.

The characteristic curves for the N-channel MOSFET are shown in *Figure 4-13*. The transistor will be used in the enhancement mode to design a small-signal amplifier similar to the JFET amplifier. Again, the amplifier will be used in an automotive application so the power supply voltage is +12V. The operating point is set as point A, again in the linear region. At point A, $V_{DS} = 8V$, $I_{DS} = 3.3$ mA, $V_{GS} = 6V$ and $V_t = 2V$. The dotted parabolic curve is the locus of points where $V_{DS} = V$, the component of V_{GS} above V_t. The points on the V_{GS} curves where the V_{DS} curve intersects are the points where the channel goes into pinch off. Operation to the right of the dotted-line curve is in pinch off; operation to the left is in the triode region. Small-signal linear amplifiers must operate in the pinch-off region. The load line is plotted for $R_L = 1.2$ kΩ just like the JFET design.

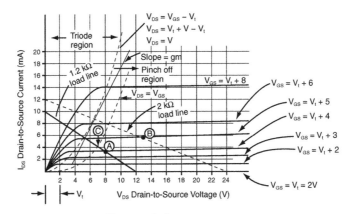

a. Characteristic curve

Point A	$V_t = 2V$	$R_L = 1.2K$			Point B	$V_t = 2V$	$R_L = 2K$	
V_{GS}	I_{DS}	V_{DS}			V_{GS}	I_{DS}	V_{DS}	
$V_t + 5V = 7V$	5.5 mA	5.5V	$gm = \dfrac{3\ mA}{2V}$		$V_t + 4 = 6V$	3.5 mA	17.0V	$gm = \dfrac{4.5\ mA}{2V}$
$V_t + 4V = 6V$	3.3 mA	8.0V			$V_t + 5 = 7V$	5.5 mA	13V	
$V_t + 3V = 5V$	2.5 mA	9.0V	$= 1500$		$V_t + 6 = 8V$	8 mA	8V	$= 2250$
Δ = 2V	3 mA	3.5V	μmhos		Δ = 2V	4.5 mA	9.5V	μmhos

b. Gain and gm

Figure 4-13: An N-channel enhancement-mode MOSFET

Fixed and Self-Bias

The MOSFET N-channel can be biased, as shown in *Figure 4-14a*, just like the JFET. The V_{DS} voltage is somewhat larger because the steady-state V_S can be much smaller due to the fact that all V_{GS} voltages are positive. In *Figure 4-14a*, $R_L = 1.2$ kΩ and $I_{DS} = 3.3$ mA; therefore, since $I_S = I_D$,

$$R_S = 0.5V/3.3 \text{ mA} = 0.152 \times 10^3 = 152 \ \Omega$$

A standard value is 150 Ω so $R_S = 150$ Ω.

To have $V_{GS} = 6V$, the gate must be at +6.5V since the source is at +0.5V; therefore,

$$R_2/(R_1 + R_2) \times 12V = 6.5V$$
$$6.5R_1 = (12 - 6.5)/6.5 \times R_2$$
$$R_1 = 0.846 \ R_2$$

R_2 is chosen as a standard value of 470 kΩ, which results in a standard value of 390 kΩ for R_1. The designed stage is shown in *Figure 4-14a*.

Drain-to-Gate Bias

Another biasing arrangement that also provides negative feedback is shown in *Figure 4-14b*. In this arrangement, since gate current is zero, $V_{DS} = V_{GS}$, and a small increase in drain current causes a small reduction in V_{DS}. The small reduction in V_{DS} is fed back to cause a small reduction in V_{GS}, which compensates for the original increase in drain current. Now, since

$$V_{DS} = V_{GS} - V_t$$

It follows that adding a V_t voltage to V_{DS},

$$V_{DS} = V_{GS} - V_t + V_t = V_{GS}$$

As a result, a new parabolic locus of points is drawn on the characteristic curves that represents $V_{DS} = V_{GS}$, as shown in *Figure 4-13a*. In other words, the curve is displaced to the right by the value of $V_t = 2V$. Because of this, the operating point shifts to point C on the load line, and $I_{DS} = 4.33$ mA and $V_{GS} = V_{DS} = 6.8V$.

Gain and g$_m$

Figure 4-13b provides some large-signal voltage gain and gm values for the circuit of *Figure 4-14a*. The voltage gain is 1.75V per V and gm is 1500 micromhos. Calculating the gain by using $-gmR_L$, the voltage gain is

$$A_V = -1500 \times 10^{-6} \times 1.2 \times 10^3 = -1.8$$

The minus sign, of course, meaning a change in phase of the signal. It was pointed out for the amplifier of *Figure 4-12* that using a larger load resistor would increase the stage voltage gain. A new load line using a 2 kΩ resistor is drawn on the characteristic curves of *Figure 4-13*. The operating point is now point B and $A_V = 4.25V$ per V, gm = 2250 micromhos and $-gmR_L = 2250 \times 10^{-6} \times 2 \times 10_3 = 4.25$. The concern with this design is the increased power supply voltage to +24V and increased power dissipation when the operating point is point B where $I_{DS} = 5.5$ mA and $V_{DS} = +13V$.

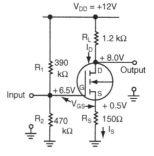

a. Self and fixed bias

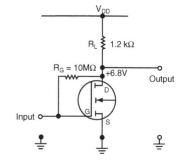

b. Drain-to-gate bias

Figure 4-14: N-channel MOSFET small-signal amplifiers

Operational Amplifiers

Integrated circuit manufacturers have provided an excellent product—the operational amplifier (op amp)—to designers of electronic circuits for signal conditioning sensor signals. Many types and varieties are available for a wide spectrum of applications. System designers that need amplification in their design need not design an individual amplifier circuit but can use an op amp instead.

The term "op amp" refers to a direct-coupled amplifier that was used initially in analog computers to perform mathematical computations, while solving real-time control system problems. Op amps are DC amplifiers that have high gain, high input impedance, low output impedance, and wide bandwidth. Another significant advantage is that the amplifier's characteristics can be varied using external components.

Figure 4-15 describes a general-purpose op amp. The amplifier has two inputs and one output. The amplifier output is normally a linear output voltage, V_O, that is proportional to the difference of the voltage between the two inputs. Thus, it is classified as a differential amplifier. The two inputs

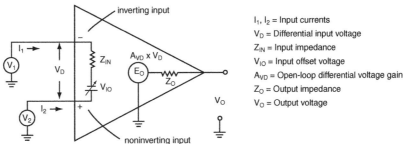

I_1, I_2 = Input currents
V_D = Differential input voltage
Z_{IN} = Input impedance
V_{IO} = Input offset voltage
A_{VD} = Open-loop differential voltage gain
Z_O = Output impedance
V_O = Output voltage

Figure 4-15: General-purpose operational amplifier

are identified with a minus and a plus sign. The input with the *minus sign* is called the *inverting input*; the input with the *plus sign* is the *noninverting input*. If the noninverting input is more positive than the inverting input, the output voltage, V_O, is positive with respect to ground. Conversely, if the inverting input is more positive than the noninverting input, V_O will be negative with respect to ground. When both inputs are referenced to ground, and the inverting input is more positive, V_O swings negative; when the non-inverting input is more positive, V_O swings positive.

Characteristics

Return to *Figure 4-15*. The output, V_O, can be represented by a generator, $E_O = A_{VD} \times V_D$, fed to the output through the output impedance, Z_O. E_O is the input differential signal, V_D, amplified by the open-loop differential gain, A_{VD}. Z_{IN} is the input impedance, and V_{IO} is the input offset voltage that causes the output voltage to be displaced from zero volts when there is no differential input signal. A_{VD} is usually a very large number (>20,000) in most modern day op amps; therefore, even a very small input signal drives the output into saturation. This is a distorted output as shown previously in *Figure 4-8*. As a result, normal operation is with feedback from output to input to set the gain of the op amp at a particular value.

Setting Gain

Look at *Figure 4-16*. A resistor, R_f, is connected from the output back to the inverting input to control the gain of the op amp with negative feedback.

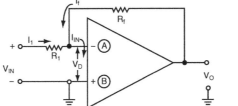

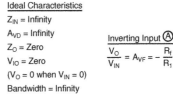

Ideal Characteristics
Z_{IN} = Infinity
A_{VD} = Infinity
Z_O = Zero
V_{IO} = Zero
(V_O = 0 when V_{IN} = 0)
Bandwidth = Infinity

Inverting Input Ⓐ
$$\frac{V_O}{V_{IN}} = A_{VF} = -\frac{R_f}{R_1}$$

Figure 4-16: Op amp with negative feedback and signal to inverting input

If the output goes positive, as it would for an input signal on A going negative, a portion of the positive output signal is fed back to the input to cancel part of the input signal. In *Figure 4-16*, the ideal op amp characteristics are listed. One of these is Z_{IN} = infinity. As a result, $I_{IN} = 0$; therefore, from *Figure 4-16*,

$$I_1 + I_f = 0$$

Since

$$I_1 = (V_{IN} - V_D)/R_1 \text{ and since } V_D = 0 \text{ because } I_{IN} = 0, \text{ then}$$
$$I_1 = V_{IN}/R_1$$

and
$$I_f = V_O/R_f \text{ because } V_D = 0, \text{ then}$$
$$V_{IN}/R_1 + V_O/R_f = 0$$
$$V_O/R_f = -V_{IN}/R_1$$

and
$$V_O/V_{IN} = -R_f/R_1$$

or
$$A_{Vf} = -R_f/R_1$$

The gain of the inverting operational amplifier with feedback, A_{Vf}, is determined by the ratio R_f/R_1, both external components to the amplifier itself.

In *Figure 4-17*, V_{IN} is applied to the noninverting input; therefore,

$$V_{R1} = V_{IN} - V_O \text{, but because } V_O = 0 \text{ since } I_{IN} = 0,$$
$$V_{R1} = V_{IN}$$

Therefore,

$$I_1 = V_{IN}/R_1$$

Since,

$$I_1 + I_f = 0, \ \ I_f = I_1$$

Now,

$$V_O = V_{R1} + V_{Rf} = V_{R1} + I_f R_f$$

Now, with substitution for V_{R1} and I_f,

$$V_O = V_{IN} + I_1 R_f$$

Therefore, since $I_1 = V_{IN}/R_1$

$$V_O = V_{IN} + (V_{IN}/R_1) \times R_f$$
$$V_O = V_{IN} (1 + R_f/R_1)$$

Or
$$V_O/V_{IN} = 1 + R_f/R_1$$

And thus, $A_{Vf} = 1 + R_f/R_1$

The feedback gain, A_{Vf}, for an op amp with the signal on the noninverting input is one plus the feedback gain of an op amp with the signal on the inverting input. The output is out of phase for the inverting input and in phase for the noninverting input.

In summary,

For inverting input: $\qquad A_{Vf} = -R_f/R_1$

For noninverting input: $\qquad A_{Vf} = 1 + R_f/R_1$

Even though the manufactured op amps are not ideal, the parameters are such that making the ideal amplifier assumptions cause very small errors (<0.5%). The above equations, when used for amplifier designs, will provide circuit performance that is well within the accuracy of other components used.

Example 4. Calculating Op Amp Gain

Calculate the op amp gain indicated using the values of R_f and R_1 as shown.

Solution:

Rf	R1	Inverting	Noninverting
1 MΩ	10 kΩ	$1 \times 10^6/10 \times 10^3 = 100$	101
870 kΩ	56 kΩ	$870 \times 10^3/56 \times 10^3 = 15.5$	16.5
330 kΩ	4.7 kΩ	$330 \times 10^3/4.7 \times 10^3 = 70.2$	71.2
100 kΩ	3.3 kΩ	$100 \times 10^3/3.3 \times 10^3 = 30.3$	31.3

Op-Amp Power Supplies

Notice that in *Figure 4-16* and *Figure 4-17* there are no power supply connections. They were omitted for clarity. Most op amps operate from plus and minus power supplies, but many manufacturers now have units that operate from a single supply. Units that operate from dual power supplies, use plus and minus voltages of equal value. The manufacturers recommend operating

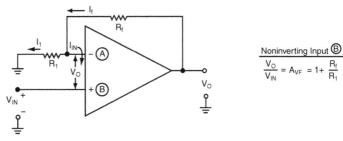

Figure 4-17: Op amp with negative feedback and signal to noninverting input

voltages, but most operate over a range of supply voltages. Parameters are tested for guaranteed values shown on data sheets at a particular power supply voltage.

Op-Amp Offset Correction

Many op amps in the past provided external package pins to aid in adjusting the amplifier output voltage for any offset voltage that might cause the output voltage to be other than zero when the input voltage is zero. The package pins are shown in *Figure 4-18*. Pin #1 and pin #5 are used to inject a current into the input stage and adjust V_O to zero when V_{IN} is zero. A high-value variable resistance is placed across the pins and the variable contact is fed by a current source. The resistance is adjusted until the output voltage is zero.

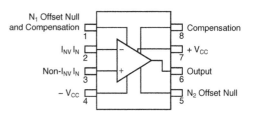

Figure 4-18: Op amp with offset adjustment and frequency compensation

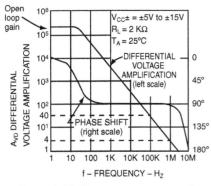

Figure 4-19: Frequency response of general purpose op amp

Frequency Response

Figure 4-19 shows the typical frequency response of a general-purpose op amp like the 741. The maximum open-loop gain of 200,000 is from DC to about 20 Hz and then the gain declines on a straight line until the gain equals 1. For this case, R_f is infinity so the feedback loop is open. When feedback is added, the gain is reduced and the gain vs. frequency curve fits under the open-loop gain response curve. For example, if feedback is added so the amplifier has a gain of 40—shown by the dotted line in *Figure 4-19*—then the frequency response is flat out to 100 kHz before it starts to roll off. If the gain is reduced to 4, the frequency response stays flat until 1 MHz before it starts to roll off.

The phase shift of the output signal is also shown in *Figure 4-19*. If the amplifier gain were greater than 1 at any frequency and the phase shift were greater than 180°, the amplifier circuit would oscillate at that frequency. One of the parameters where commercial op amps definitely deviate from the ideal specifications is in the bandwidth. Available op amps are definitely limited to a finite bandwidth.

Frequency Compensation

Frequency compensation is the act of adding external components to stabilize the op amp and keep it from oscillating. Some op amps provide pin connections to which external capacitance can be connected to stabilize the op amp. This feature is shown for the package of *Figure 4-18*. Capacitance is added across the pins marked "compensation." When the application demands more and more frequency response and higher frequency operation, more attention must be paid to circuit layout and lead lengths to keep the op amp from oscillating. For this reason, op amps with external connection pins for compensation may be a necessity.

Conditioning the Output of a Pressure Sensor

It is necessary to amplify the output signal from a pressure sensor in order to have a voltage great enough to input it to an analog-to-digital converter. In *Chapter 3*, the construction of a pressure sensor was shown in *Figure 3-12*. Such a sensor is connected to an op amp in *Figure 4-20* that is acting as a difference amplifier. It is amplifying the difference voltage $V_2 - V_1$ identified as V_{IN} in *Figure 4-20*. There are specific requirements for the op amp to be a difference amplifier. The ratio R_3/R_2 must be equal to the ratio R_f/R_1. R_f/R_1 determines the differential gain. With R_f/R_1 equal to 20 for the circuit, R_3/R_2 must be 20. R_2 is 2 kΩ; therefore, R_3 is 40 kΩ. The reason that R_2 is 2 kΩ is to have the input impedance to the sensing circuit, which is $R_1 + R_2$, as high as possible to keep from loading the pressure gauge bridge and causing inaccuracies. R_1 tends to be small to allow the gain to be high, but this keeps the input impedance low, which would load the circuit. There is a compromise here.

The signal from the sensor is normally in millivolts (mV), and the input to the analog-to-digital converter needs to be 1V or more. If the input voltage is 2 mV and the difference amplifier has a gain of 20, the output voltage will be $V_O = 20 \times 2 \times 10^{-3} = 40$ mV. More amplification is needed. Another op amp with the signal fed to the noninverting input is added with a gain of 41. Its output voltage will be $V_O = 41 \times 40 \times 10^{-3} = 1.64$V. Op amps are manufactured with dual circuits in a package; therefore, only one IC is used. The power supply voltages (plus and minus) are chosen for the convenience of the application. There are many op amps that operate from 3–4 volts to 20 volts.

Common-Mode Rejection

One problem with instrumentation amplifier circuits such as *Figure 4-20* is that the signal output from the sensor is very small, but the noise voltage picked up on the leads connecting the sensor to the amplifier may be 100 to 1000 times greater.

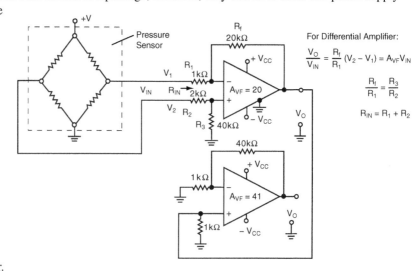

For Differential Amplifier:

$$\frac{V_O}{V_{IN}} = \frac{R_f}{R_1}(V_2 - V_1) = A_{VF}V_{IN}$$

$$\frac{R_f}{R_1} = \frac{R_3}{R_2}$$

$$R_{IN} = R_1 + R_2$$

Figure 4-20: Amplifying a pressure-sensor output

An op amp with a high common-mode rejection must be chosen for such applications. Common-mode rejection means that any signal appearing on both inputs at the same time will not appear at the output. Only the differential signals will be amplified. Common-mode rejection ratio is:

$$CMRR = A_{VD}/A_{CM}$$

where A_{VD} is the differential gain and A_{CM} the common-mode gain. In decibels,

$$CMRR_{dB} = 20\log_{10}A_{VD}/A_{CM}$$

Example 5. Common-Mode Rejection

If an op amp has an $A_{VD} = 100$ dB and a CMRR = 80 dB, what is the common-mode gain?

Solution:

$$
\begin{aligned}
CMRR_{dB} &= 20\log_{10}A_{VD}/A_{CM} \\
80 &= 20\log_{10}A_{VD}/A_{CM} \\
4 &= \log 10 A_{VD}/A_{CM} \\
10^4 &= A_{VD}/A_{CM} \\
A_{CM} &= A_{VD}/10^4
\end{aligned}
$$

$$
\begin{aligned}
A_{VDdB} &= 20\log_{10}V_O/V_{IN} \\
100 &= 20\log_{10}V_O/V_{IN} \\
5 &= \log_{10}V_O/V_{IN} \\
10^5 &= V_O/V_{IN} = A_{VD}
\end{aligned}
$$

therefore, $A_{CM} = 10^5/10^4 = 10$

A More Sophisticated Pressure Sensor Amplifier

Burr-Brown, manufacturers of op amps, in one of their application notes[3], present a much more sophisticated pressure sensor circuit. It is shown in *Figure 4-21*. It is presented here to demonstrate several other uses of op amps that are very important to signal conditioning circuitry, and, in addition, the circuit is designed using only one power supply. The circuit uses four op amps, two in each package. In the circuit, A_3 and A_4 constitute a two-op-amp instrumentation amplifier whose output voltage is:

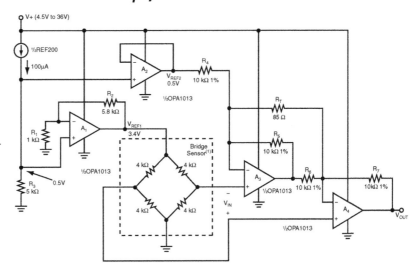

Figure 4-21: A more sophisticated pressure-sensor amplifier

$$V_{OUT} = V_{IN}(2(1 + R/R_T)) + V_{OUT1}$$

V_{OUT1} is the offset voltage at the output when $V_{IN} = 0$, or when there is zero pressure on the sensor. For the circuit shown, because the design required an output from the instrumentation amplifier when the input voltage was zero, V_{OUT1} was set at +0.5V.

---

[3] Burr-Brown, AB-033A Application Note, © 1991, Burr-Brown Corporation.

Voltage Follower

The +0.5V is set using what is called a voltage follower. The op amp A_2 is a non-inverting amplifier with $R_f = 0$ and R_1 equal to infinity; therefore, its gain, A_{VD} is:

$$A_{VD} = 1 + R_f/R_1 = 1$$

and $\quad V_O = V_{IN}$

It is a unity gain amplifier, and its configuration is as shown in *Figure 4-22*. Its output voltage equals its input voltage, thus, the name voltage follower. It has a very high input impedance and a low output impedance so it isolates output from input. It has +0.5V applied to its noninverting input so it pro-duces +0.5V at its output.

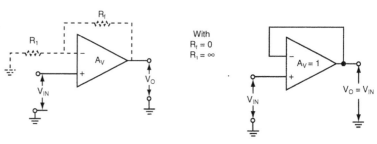

Figure 4-22: Unity gain amplifier or voltage follower

The A_1 op amp is a similar stage but has a finite gain set by $R_f = 5.8$ kΩ and $R_1 = 1$ kΩ, so that its gain is:

$$A_{VD} = 1 + 5.8 = 6.8$$

It also has +0.5V on its noninverting input, so that it produces an output voltage of $6.8 \times 0.5 = 3.4$V. This output voltage is used as a reference voltage for the bridge sensor and will remain very stable. Op amps make excellent voltage followers and produce outputs that are isolated from the inputs because of the high input impedance and low output impedance. They do not have to be unity gain amplifiers.

The component called REF200 in the circuit is one half of a dual constant-current source. It supplies a constant current of 100 μA. A constant-current source can be a high resistance connected to a power supply; however, REF200 is made up of active devices that produces a very accurate, stable source of 100 μA. The 100 μA through a precision 5 kΩ resistor produces a very accurate +0.5V required for the circuit.

The gain of the A_3 and A_4 op amps, as stated previously, is two times the gain of a noninverting amplifier; therefore, with $R_T = 85$ Ω and $R = 10$ kΩ

$$A_{VD} = 2 (1 + 10 \times 10^3)/85 = 2(1 + 118) = 238$$

Current Mirror

One other circuit that is useful for signal conditioning is a current mirror shown in *Figure 4-23*. It can be made using MOS or bipolar transistors, and there are more elaborate circuits than those shown in *Figure 4-23*. In each of these circuits, the output current is equal to a constant factor times the input current. When transistors are made in integrated circuit form and used for current mirrors, they are next to each other, and, therefore, have identical characteristics.

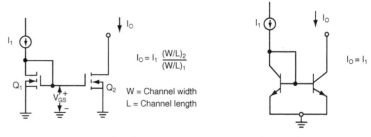

$$I_O = I_1 \frac{(W/L)_2}{(W/L)_1}$$

W = Channel width
L = Channel length

a. MOS circuit $I_O = I_1$ **b. Bipolar circuit**

Figure 4-23: Current mirrors

As a result, the current mirrors produce very accurate outputs vs. inputs. They are used to supply accurate currents in themselves or to establish accurate voltage references by developing a voltage across precision resistors. The output current for the MOS transistors turns out to be the input current times the ratio of the channel width, W, to the channel length, L, of the MOS transistors used in the design.

Example 6. W/L Ratio of Current Mirrors

In *Figure 4-23a*, if $I_1 = 50 \mu A$ and $(W/L)_2 = 4$ and $(W/L)_1 = 2$, what is the value of I_0 for the current mirror?

Solution:

$I_0 = I_1 (W/L)_2/(W/L)_1$ where W is channel width and L channel length

$I_0 = 50 \mu A \times 4/2 = 100 \mu A$ for the MOS transistors used in the design.

Applications of Op Amps

Thermocouple Amplifiers

Figure 4-24a shows an amplifier for use with a thermocouple. The thermocouple is connected to the non-inverting input. A 1 MΩ variable resistor is used for R_f so that the gain and sensitivity can be adjusted to the requirements of the application. Short lead lengths and bypass capacitors on the power supply leads may be required at the highest gain settings.

Solar Cell or Photodiode Amplifier

Figure 4-24b is an amplifier for use with a photodiode or a solar cell. Again the gain is adjustable to fit the application. Short lead lengths and component placement are important in both of these applications to make sure the circuit does not oscillate.

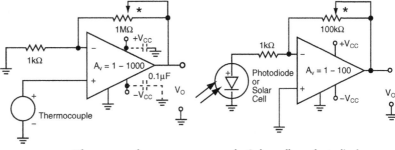

a. Thermocouple   b. Solar cell or photodiode

Thermistor Amplifier

Figure 4-24c is an amplifier to use with a thermistor temperature sensor. Beside a gain adjustment, there is a level adjustment to set the output at a particular temperature. 25°C might be the temperature for many applications.

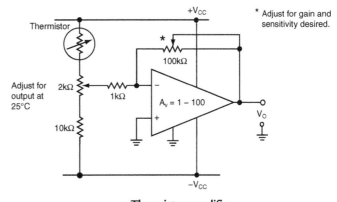

c. Thermistor amplifier

Figure 4-24: Sensor amplifiers using op amps

Oscillators

Recall that with an op amp, as shown in *Figure 4-25b*, when a signal, v_i, is applied to the noninverting input that the output, v_o, is in phase with the input signal. If now the output is fed back to the noninverting input through a feedback circuit, as shown by the dotted box, the output signal reinforces the input and the circuit is turned into an oscillator. An oscillator is a circuit that outputs a continuous signal, usually at a constant frequency, with $v_i = 0$. In other words, the oscillator puts out a continuous signal without having an input. In *Figure 4-25*, β is the gain of the feedback network shown in dotted lines. It either increases or decreases v_O as it feeds back the signal from output to input; therefore,

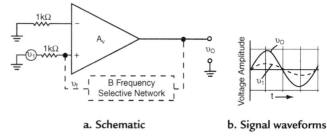

a. Schematic   b. Signal waveforms

Figure 4-25: An oscillator

$$v_f = \beta v_O$$

Since $v_i = 0$, v_f when amplified by A_v produces v_O, and

$$v_O = A_v \beta v_O$$

or, cancelling out v_O,

$$A_v \beta = 1$$

The oscillator will maintain an output signal without an input signal when the loop gain is equal to or greater than 1 and will oscillate at a frequency where the phase from input back to input is 360°.

It so happens, when many amplifiers are built, not enough precaution is taken with component layout so that part of the output signal is fed back to the input. The circuit oscillates at a sine-wave frequency where the loop gain is greater than 1 and the phase shift through the feedback loop is 180° or greater. This assumes that the phase shift through the amplifier stage is 180°. There are many types of oscillators that output waveforms other than sine-waves—square-wave, triangular waves, pulses of constant width, or pulses that vary in width, but the principle is the same—the phase shift from input to output to input is 360°, and the gain is greater than 1 through the feedback loop.

Power Amplifiers

Class A Amplifiers

The small-signal amplifiers that have been discussed are Class A amplifiers—there is current in the transistors, as discussed in *Figure 4-9*, for 360° of the signal cycle. When Class A amplifiers are used for power amplifiers, where high current and high voltage swings are required at the same time, there is a large amount of power dissipated in the amplifier stage itself rather than being delivered to the load. The efficiency of power transfer to the load is low. The maximum efficiency is 25%, but in practical applications it usually is only 10% to 20%. As a result, Class AB, Class B and Class C amplifiers are used for power amplifiers. Because current in Class C amplifiers flows only for a small portion of an input signal cycle, current pulses are applied to a tuned circuit load resonant at a particular frequency, or over a narrow band of frequencies. Thus, Class C amplifiers are used for frequency-selective power amplifiers, and most applications of power amplifier circuits in this book will use Class AB or Class B amplifiers.

Class B

Figure 4-26a is the circuit for a simple complementary bipolar transistor power amplifier. Q_1 is a NPN power transistor whose collector is connected to $+V_{CC}$. Q_2 is a PNP power transistor whose collector is connected to $-V_{CC}$. The bases of the two transistors are connected together to V_i, the input voltage. When $V_i = 0$, both transistors are off and there is no current through R_L, the load, so $V_O = 0$. As V_i increases positively, when V_{BEQ1} is exceeded, Q_1 conducts and becomes an emitter-follower with a voltage gain of 1.

$V_O = V_i - V_{BEQ1}$ results as shown in the transfer characteristics of *Figure 4-26b*. As V_i returns to zero and increases negatively, when V_{BEQ2} is exceeded, Q_2 conducts and it becomes an emitter-follower. $V_O = -(V_i - V_{BEQ2})$ as shown in the transfer characteristics of *Figure 4-26b*.

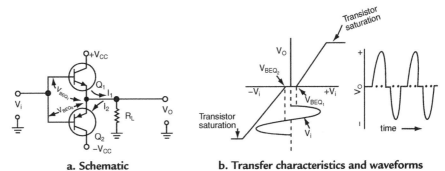

a. Schematic

b. Transfer characteristics and waveforms

Figure 4-26: Simple complementary bipolar transistor Class B power amplifier

Q_2 is cutoff while Q_1 conducts; Q_1 is cutoff while Q_2 conducts. There is no voltage gain but significant current gain to provide the power amplification. Each transistor is biased Class B with conduction only over 180° of the input signal cycle. Note, however, as shown in *Figure 4-26b*, that there is quite a bit of distortion, called crossover distortion, especially when the signal amplitude is small.

Figure 4-27 shows a Class B complementary transistor power amplifier that has the crossover distortion eliminated. It is eliminated because the transistors are operating in Class AB where each has a small quiescent current through it. I_1 is the current through Q_1 and I_2 is the current through Q_2 when $V_i = 0$. A biasing resistor, R_1, supplies a bias current to the diodes D_1 and D_2 to provide a two-diode constant voltage between the bases of Q_1 and Q_2. Q_1 and Q_2 are matched to have the same V_{BE}. As a result, $I_1 = I_2$.

When the diode voltage drop is matched to the V_{BE} of Q_1 and Q_2, point A in the circuit of *Figure 4-27a* will be at V_{BE}, point B will be at $2V_{BE}$ and point C will be at the same voltage as point A, V_{BE}. *Figure 4-27b* shows the transfer characteristics. When $V_i = 0$, $V_O = V_{BE}$. The transfer characteristics are displaced by V_{BE} and there is no crossover distortion.

The circuit operates as follows: V_i increases positively to cause point B to increase to $2V_{BE} + V_i$ and for V_O to increase by V_i. Since the base of Q_2 is at V_i and its emitter is at $V_O = V_{BE} + V_i$, Q_2 does not change in emitter-base voltage so I_2 remains constant, but the current from Q_1 flows through R_L to produce V_O across R_L as long as the input voltage is positive. When V_i increases negatively it causes Q_2 to conduct I_2 through R_L. At the same time, point B is pulled to $2V_{BE} - V_i$ and since the output is at $V_{BE} - V_i$, the V_{BE} of Q_1 remains constant and I_1 remains constant. The output follows V_i due to I_2.

The output voltage cannot go any higher than about $+V_{CC} - 2V$ for the positive swing, or $-V_{CC} + 2V$ for the negative swing otherwise the transistors will go into saturation and the output signal will be distorted. Thus, the maximum load current, I_L, drawn by

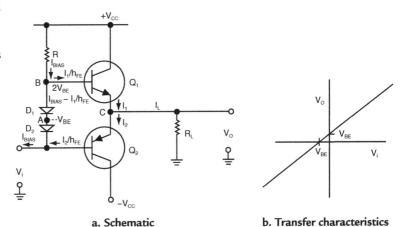

a. Schematic

b. Transfer characteristics

Figure 4-27: Class AB power amplifier

either transistor will be $(V_{CC} - 2V)/R_L$, and it determines the power output of the amplifier. I_L divided by h_{FE} determines the maximum base current that will need to be supplied to the bases when the respective transistor is driving the output. I_{BIAS} of *Figure 4-27a* must always be larger than the maximum base current required to drive the load in order to have a proper design; therefore, using the maximum base current, R can be determined. Where the efficiency of the Class A amplifier is 10%–20%, Class B amplifiers can have efficiencies as high as 78.5%. Usually they average about 60%.

Class B Audio Power Amplifier

A very successful Class B power amplifier used for signals with frequencies in the audio range is shown in *Figure 4-28*. Q_1 and Q_2 are biased just into conduction with a constant current supplied by R_1 from $+V_{CC}$. This design condition eliminates crossover distortion. The circuit operates as follows: V_i is applied to the primary of the input transformer, T_1. It has a center-tapped secondary each side of which feeds a base of Q_1 or Q_2. The positive-going alternation of the input signal produces

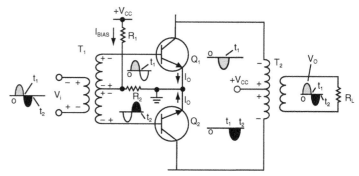

Figure 4-28: Class B transformer coupled audio power amplifier

a positive-going signal on the base of Q_1. The resulting base current produces amplified collector current from Q_1 in the upper half of the primary of the output transformer, T_2. A positive-going output signal appears across the load resistor, R_L, connected across the secondary of the output transformer.

The negative-going alternation of V_i, through the secondary connection of T_1, produces a positive-going signal on the base of Q_2. Amplified Q_2 collector current is produced in the lower half of the primary of T_2. A negative-going output signal appears across R_L. Q_1 conducts the power transferring collector current on the positive alternation of V_i, and Q_2 conducts similar power transferring collector current on the negative alternation of V_i. Through transformer action, a positive-going signal appears across R_L when Q_1 conducts, and a negative-going signal when Q_2 conducts. Both transistors operate in Class B because they only conduct for 180° of the input signal. This power amplifier does not amplify DC. It must have time-varying signals because of the transformer action.

Q_1 and Q_2 should be matched transistors so that there is the same quiescent current, I_Q, through each transistor, and so that they have the same h_{FE}. I_Q should be less than 1mA, but with $I_Q = 1$ mA, and a minimum $h_{FE} = 40$,

$$I_{BIAS} = 2I_Q/h_{FE} = 2 \text{ mA}/40 = 0.5 \times 10^{-4} = 50 \text{ μA}$$

With I_{BIAS}, R_1 can be calculated for a given V_{CC}. The collector-to-emitter breakdown voltage must be greater than $2V_{CC}$, and the turns ratio, N_S/N_P, must be chosen to match the impedance of the load, R_L, to the output load required on the transistors, Q_1 and Q_2, at the power level desired.

Special Signals

Square Waves from Sine Waves

There are three other special signal-conditioning circuits that are important. The first is a circuit that produces pulses from a time-varying input signal. As shown in *Figure 4-29a*, an easy way of producing square waves from sine waves is to use diodes, either standard diodes or zener diodes. The amplitude of the square waves is determined by the diode drop. The higher the amplitude of the input signal the squarer the output waveform.

The use of a comparator is shown in *Figure 4-29b*. A comparator is essentially an op amp without any feedback. Since op amps have very high open-loop gain, any small signal at either input will drive the output into saturation. A positive signal drives the output to $+V_{CC} - V_{SAT}$, and a negative signal drives the output to $-V_{CC} + V_{SAT}$. If a reference voltage, V_R is placed on one input, the switching point will move to V_R. Control of the pulse width can be provided by the variation of V_R. The repetition of the square waves will be at the frequency of the input.

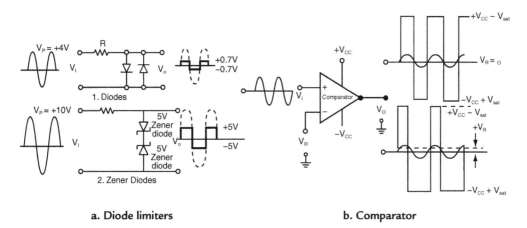

a. Diode limiters   b. Comparator

Figure 4-29: Square waves from sine waves

Trigger Pulses from Square Waves

In many applications there is a real need to provide accurate trigger pulses at particular times. One of the simplest circuits for producing sharp timing pulses is the differentiating circuit formed with

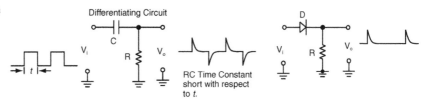

Figure 4-30: Producing timing pulses from square pulses

a capacitor and resistor as shown in *Figure 4-30*. A pulse with a fast rising leading edge passes through C and appears across R as a sharp rising pulse. The time constant of RC is short compared to the pulse width t. C rapidly charges and V_O goes to zero while the maximum amplitude of the input pulse continues. When the V_i pulse returns to zero, the capacitor again transmits the fast falling trailing edge of the pulse across R, producing a negative sharp falling pulse. Out of the rectangular pulse of V_i both a positive and a negative timing pulse is obtained. *Figure 4-30* shows how the positive pulse is recovered. If the diode, D, is reversed, the negative pulse is recovered.

In the circuit of *Figure 4-31*, the R and C are reversed. Now the circuit is called an integrating circuit. When V_i is a series of pulses, an integrating circuit can produce a DC voltage from the pulses. The resistor now is in series with the input signal and the capacitor across it. The RC time constant in this case is very large compared to t. As the fast rising leading edge of the V_i pulse is applied, since the RC time constant is large, C changes slowly and doesn't reach full charge until the trailing edge of the input pulse. When the trailing edge of the pulse appears, C tries to discharge, but because of the large RC, C discharges very little.

The charge and discharge repeats itself and a waveform with a ripple appears as V_O. The V_i pulses have been converted to a DC output voltage.

Op amps can be connected as integrators and differentiators but are much more complicated than the simple RC circuits. They are used in very sophisticated circuits when needed.

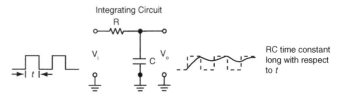

Figure 4-31: Integrating pulses into DC signals

RC Time Constants

The concept of a time constant used in the differentiating and integrating circuits is shown in *Figure 4-32*. When a time varying voltage, in this case a pulse with a sharp rising leading and trailing edge, is applied to a capacitor through a resistor, the charge on the capacitor can only change as rapidly as the current through will allow. For example, for the integration schematic, the voltage across the capacitor cannot change as fast as the input voltage pulse. It is restricted because the resistor limits the current. The charge on the capacitor builds up at a predictable rate as shown by the curve A in *Figure 4-32b*. Curve A is the voltage, V_C, across the capacitor plotted against τ. τ is called the time constant and is equal to the ohms resistance in the circuit times the farads of capacitance in the circuit. As shown by curve A, after one time constant, the voltage across the capacitor has obtained a value 63.2% of its final value. It takes at least five times the time constant for the capacitor to be charged to full value. The value of voltage on the capacitor in a given time can be estimated by knowing the relationship of the time constant to the pulse width t. For integration, τ should be at least five times t. For differentiation, where the capacitor must charge rapidly with respect to t, τ should be only one-fifth t, or even smaller.

On the trailing edge of the pulse applied to the integration circuit, the capacitor is trying to discharge. Its voltage can change only as rapidly as the time constant will allow. Its discharge is described by curve B in *Figure 4-32b*. In one time constant, the voltage has reduced to 36.8% of its value when the trailing edge of the input pulse changed. It takes at least five time constants for the capacitor to completely discharge. Applying the curve A

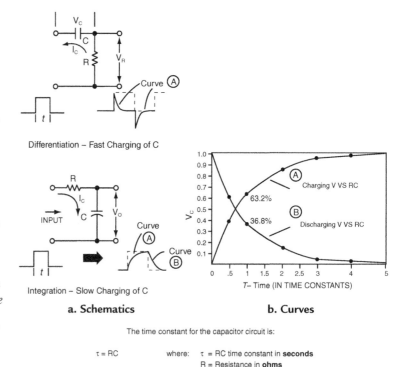

Differentiation – Fast Charging of C

Integration – Slow Charging of C

a. Schematics

b. Curves

The time constant for the capacitor circuit is:

$$\tau = RC$$

where: τ = RC time constant in **seconds**
R = Resistance in **ohms**
C = Capacitance in **farads**

Figure 4-32: RC time constants
Courtesy of Master Publishing, Inc.

to the voltage across the capacitor when the time constant is such that the capacitor charges rapidly, the differentiator circuit of *Figure 4-30* results. Here the time constant should be such that the capacitor fully charges in one-fifth of the pulse width. Thus, 5τ should be equal to one-fifth the pulse width to allow the capacitor to charge fully.

By applying curve A and B to a variety of circuits, it is possible to visualize many different signal shapes dependent upon the relationship of the time constant to the time of pulses or time varying signal edges applied.

Example 7. RC Time Constant

a. Using *Figure 4-30*, for differentiation, 5τ must be no greater than one-fifth the pulse width t so that C charges rapidly. If $t = 10\ \mu S$, what is the value of R if $C = 0.001\ \mu F$? Remember $\tau = RC$.

b. Using *Figure 4-31*, for integration, it should take at least 5τ to charge the capacitor in the pulse width t. C charges very slowly. Using the same values of t, and C, what is the value of R? Remember $\tau = RC$.

Solutions:

a.
$$5\tau = t/5$$
$$25\tau = t$$
$$25RC = t$$
$$R = t/25C$$
$$R = 10 \times 10^{-6}/25 \times 1 \times 10^{-9}$$
$$R = 0.4 \times 10^3 = 400\ \Omega$$

b.
$$5\tau = t$$
$$5RC = 10 \times 10^{-6}$$
$$R = 10 \times 10^{-6}/5C$$
$$R = 10 \times 10^{-6}/5 \times 1 \times 10^{-9}$$
$$R = 2 \times 10^3$$
$$R = 2,000\ \Omega$$

Frequency Selection

There is signal conditioning that really doesn't amplify or change the waveform of the AC signal, but rather it is a frequency selection process. Many analog signals contain a composite of signals with different frequencies. In fact as discussed in *Basic Communications Electronics*,[4] signals of different frequencies are mixed together to form a resultant signal that is the sum of the two frequencies or the difference of the two frequencies. Each of these signals can be processed separately to communicate the information in the original signals. As a result, there is a need to select out the desired signal by its frequency from companion signals of different frequencies. Frequency selection signal conditioning is required for this.

Band-Pass Filters

Look at *Figure 4-33a*. It is the familiar frequency response curve that was discussed for amplifiers. The mid-band gain of the curve is A, and that gain is maintained out to a frequency of f_H, the high-frequency cutoff point, and down to a frequency of f_L, the low-frequency cutoff point. At f_H and f_L the gain is reduced to 0.707A. Since at the cutoff points the gain is down -3 dB from the mid-band gain, they are called the "minus 3 dB" points.

A signal with a frequency much higher than f_H, as the curve shows, will be attenuated, that is, it will have a much lower amplitude; the higher the frequency the larger the attenuation. Similarly, signal frequencies below f_L will be attenuated in amplitude; the lower the frequency the larger the attenuation. Frequencies from f_L to f_H will have constant gain and not be attenuated. A circuit that has such a frequency response curve can be considered a *band-pass* filter. It passes signals with frequencies in the band from f_L to f_H and attenuates the others.

It can be considered as a circuit that selects signals with frequencies within the band and rejects signals with frequencies outside the band.

---

[4] *Basic Communications Electronics*, J.W. Hudson, G. Luecke, ©1999, Master Publishing, Inc., Lincolnwood, IL.

Low-Pass Filter

Look at *Figure 4-33b*. Here the frequency response curve extends down to zero frequency—DC. The frequency response of a DC amplifier was shown previously. Its frequency response curve looks just like this. As the signal frequency is increased from zero, a frequency f_C, the cutoff frequency, is reached where the signal begins to be attenuated; further increases in signal frequency increases the attenuation. Circuits that have such a signal frequency response are called *low-pass* filters. Signal with frequencies below f_C pass without attenuation; signals with frequencies above f_C are attenuated.

High-Pass Filter

Circuits with a frequency response as shown in *Figure 4-33c* are called *high-pass* filters. Signals with frequencies below f_C, the cutoff frequency, are "not passed" (attenuated), while signals with frequencies above f_C are "passed" without attenuation.

The filters of *Figure 4-33a,b,c* can have amplification built in or they may be circuits built with passive components that have no amplification. When they are used in combination with other signal conditioning circuits, they can select signal frequencies that cover a band of frequencies above f_L and below f_H, or they can pass only signal frequencies below a frequency f_C, or they can pass only signal frequencies above a frequency f_C. The rate at which the circuits attenuate is determined by the design of the filter.

a. Band-pass filter

b. Low-pass filter

c. High-pass filter

Figure 4-33: Filters are used for frequency selection

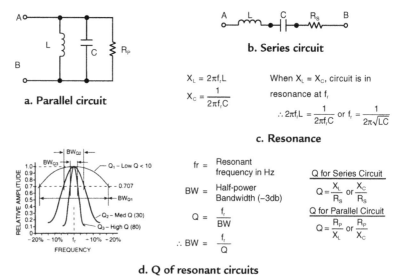

a. Parallel circuit

b. Series circuit

$$X_L = 2\pi f_r L$$
$$X_C = \frac{1}{2\pi f_r C}$$

When $X_L = X_C$, circuit is in resonance at f_r

$$\therefore 2\pi f_r L = \frac{1}{2\pi f_r C} \text{ or } f_r = \frac{1}{2\pi\sqrt{LC}}$$

c. Resonance

$f_r = $ Resonant frequency in Hz

$BW = $ Half-power Bandwidth (−3db)

$$Q = \frac{f_r}{BW}$$

$$\therefore BW = \frac{f_r}{Q}$$

Q for Series Circuit

$$Q = \frac{X_L}{R_s} \text{ or } \frac{X_C}{R_s}$$

Q for Parallel Circuit

$$Q = \frac{R_P}{X_L} \text{ or } \frac{R_P}{X_C}$$

d. Q of resonant circuits

Figure 4-34: Tuned-circuit filters

Courtesy of Master Publishing, Inc.

Tuned-Circuit Filters

A special kind of band-pass filter used extensively in communications circuits is called a tuned-circuit filter. It is used to select out a narrow band of signal frequencies. The tuned-circuit filter is formed by a combination of an inductance, L, a capacitance, C, and a resistance, R. It is designed at a particular frequency, f_r, called the resonant frequency. The resonant frequency is special because at f_r the inductive reactance ($X_L = 2\pi fL$) is equal to the capacitive reactance ($X_C = 1/2\pi fC$). When these reactances are equal, as shown in *Figure 4-34*, the resonant frequency is given by:

$$f_r = 1/2\pi\sqrt{LC} \text{ where L is in henries and C is in farads}$$

The band pass characteristics are shown in *Figure 4-34d*.

Bandwidth and Q

The frequency response of the band-pass filter is plotted in *Figure 4-34d* similar to the frequency response of *Figure 4-33*, except the frequency separation between f_L and f_H is very small, within a 20% variation on each side of f_r. The 0.707 attenuation points for f_L and f_H are shown. The frequency band between f_L and f_H is called the *bandwidth*, BW. Bandwidth is defined as the half-power bandwidth—the frequency band between the –3 dB attenuation points on the response curve.

A means of describing the narrowness of the band-pass filter response is to use a quality factor called Q. Q = f_r/BW and is shown in *Figure 4-34d*. Note that in the equation, if the bandwidth is very narrow around the resonant frequency f_r, Q will be large. The wider the bandwidth, the smaller Q will be. *Figure 4-34d* shows how the response curve varies as Q varies. If the Q of the band-pass filter is known, and f_r is known then the bandwidth can be calculated using BW = f_r/Q.

Either Parallel or Series Resonant Circuits

A resonant circuit can be either a parallel resonant circuit or a series resonant circuit. Both are shown in *Figure 4-34*. *Figure 4-34a* is the parallel circuit; *Figure 4-34b* is the series circuit. The equation for f_r is the same but calculating BW is different. Here are the equations for Q:

Series Circuit	Parallel Circuit
$Q = X_L/R_S$ or $Q = X_C/R_S$	$Q = R_P/X_L$ or $Q = R_P/X_C$

R_S and R_P are as shown in *Figures 4-34a,b* respectively. If the circuit values for L, C, and R are known, fr, Q, and BW can be calculated. Or if the bandwidth that is desired is known around a frequency f_r, then Q, and the $\sqrt{LC}$ can be calculated. Then L or C can be can be calculated after either one is chosen. Following L and C, R_P and/or R_S can be calculated.

Typical Application of Filters

The typical application of filters is shown in *Figure 4-35*. A broad bandwidth signal is amplified to increase its amplitude. Then a selection of particular signal frequencies is accomplished by passing the signal through a frequency selection filter. Some circuits will require a low-pass filter, others a band-pass filter, others a high-pass filter, and yet others a tuned-circuit filter.

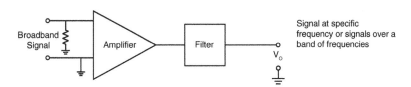

Figure 4-35: Use of filters

Figure 4-36 is a tuned-circuit band-pass filter amplifier. It uses an N-channel depletion mode JFET as the active device. The tuned circuit in the drain results in the amplification of only a narrow band of frequencies. The f_r and Q of the circuit can be adjusted to a wide range of frequencies and bandwidth limited, of course, by the frequency response of the JFET itself. The circuit is a combination of a tuned circuit to get the band pass desired and an amplifier to increase the amplitude of the signal.

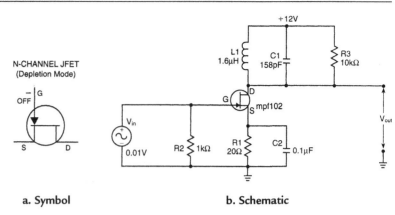

a. Symbol

b. Schematic

Figure 4-36: RF-tuned amplifier
Courtesy of Master Publishing, Inc.

Example 8. Bandwidth and Q

If the tuned circuit of *Figure 4-36* has $X_L = 100\ \Omega$, $R_P = 10,000\ \Omega$ and $f_r = 10$ MHz, what is the Q of the circuit and the bandwidth?

Solution:

$Q = R_P/X_L = 10000/100 = 100$

$BW = f_r/Q = 10 \times 10^6/1 \times 10^2 = 100$ kHz

The resonant frequency of the circuit is 10 MHz. At resonance the load is R3 = 10 kΩ in parallel with QX_L. The Q of the circuit is 100, the bandwidth is 100 kHz and the voltage gain is 30 dB.

Summary

Signals from sensors need signal conditioning. The prime signal conditioning is amplification. In this chapter, individual bipolar and MOS transistor amplifiers have been explained, followed by op amps and power amplifiers. Several special signal-conditioning circuits conclude the chapter. In the next chapter, analog-to-digital and digital-to-analog converters will be discussed.

Chapter 4 Quiz

1. Signal conditioning:
    a. leaves the signal unchanged.
    b. means to modify the signal to adjust it to the application.
    c. does not include amplification.
    d. doesn't occur in the A-to-D or D-to-A chain of functions.
2. Amplification:
    a. is a signal conditioning function.
    b. is performed by bipolar transistor circuits.
    c. is performed by field-effect transistor circuits.
    d. a, b, c above.
    e. none of above.
    f. a only above.

3. The amplifier circuit of *Figure 4-4:*
   a. has no linear operating range.
   b. has a linear operating range of I_B from 0.02 mA to 0.1 mA.
   c. has a linear operating range of V_{CE} from 2 to 10 volts.
   d. b and c above.
   e. none of the above.

4. In a common-emitter bipolar transistor amplifier:
   a. the base-emitter junction is forward biased and the collector-base junction is forward biased.
   b. the base-emitter junction is reverse-biased and the collector-base junction is forward-biased.
   c. the base-emitter junction is reverse-biased and the collector-base junction is reverse-biased.
   d. the base-emitter junction is forward-biased and the collector-base junction is reverse-biased.

5. In a common-emitter bipolar transistor amplifier:
   a. the forward-biased base-to-emitter voltage is approximately 0.7V.
   b. the forward-biased base-to-emitter voltage is greater than 10V.
   c. the reverse-biased collector-to-base junction is approximately 0.7V.
   d. the collector is tied to ground.

6. Common ways of biasing a bipolar transistor amplifier circuit are:
   a. fixed-current I_B bias.
   b. voltage-divider bias.
   c. collector-feedback bias.
   d. a, b, c above.
   e. none of the above.
   f. b only above.

7. The voltage gain, A_v, of a common-emitter bipolar transistor amplifier is, where R_L = total load resistance and I_E is DC emitter current in mA, and I_C is DC collector current in mA:
   a. $R_L \times I_E$.
   b. $R_L \times I_E \times I_C$.
   c. $R_L I_E$ divided by 0.026.
   d. $V_{BE} + V_{CE}$.

8. The voltage gain, A_V, expressed in dB is:
   a. $A_V = 20 \text{ dB} \times A_V$.
   b. $A_V = 20\log_{10}A_V$.
   c. $A_V = 20\log_{10}I_C$.
   d. $A_V = 10\log_{10}A_V$.

9. The capability of an amplifier to handle signals over a frequency range is:
   a. called its frequency response.
   b. called its amplitude or gain.
   c. called its single-frequency gain.
   d. called its linearity.

10. Cascaded amplifiers:
    a. are amplifiers coupled together to increase the overall gain.
    b. can use different means of coupling between stages.
    c. the dB gain can be added to arrive at overall gain.
    d. b only above.
    e. a, b, c above.
    f. none of above.

11. With AC coupling between stages of a cascaded amplifier:
    a. the frequency response does not go down to zero frequency.
    b. the frequency response goes down to zero frequency.
    c. the capacitance coupling doubles the high-frequency response.
    d. the inductance coupling limits the low-frequency response.
12. The dynamic range of an amplifier is:
    a. the signal range that is twice where distortion begins.
    b. the signal range that extends beyond distortion.
    c. the signal range from small-signal to where distortion begins.
    d. the signal range that is 10 dB below the distortion point.
13. A class B amplifier:
    a. operates only for 30° of the input signal cycle.
    b. operates only for 180° of the input signal cycle.
    c. operates only for 10° of the input signal cycle.
    d. operates for 360° of the input signal cycle.
14. For field-effect transistor amplifiers:
    a. a change in voltage from gate to source causes a change in voltage from source to source.
    b. a change in current from gate to source causes a change in voltage from drain to source.
    c. a change in voltage from gate to source causes a change in current from drain to source.
    d. a change in current from gate to source causes a change in current from drain to source.
15. Transconductance for FETs is defined as:
    a. a change in a current as a result of a change in a voltage.
    b. a change in a voltage as a result of a change in a current.
    c. a change in a voltage as a result of a change in a voltage.
    d. none of the above.
16. Field-effect transistors operate:
    a. in the enhancement mode and depletion mode at the same time.
    b. in the enhancement mode.
    c. in the depletion mode.
    d. b and c above.
    e. a only above.
    f. none of above.
17. Ideal operational amplifiers are amplifiers with:
    a. zero Z_{IN}, infinite gain, zero Z_O, infinite bandwidth and zero offset.
    b. infinite Z_{IN}, infinite gain, zero Z_O, infinite bandwidth and zero offset.
    c. infinite Z_{IN}, zero gain, zero Z_O, infinite bandwidth and zero offset.
    d. infinite Z_{IN}, infinite gain, infinite Z_O, zero bandwidth, and zero offset.
18. The frequency response of a general-purpose op amp:
    a. increases as the overall gain is reduced to one.
    b. stays the same if overall gain is reduced.
    c. reduces as the overall gain reduces.
    d. none of the above.
19. When $R_f = 0$ and $R_1 = $ infinity, an op amp becomes:
    a. an amplifier with gain equal to infinity.
    b. an amplifier whose output voltage equals its input voltage.
    c. a unity-gain amplifier.

    d.   a only above.

    e.   b and c above.

    f.   none of above.

20. An oscillator maintains an output signal without an external input signal:

    a.   when its internal gain is at least 1 and the phase of the output feedback to its input equals 360°.

    b.   when its internal gain is 0 and the phase of the output feedback to its input equals 360°.

    c.   when its internal gain is at least 1 and the phase of the output feedback to its input equals 180°.

    d.   when its internal gain is at least 1 and the phase of the output feedback to its input equals 90°.

21. A Class B complementary bipolar transistor amplifier:

    a.   has no crossover distortion.

    b.   has crossover distortion.

    c.   cannot have a circuit that eliminates crossover distortion.

    d.   none of the above.

22. Diodes are very important:

    a.   in forming special signal shapes or timing signals.

    b.   in circuits conducting current in both directions.

    c.   because they have the same characteristics in the forward and reverse direction.

    d.   c only above.

    e.   none of the above.

23. RC time constants are:

    a.   combinations of inductance and capacitance in circuits.

    b.   very important in integrating and differentiating circuits.

    c.   combinations of inductance and resistance in circuits.

    d.   not used extensively in electronic circuits.

24. A band-pass filter has:

    a.   a low-frequency and high-frequency cutoff point.

    b.   only a low-frequency cutoff point.

    c.   only a high-frequency cutoff point.

    d.   frequency response down to zero frequency.

25. The Q of a band-pass filter is:

    a.   equal to X_L/R_s or X_C/R_s for a series resonant circuit.

    b.   equal to R_p/X_L or R_p/X_C for a parallel resonant circuit.

    c.   equal to f_r/BW.

    d.   all of the above.

    e.   none of the above.

    f.   c only above.

Answers: 1.b, 2.d, 3.d, 4.d, 5.a, 6.d, 7.c, 8.b, 9.a, 10.e, 11.a, 12.c, 13.b, 14.c, 15.a, 16.d, 17.b, 18.a, 19.e, 20.a, 21.b, 22.a, 23.b, 24.a, 25.d.

Analog-to-Digital and Digital-to-Analog Conversions

Introduction

As this chapter begins to develop an understanding of converting an input analog signal to digital codes or converting digital codes to analog signals, let's look again at the binary numbering system as illustrated in *Figure 5-1a*. This same illustration was shown in *Figure 1-5*. It is repeated here to emphasize again the digit position weighted value in a binary numbering system. Recall that the binary number is made up of **bi**nary dig**its** (**bits**) in each digit position of the binary number. Each bit can only have two values, 0 or 1. Each digit position has a weighted value that is the binary digit value of 1 or 0 multiplied by the weighted value of the digit position. If the bit is a 1 the digit position has the weighted position value; if the bit is a 0, the weighted position value is 0. The total value of the binary number is the sum of all the weighted position values.

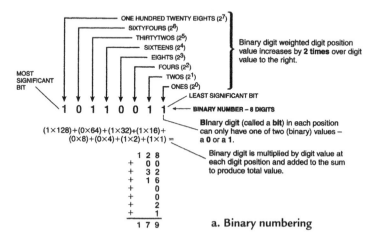

a. Binary numbering

Digit Position	Digit Position Value 2^{n-1}		Binary Number	Decimal Equivalent
1	$2^0 =$	1	$\times$ 1	1
2	$2^1 =$	2	$\times$ 1	2
3	$2^2 =$	4	$\times$ 0	0
4	$2^3 =$	8	$\times$ 0	0
5	$2^4 =$	16	$\times$ 1	16
6	$2^5 =$	32	$\times$ 1	32
7	$2^6 =$	64	$\times$ 0	0
8	$2^7 =$	128	$\times$ 1	128

b. Equivalent decimal

Figure 5-1: Binary number and equivalent decimal

As shown in *Figure 5-1a*, the binary digit weighted digit position value increases by 2 times over the digit value to the right. This is very important to the design of digital-to-analog converters (DACs) and analog-to-digital converters (ADCs).

For example, the most significant bit (MSB) of the 8-bit binary number shown in *Figure 5-1a* has a weighted digit position value of 128. This is one-half the total value of 256 of the 8-bit binary number. Note also that the weighted digit position value for the next digit to the right (the 7th bit) is 64, or one-half the MSB value. This reduction by one-half in weighted digit position value as the bit position is moved to the right continues down to the least significant bit (LSB). The design of DACs and ADCs is based on testing the value of the input quantity to see if it is greater than the MSB value; if it is, is it greater than the MSB value *plus* the weighted digit position value of the next bit to the right? If it is, is it greater than a total of the previous bit values plus the weighted digit position value of the next bit to the right? The process continues until the input is less than the sum of the weighted values. Then the last digit weighted position value is not added but made equal to zero and a weighted position value of a next bit to the right is added and the total tested again. This process continues until the value is determined or the LSB's value is included which indicates that the evaluation is complete. In other words, as the input value is tested, the digit values are added or they are set at zero as the digit positions from MSB to LSB are evaluated, and when the LSB is reached it is the end of the evaluation.

Decimal Equivalent of a Binary Number

It is important to the A-to-D and D-to-A process to know the decimal equivalent of a binary number. *Figure 5-1b* summarizes the evaluation process. It shows how the binary digit weighted position value is multiplied by the bit value at each bit position and the total of all bit values summed to arrive at the decimal value.

Example 1. **Converting a Decimal Number to Binary**

Convert the number 4311 to a binary number.

Solution: Binary Number: **1000011010111**

		Check:		
4311/2	= 2155 with a remainder of **1**	**1 × 4096**	**=**	**4096**
2155/2	= 1077 with a remainder of **1**	**0 × 2048**	**=**	**0**
1077/2	= 538 with a remainder of **1**	**0 × 1024**	**=**	**0**
538/2	= 269 with a remainder of **0**	**0 × 512**	**=**	**0**
269/2	= 134 with a remainder of **1**	**0 × 256**	**=**	**0**
134/2	= 67 with a remainder of **0**	**1 × 128**	**=**	**128**
67/2	= 33 with a remainder of **1**	**1 × 64**	**=**	**64**
33/2	= 16 with a remainder of **1**	**0 × 32**	**=**	**0**
16/2	= 8 with a remainder of **0**	**1 × 16**	**=**	**16**
8/2	= 4 with a remainder of **0**	**0 × 8**	**=**	**0**
4/2	= 2 with a remainder of **0**	**1 × 4**	**=**	**4**
2/2	= 1 with a remainder of **0**	**1 × 2**	**=**	**2**
½	= 0 with a remainder of **1**	**1 × 1**	**=**	**1**
				4311

Digital Codes of ADC

The discussion of ADCs and DACs starts by examining the codes generated by an ADC as a result of an analog input signal. *Figure 5-2b* shows the digital codes generated by a 4-bit analog-to-digital converter which has 16 codes of four bits each that are generated as an analog signal increases from 0 to 15/16 of full scale. As the signal increases 1/16 of full scale, the code changes by a digital bit. As the analog signal, shown in *Figure 5-2a*, varies in amplitude with time, the digital code generated by the ADC changes to represent the amplitude of the analog signal at the time the signal was sampled. This is demonstrated by superimposing the analog signal of *Figure 5-2a*, onto the ADC transfer curve shown in *Figure 5-2b*. The points of sampling are shown and numbered from 1 through 16, and correspond to the sampling points versus time shown in *Figure 5-2a*.

The digital codes generated at each sampling point in *Figure 5-2b* are listed in *Figure 5-2c*. The digital code generated at a particular sample is the code *nearest the amplitude just exceeded by the signal* but not large enough to generate the next code step. These codes from the sampling points appear in sequence at the output of the ADC to describe the analog signal. Depending on the ADC, the digital codes may be presented a bit at a time in series, or all bits together in parallel at specific times determined by a timing network.

As shown in *Figure 5-3*, the digital data from the ADC, represented in codes, is manipulated by computing networks to alter, modify and redefine the data, but it emerges from the computing networks again as a series of digital codes, again timed by the timing network. The codes are presented to the DAC to be converted back to an analog signal. The circuit discussion begins with a DAC.

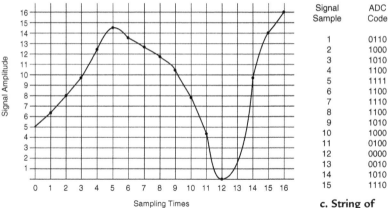

a. Analog signal versus time

Signal Sample	ADC Code
1	0110
2	1000
3	1010
4	1100
5	1111
6	1100
7	1110
8	1100
9	1010
10	1000
11	0100
12	0000
13	0010
14	1010
15	1110

c. String of codes from ADC

A Resistor Network DAC

Recall that in a digital code, the MSB's weighted binary digit position value is equal to one-half the value of the full code value, and that the next least significant bit is one-half the MSB's digit position value. This principle is used to design the DAC shown in *Figure 5-4*. It is called a R/2R ladder DAC. The circuit, shown in *Figure 5-4a*, is a resistor network with a particular combination of resistor values. From a reference voltage of V_{REF} to ground, there are resistors with 2R values for each bit separated by resistors with R values, and terminated in a resistor to ground with a 2R value. The circuit is for a 4-bit DAC. A switch at the end of each bit resistor of 2R value either

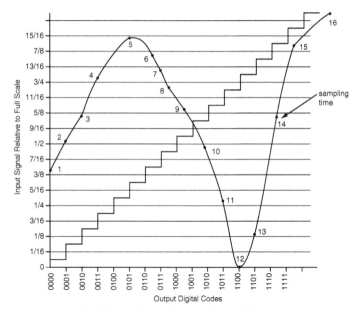

b. AC signal superimposed on ADC steps

Figure 5-2: Converting AC signals to digital codes—A-to-D conversion

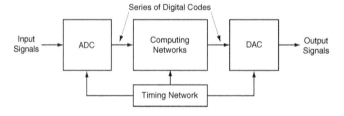

Figure 5-3: Computing network manipulates digital data

connects to ground or to the input of an operational amplifier used as a summing amplifier. If each bit in the code to be converted is a 0, then each 2R resistor is connected to ground and there is zero current into the summing amplifier.

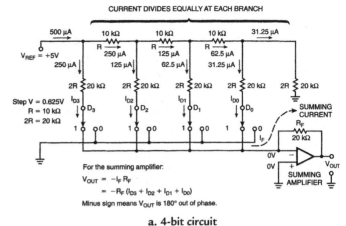

a. 4-bit circuit

The Equivalent Resistance of the Network

Looking at the right end of the network at the LSB leg, the equivalent resistance of the 2R resistors in parallel is R. This equivalent resistance R in series with the R between the second LSB leg equals 2R. This 2R parallels the 2R of the second LSB leg to make an equivalent resistance of R. This process continues so that the equivalent resistance of the network between V_{REF} and ground is R.

The Digit-Position Currents

When all bits are zero, all bit resistors are connected to ground; with R = 10 kΩ and V_{REF} = 5V, the current into the network is 500 μA. The current into the MSB leg is 250 μA and the current into the remaining network is 250 μA. At the next lower significant bit, the 250 μA divides into

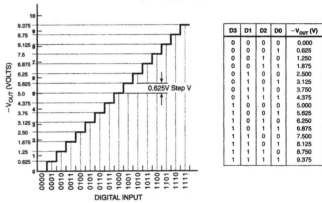

b. V_{OUT} vs. digital input (4-bit conversion)

Figure 5-4: 4-bit R/2R ladder DAC

D3	D1	D2	D0	$-V_{OUT}$ (V)
0	0	0	0	0.000
0	0	0	1	0.625
0	0	1	0	1.250
0	0	1	1	1.875
0	1	0	0	2.500
0	1	0	1	3.125
0	1	1	0	3.750
0	1	1	1	4.375
1	0	0	0	5.000
1	0	0	1	5.625
1	0	1	0	6.250
1	0	1	1	6.875
1	1	0	0	7.500
1	1	0	1	8.125
1	1	1	0	8.750
1	1	1	1	9.375

125 μA down the next lower significant bit leg and 125 μA to the remaining network. The 125 μA divides into 62.5 μA down the second LSB leg and 62.5 μA to the remaining network. The 62.5 μA divides to 31.25 μA down the LSB leg and 31.25 μA to the remaining network. Therefore, each current in the bit legs is one half of the current in the bit position to the left, just like the digit position values in a binary number. Thus, the summing of the currents in the digit position legs results in the value of the binary number.

The Summing Amplifier

Refer again to *Figure 5-4a*, when the code bit is equal to 1, the bit leg current is connected to a summing amplifier. For the summing amplifier:

$$V_{OUT} = -I_F R_F$$

Where I_F is the current into the inverting input of the operational amplifier, and R_F is the feedback resistor from output to input. The minus sign means the output is 180° out of phase from the input.

Since,

$$I_F = I_{D3} + I_{D2} + I_{D1} + I_{D0}$$

then

$$V_{OUT} = -(I_{D3} + I_{D2} + I_{D1} + I_{D0})\ R_F$$

When the code is 0101, then

$$V_{OUT} = -(I_{D2} + I_{D0})\ R_F$$

If $I_{D2} = 125$ µa and $I_{D0} = 31.25$ µa, then, with $R_F = 20$ kΩ

$$V_{OUT} = -20(156.25) \times 10^{-3}$$
$$= -3.125V$$

Here are two more examples:

A. the code 0001 results in:

$$V_{OUT} = -(I_{D0})\ R_F = -(31.25\ \text{µa}) \times 20\ \text{k}\Omega$$
$$= -625 \times 10^{-3}$$
$$= -0.625V$$

B. the code 1111 results in:

$$V_{OUT} = -(I_{D3} + I_{D2} + I_{D1} + I_{D0})\ R_F$$
$$= -(250\ \text{µa} + 125\ \text{µa} + 62.5\ \text{µa} + 31.25\ \text{µa})\ 20\ \text{k}\Omega$$
$$= -9375 \times 10^{-3}$$
$$= -9.375V$$

The codes and the output voltage at each step are shown in *Figure 5-4b*.

No Change in Current for Bits of 1 or 0

The current in the digit position legs remains the same whether the bit is a 1 or a 0. Even when the current leg is connected to the summing amplifier inverting input, because of the high input impedance of the summing amplifier, there is no current from the inverting input to the noninverting input. As a result, The inverting input is at the same potential as the noninverting input, which is ground. There is no change in the currents because in both cases, whether a 1 or a 0, the terminating point is at ground.

Example 2. **Output Voltage of R/2R Ladder DAC**

The output voltage from a R/2R ladder DAC for n bits can be expressed as:

$$V_{OUT} = -(\text{decimal equivalent of binary number})\ \frac{V_{REF}}{2^n}$$

What is the output voltage of a 4-bit R/2R DAC with an input code of 1010 and a reference voltage of 10 volts?

Solution:

The decimal equivalent of 1010 is $8 + 2 = 10$ and $2^4 = 16$, therefore,

$$V_{OUT} = -(10)10/16 = 6.25V$$

Check your answer with code and voltage given in *Figure 5-4b*.

A Simple Resistor-String DAC

One of the simplest DACs, from a circuit standpoint, is the resistor-string DAC. $2n - 1$ resistors of equal value are interconnected from a reference voltage, V_{REF}, to ground. The outputs from the resistor string are fed to a decoder. The decoder closes the appropriate switch dictated by the input digital code. A resistor-string for a 4-bit DAC is shown in *Figure 5-5a* and given in more detail in *Figure 5-6a*.

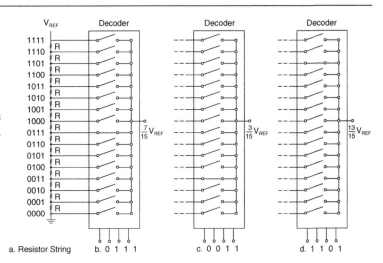

a. Resistor String   b. 0 1 1 1   c. 0 0 1 1   d. 1 1 0 1

Figure 5-5: Decoder for resistor string

The first position of the string is for zero volts. The next position has a voltage

$$V = (R/15R) \times V_{REF}$$

or

$$V = V_{REF}/15$$

or for any string, since it is the number 1 position

$$V_1 = 1 \times V_{REF}/2^n - 1$$

where n = bits in digital code.

The next position, position 2, has a voltage

$$V_2 = (2R/15R) \times V_{REF}$$

or

$$V_2 = 2V_{REF}/15$$

or for position 2 for any string

$$V_2 = 2 \times V_{REF}/2^n - 1.$$

The position one removed from V_{REF} has a voltage

$$V_{14} = 14 \times V_{REF}/2^n - 1.$$

The Decoder

Figure 5-5b,c and d shows the details of the decoder as different input codes are received to identify the analog voltage and produce an analog voltage equivalent.

Figure 5-5b is for an input code of 0111; *Figure 5-6c* is for a code of 0011, and *Figure 5-6d* is for a code of 1101.

Accuracy and Increments

The number of increments or steps in a digital code with n bits is shown in *Figure 5-6b*. If n = 4, there are 16 increments; if n = 10, there are 1024 increments. The n corresponds to the resolution for the ADC and DAC systems. If a system needs to have an accuracy of 0.5%, the measurement must be made to 1 part in

200; therefore, an 8-bit system which has 256 increments must be used. A 10-bit system must be used for an accuracy of 0.1% (1 part in 1000). With a full-scale range (FSR) set at a particular voltage, then the voltage value of each increment is FSR/2^n. The voltage increment for an FSR = 10V is 10/1024 or about 10 millivolts (9.77 to be exact).

General System Increments

Here are the increments for a general system. The first increment above zero in the resistor string is equal to:

$$V_{REF}/(2^n - 1)$$

which is equal to FSR/2^n; therefore,

$$V_{REF}/(2^n - 1) = FSR/2^n$$

or $\qquad V_{REF} = (2^n - 1)/2^n \times FSR$

The increment is then,

$$\text{Incremental Voltage} = \frac{(2^n - 1)/2^n \times FSR}{2^n - 1} = FSR/2^n$$

For the eighth code (the 7th increment) of a 4-bit resistor string, the voltage is:

$$7\text{th Incremental voltage} = 7 \times FSR/16$$

A Simple Current-Steering DAC

A DAC similar to the resistor-string DAC can be designed by decoding a binary code and switching binary-weighted currents into a current summing amplifier. Its design is based on the same principles used for the R/2R ladder DAC of *Figure 5-4*. *Figure 5-7* shows a very simplified version of such a DAC. A binary input code is decoded and the appropriate binary-weighted constant current is routed to a current summing amplifier to produce a proportional output voltage. If it is a 10-bit DAC, the MSB constant-current source is 512 times the LSB constant-current source. Summing the constant currents from the bit positions that have a value of 1 produces the proportional analog output voltage.

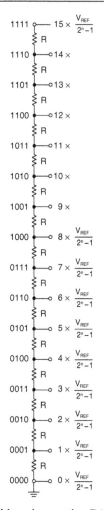

a. 4-bit resistor-string DAC

n =	4	6	8	10	12
INCR	16	64	256	1024	4096

b. Increments in N-bit code

Figure 5-6: Resistor-string DAC

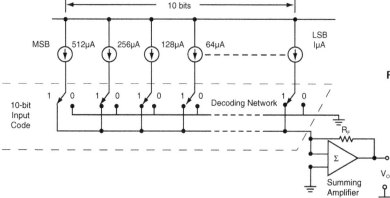

Figure 5-7: Simple switched constant-current 10-bit DAC

An excellent application in integrated circuits is to produce the binary-weighted currents using current mirrors of the type shown in *Figure 4-23*. The currents produced are precise and uniform because of the side-by-side processing on the IC chip and wafer or slice.

Example 3. Summing Constant-Current DAC Output

Using the constant-current DAC circuit of *Figure 5-7* and a 10-bit code, what is the final summed current for the binary input of 0110011101 when the LSB current is 20 µA?

Solution:

MSB									LSB
0	1	1	0	0	1	1	1	0	1
	5120	2560			320	160	80		20 = 8,260 µA = 8.26 mA

Analog-to-Digital Converters (ADC)

The input portion of *Figure 5-3* is an ADC, an analog-to-digital converter. One of the earliest ADCs was the counting ADC shown in *Figure 5-8*. It is made up of a binary counter that counts pulses from a central clock. The counters binary output is fed to two units—a DAC and a latch. Each unit has the number of input or output bit lines to cover the number of bits required from the ADC. Notice the DAC in the loop. This is the reason that the discussion of the DAC came first. The binary code input to the DAC produces an analog voltage that feeds one input of a comparator. The analog input volt-

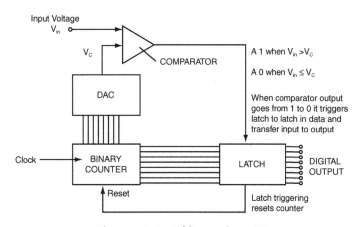

Figure 5-8: An 8-bit counting ADC

age to be converted to a digital output is the other comparator input. When the input from the DAC is lower than the analog input, the comparator will be a high voltage (a digital 1); when the input from the DAC is equal to or greater than the analog input, the comparator output is a low voltage (a digital 0). When the comparator output changes from a high voltage to a low voltage, it triggers the latch to latch in the binary values from the bit lines of the counter. Thus, the output of the latch is the binary code matching the value of the input analog voltage.

The A to D process works like this. The counter is reset to a count of zero. The DAC output is zero as a result. If the analog input voltage, V_{in}, is some positive value, the comparator output will be a 1. As the clock increments the counter, the output of the DAC will increase in steps, each a small positive voltage. If the DAC output is a lower positive voltage than V_{in}, the counter continues to count and increases the DAC output voltage until it is greater than V_{in}. This triggers the comparator, its output goes to 0 to latch in the binary code at the output to the ADC and reset the counter. Resetting the counter to zero causes the comparator output to go to a 1 and the ADC is ready for another conversion. One of the disadvantages of the counting ADC is the time for conversion. The conversion time can be as great as $2^n - 1$ clock cycles, where n is the number of bits of the binary output of the ADC.

Successive Approximation Register (SAR) ADC

An improvement in conversion time results when using a Successive Approximation Register (SAR) ADC. As shown in *Figure 5-9*, the counter of *Figure 5-8* is replaced with logic, register and latch circuits to make up the SAR, one of the most popular ADCs. The SAR can have conversion times from 100 µS to 1 µS and up to 16 bits in resolution. Semiconductor technologies of bipolar, CMOS and combinations of both have been used to design the SAR. The SAR seems to be the design of choice for the conversion time required because the desired performance can be obtained at a reasonable cost. In addition, system throughput (speed) can be traded for accuracy—increasing speed decreases accuracy.

The SAR gets its name from successively comparing the input analog voltage to the output of a DAC that has a binary-weighted code at its input. The conversion process begins by setting the MSB of the input to the DAC from the SAR to a 1. All the other bits are set to 0. This produces an analog voltage at the DAC output equal to one-half the full-scale range of the DAC. At the comparator, as with the counting ADC, the DAC output is compared to the input analog voltage. If the input voltage is greater than the DAC voltage,

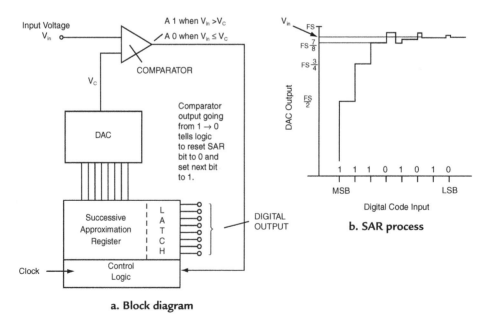

a. Block diagram

Figure 5-9: Successive approximation ADC

74

the comparator output is a 1 and the SAR MSB is left at a 1, and the next most significant bit input to the DAC is set to a 1. With the MSB and next significant bit set to a 1, the output from the DAC will now be one-half plus another one-quarter to equal three-quarters of the full-scale range of the DAC. The sequence is shown in *Figure 5-9b*.

The sequence continues to set the next most significant bit to a 1 (all other bits are zero) as long as the comparator output is a 1. Each time a binary-weighted voltage is added by the DAC to its output—one eighth, one sixteenth, one thirty-secondth, and so on—the comparator output will be a 1 as long as the input voltage is greater than the output of the DAC. When setting the next significant bit to a 1 causes the input voltage to be less than the DAC output, the comparator output goes to 0. This results in setting the last significant bit back to a 0 from a 1, reducing the DAC output below the input voltage. But at the same time the next most significant bit is set to a 1 and the DAC output increased again; however, this time only say one thirty-secondth of an increment of voltage is added instead of the one-sixteenth that was added at the bit before. This is shown in *Figure 5-9b*. The successive approximation continues until all bits are tested and the closest approximation is obtained. The result is that the SAR output bit either is set to a 1 or a 0 depending on the result of the comparison of the output of the DAC and the input voltage. The final digital code for Figure 5-9b is 11101010.

The time to convert the input analog voltage to a digital output is n clock cycles, much less than the counting ADC. *Figure 5-9b* shows that after n clock cycles all the bits have been tested and set and the SAR output will be the digital output code. The output can be taken in parallel or shifted out as each comparison is made. This is an additional advantage of the SAR ADC.

Example 5. Maximum Conversion Time for SAR DAC
Repeat the calculation of the maximum conversion time of an 8-bit, 12-bit and 16-bit SAR ADC. The clock frequency is 1 MHz whose period is 1 µS.
Solution:

n	Max. Conversion Time (n clock cycles)
8	8 µS
12	12 µS
16	16 µS

Capacitor Charge-Redistribution ADC

A block diagram of a hybrid resistor-tree, capacitor charge-redistribution ADC is shown in *Figure 5-10*. It consists of a resistor-tree conversion circuit that handles M bits of the ADC output and a charge-redistribution capacitor bank conversion circuit that handles K bits of the ADC output. The control logic, under synchronization by the clock, provides the switching logic for setting the bits in the SAR, the switch settings for the resistor tree and the switch settings for the capacitor bank. After comparison of a bit by the comparator, similar to the process previously described for the SAR ADC, the bit evaluation by the comparator is fed to the SAR to set the bit for the ADC output. The hybrid DAC is a compromise between using an all capacitor charge-redistribution circuit and an all resistor-tree circuit. The capacitor-charge redistribution has slow conversion times;

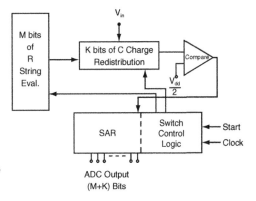

Figure 5-10: Hybrid R-tree capacitor charge-redistribution ADC

the resistor-tree circuit has faster conversion times but uses large IC real estate, especially as the bits in the ADC output increase. In integrated circuits, resistors use more area than capacitors.

The ADC Operation

In the hybrid ADC, the input analog voltage is captured as an amount of charge on a bank of capacitors. The capacitors are binary-weighted and handle a certain number of bits (equal to K) of the digital code to be converted. The remaining bits (equal to M) are converted through a resistor tree conversion. The charge on the capacitors, which remains constant during the conversion, plays an important part, not only in the K bit conversions, but also in the conversion of the M bits using the resistor tree. For example, in *Figure 5-11*, M = 5 and K = 3 so that five bits are converted using the resistor tree and three bits are converted using the binary-weighted capacitors. The M bits are the five most significant and the K bits are the three least significant.

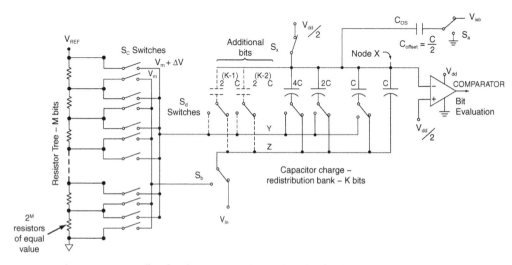

Figure 5-11: Details of resistor tree and capacitor bank (data acquisition period)

Converting the M Bits

The conversion process starts with the data acquisition period shown in *Figure 5-11*. Switch S_b is connected to the input analog voltage, V_{in}, switch S_x is connected to $V_{DD}/2$ and S_a is connected to V_{lsb}. The binary-weighted capacitor bank charges to $V_{DD}/2 - V_{in}$ because the lower end of the capacitors are connected to Node Z. The total capacitance in the bank is $2^K C$. The offset capacitor C_{OS}, equal to $C/2$, is charged to $V_{DD}/2 - V_{lsb}$. Both inputs to the comparator are connected to $V_{DD}/2$ at this time, thus, no comparison.

At the completion of the data acquisition period S_x opens, S_b switches from V_{in} and is connected to the V_m line of the resistor tree, as shown in *Figure 5-12*, and S_a is connected to ground. The voltage at Node X is the important voltage in all of the conversions for it feeds the minus input to the comparator. The plus input of the comparator is connected to $V_{DD}/2$. If the value of Node X is less than $V_{DD}/2$, the comparator output will be a 1; if it is greater than $V_{DD}/2$, the output will be a 0.

With S_b connected to the resistor tree, the control logic sets the MSB of the output digital code to a 1 and selects the tap from the resistor tree that represents one-half of full-scale range for V_m just as in the SAR ADC. As a result, the voltage at Node X is evaluated against $V_{DD}/2$. If the voltage of Node X is less than $V_{DD}/2$, the MSB of the output digital code from the SAR is set to 1. If the voltage at Node X is greater than $V_{DD}/2$ the output bit of the SAR is set to 0. This completes the evaluation of the MSB; it is either set to a 1 or a 0.

The control logic steps to evaluate the next significant bit. It sets the next significant bit to a 1. This along with the MSB value of a 1 or 0 will cause the SAR to select the corresponding

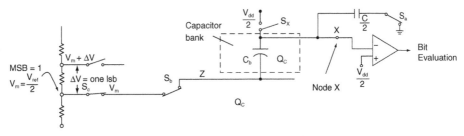

Figure 5-12: Evaluating R-tree bits

value of V_m from the resistor tree to feed the Node Z line connected to the capacitor bank. The new voltage value on the Z line causes the constant charges on the capacitor bank to redistribute and change the voltage at Node X. The new Node X voltage is compared to $V_{DD}/2$, and the bit evaluation is completed by setting the second most significant bit to a 1 or 0 depending on the result of the comparison. The bit evaluation process continues until all M bits are evaluated. This results in a set SAR code output for the M bits. At the end of the M bit evaluations, the voltage of Node X will be representative of the value of the five most significant bits in the SAR output digital code.

Converting the K Bits

The Node X voltage value is maintained as the evaluation now changes to the capacitor bank circuit to evaluate the K bits, the last three significant bits of the digital output code. The K bit evaluation is accomplished by switching the ends of the capacitors in the respective bit position, one bit position at a time, to the Y line. The Y line connects to a resistor-tree connection that is one

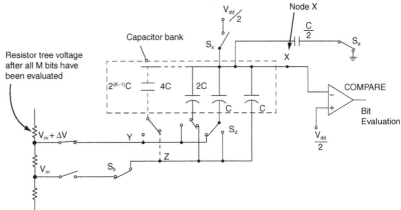

Figure 5-13: Evaluating capacitor bits

significant bit higher in voltage ($V_m + \Delta v$) than the Z line voltage.

The most significant bit of the three K least significant bits is evaluated first as shown in *Figure 5-13*. The end of its capacitor, in this case of value C, is connected to the Y line. The charge on the capacitors redistributes and changes the voltage at Node X. If Node X is greater than $V_{DD}/2$, the bit is set to a 1 and the end of the capacitor is left connected to the Y line; if it is less than $V_{DD}/2$, the bit is set to a 0 and the end of the capacitor is switched back to the Z line. With the bit set to a 1, the voltage on capacitor C is added to the resistor tree value to set the Node X voltage value.

The control logic switches the end of the next binary-weighted capacitance of the next least significant bit by changing its S_d switch and connecting it to the Y line. The charge redistributes with the new capacitor, now with a value of 2C, and the voltage at Node X changes correspondingly. The Node X voltage is compared to $V_{DD}/2$ and the bit evaluated as above and the output set to a 1 or a 0. As before, if the bit is set to a 0, the end of the capacitor is returned to the Z line with switch S_d. The process continues until all K bits are evaluated and the final SAR digital code is sent out from the SAR.

Highest Speed Conversions

The highest speed conversions are made with flash ADCs. The high speed is made possible by the use of simultaneous comparisons of the analog input voltage to references generated from a resistor string. A block diagram of a flash ADC is shown in *Figure 5-14*. For an n-bit flash converter, there are $2^n - 1$ reference voltages and $2^n - 1$ comparators required. Thus, for an 8-bit flash converter, 255 comparators are required, and for a 10-bit flash converter, 1023 comparators are required. A high price is paid for the speed advantage—high power, large silicon area for the ICs, and high cost contribute to the price that must be paid.

The conversion process is rather simple. The reference voltages are connected to the minus input of each comparator and are separated in value by one LSB. The analog input voltage is connected to the plus input of each comparator. A simultaneous comparison is made at each comparator. If the input analog voltage on the plus input is less than the reference voltage on the minus input, the output of the comparator is a 0. The comparator output will be a 1 if the input analog voltage is greater than the reference voltage. Each comparator output is presented to the decoder at the same time and the decoder's output is stored as an n-bit wide code in a latch. All the inputs of the input analog voltage that are greater than their respective resistor-string reference voltages will have comparator outputs of a 1; all the inputs that are less than their

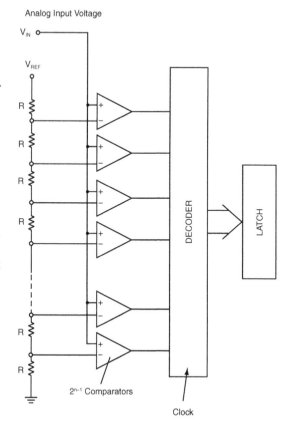

Figure 5-14: Flash converter

respective resistor-string reference voltage will have comparator outputs that are 0. The resultant digital code into the decoder results in the equivalent binary output code, for a given n-bit code, that represents the value of the input analog voltage.

Sample and Hold and Filters

Sample and Hold

There are two other functions that are associated with A to D conversions. One is sample and hold; the other is filtering. Sample and hold, as shown in *Figure 5-15*, is just what it says. The input analog signal is sampled by switch S_1 closing momentarily and charging C_1. C_1 then holds the value of the input voltage until the ADC can process the data. It probably is obvious that a capacitor that leaks its charge between samples would contribute errors to the sampling process. Likewise, switches that have variable contact resistance vary the times to charge the capacitors and contribute errors. Thus, high quality capacitors and fast switches are key to sample and hold

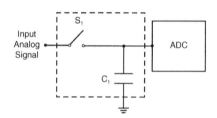

Figure 5-15: A simple sample and hold circuit

circuits. At one time, sample and hold circuits were available independently; however, most sample and hold circuits are incorporated right in the ADC. In fact, in the hybrid resistor-tree capacitor charge-distribution ADC there is no need for a sample and hold. It is built in as part of the circuit design, saving cost on providing such a circuit.

Filtering

Filtering, as shown in *Figure 5-16a*, is used to limit the bandwidth of signals. As such, it can smooth out the input signal, eliminate noise spikes, limit the high frequency response, select particular signal frequencies, and the like. *Figure 5-16b* shows their specific

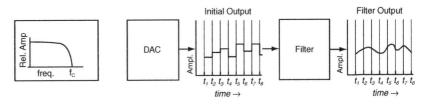

a. Bandwidth limiting

b. Filtering DAC output

Figure 5-16. Filtering

use in DAC systems. The DAC output can be a step-like signal. Filtering is used to smooth out the step nature of the signal and output a smooth analog signal. Most filters are tailored to the particular application. They are selected to control a specific need of the system; therefore, general filters are usually not the solution. The filters must be chosen specifically for the application. The example for the DAC system in *Figure 5-16b* requires that the filter be chosen for the specific system. The output signal that emerges must be a smooth continuous signal with time rather than a jagged jerky out. The result is that the input signal shown in *Figure 5-2* is reproduced very accurately after the ADC conversion and the DAC conversion.

Summary

In this chapter, DACs and ADCs have been discussed showing techniques used for each and circuits that implement the functions. In the next chapter, digital processors will be discussed. They receive the digital signals from the ADCs, modify and manipulate the digital signals, and then deliver the digital signals to the DACs.

Chapter 5 Quiz

1. In a binary number:
   a. the digit position value increases by 1 times over the digit value to the right.
   b. the digit position value increases by 2 times over the digit value to the right.
   c. the digit position value increases by 4 times over the digit value to the right.
   d. the digit position value increases by 8 times over the digit value to the right.

2. A decimal equivalent of a binary number:
   a. is the addition of all the bit position values for all the bits equal to 1.
   b. is the multiplication of all the bit position values for all the bits equal to 1.
   c. is the subtraction of all the bit position values for all the bits equal to 1.
   d. is the division by 2 of the bit position value of the LSB.

3. The principle used to design a resistor-string DAC is:
   a. the MSB is one-eighth the full value and the next bit position is one half of the MSB value.
   b. the MSB is one-fourth the full value and the next bit position is one half of the MSB value.
   c. the MSB is one-third the full value and the next bit position is one half of the MSB value.
   d. the MSB is one-half the full value and the next bit position is one half of the MSB value.

4. The equivalent resistance between V_{REF} and ground of the R/2R ladder DAC is:
   a. 4R.
   b. 2R.
   c. R.
   d. R/2

5. The digit position current in the R/2R ladder DAC is:
   a. one half the current in the bit position to the left.
   b. one eighth the current in the bit position to the left.
   c. one fourth the current in the bit position to the left.
   d. equal to the current in the bit position to the left.

6. The voltage increment from a 10-bit resistor-string DAC with 10V applied is:
   a. about 10 volts.
   b. about one volt.
   c. about 10 millivolts.
   d. about 100 millivolts.

7. A simple current-steering DAC:
   a. combines both voltage and current to produce the analog output.
   b. adds binary-weighted voltages to produce the analog output.
   c. produces the analog voltage by sensing a resistor string.
   d. adds binary-weighted constant currents to produce the analog output.

8. The counting ADC:
   a. contains a DAC whose input is the output of a counter.
   b. contains a comparator to compare the analog input to the output of a DAC.
   c. latches the counter output code when the comparator inputs are equal.
   d. all of the above.
   e. a and b only above.

9. The SAR gets its name from a process that:
   a. successively compares the input analog voltage to the output of a DAC that has a binary-weighted input code.

   b.   sums a series of binary-weighted currents.

   c.   sums current from a ladder resistor network.

   d.   sums voltages from a resistor string.

10. The maximum conversion time for a SAR DAC is:

   a.   4n clock cycles.

   b.   n clock cycles.

   c.   8n clock cycles.

   d.   n/2 clock cycles.

11. In the hybrid resistor-tree capacitor charge-redistribution ADC:

   a.   (M − K) bits are converted using an R tree and (M + K) bits using a C network.

   b.   K bits are converted using an R tree and M bits using a C network.

   c.   M bits are converted using an R tree and K bits using a C network.

   d.   (M + K) bits are converted using an R tree and (M − K) using a C network.

12. In the hybrid resistor-tree capacitor charge-distribution ADC:

   a.   the K bits are evaluated first and then the M bits.

   b.   the M bits are evaluated first and then the K bits.

   c.   the K and M bits are evaluated at the same time.

   d.   only the K bits are evaluated.

13. In Flash ADCs:

   a.   the high speed is made possible by simultaneous comparisons.

   b.   there are as many comparators as there are bits.

   c.   there are as many reference voltages as there are bits.

   d.   the basic string for comparisons is a capacitor charge-redistribution network.

14. A sample-and-hold circuit:

   a.   has a momentary switch that connects the input voltage to a capacitor long enough for the capacitor to charge.

   b.   has a resistor in series with a capacitor in series with a switch.

   c.   has a capacitor that is charged to hold the value of the input voltage.

   d.   a only above.

   e.   a and c above.

   f.   none of the above.

15. Filtering is important to DAC operation:

   a.   because it adds noise to the output signal.

   b.   because it returns the DAC output to a smooth continuous signal.

   c.   because it selects one frequency to pass on from the output.

   d.   because it acts as a very high-frequency high-pass filter.

Answers: 1.b, 2.a, 3.d, 4.c, 5.a, 6.c, 7.d, 8.d, 9.a, 10.b, 11.c, 12.b, 13.a, 14.e, 15.b.

Digital System Processing

Introduction

Previous chapters have sensed the analog signal, conditioned the signal and converted it from analog to digital. In this chapter, the processing of the digital signal to modify, calculate, manipulate, change the form of the signal or to route the signal to particular channels is discussed. All or any of these processing operations may be needed to accomplish a task predetermined by the application that is being fulfilled. The total system is designed to perform a task, and the digital processor is a very important part of the system.

Digital Processor or Digital Computer

As the name implies, the digital processor inputs, stores, performs operations and outputs digital signals. Performing logic or arithmetic computations, modifying the format of the signal, storing data temporarily or more permanently, decoding signals for display and outputting signals are some of the operations dictated by the instructions in the application program.

Figure 6-1 shows the basic structure of a digital processor, more generally called a digital computer. The main brain of the structure is the CPU (central processing unit) where the operations that are performed are decided upon and controlled. The digital signals

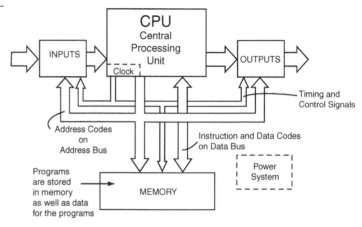

Figure 6-1: A digital processor

in the form of binary codes that tell the digital processor which operation to perform are called instructions. Each digital processor is manufactured to respond to a particular set of instructions. Each instruction in the set will cause the digital processor to do a unique operation. For example, an instruction might cause the digital processor to input a digital signal from a particular input. Or an instruction might tell the processor to take the input signal and store it temporarily, or to store it in memory more permanently. Another instruction might take a digital signal that has been operated on by the processor and output it to a particular output. Or an instruction might tell the processor to do a logical operation (for example, AND two binary numbers together), or to do an arithmetic operation like ADD two binary numbers, or maybe subtract them. The instructions, presented in sequence to the processor, are called a program.

Digital Computer Program

The arrangement of the instructions, one after another, for the digital processor to perform set operations in a particular sequence to accomplish a task is called a *program*. The set of instructions in a program is stored in memory to be recalled each time that the desired task is required. If a different task is required, then a different program is needed.

As shown in *Figure 6-1*. The instructions of a program are stored in memory at specific addresses, usually in sequence, and are moved from memory to the CPU over the data bus. It is just like a home with a particular address. The post office uses the address to deliver the mail. In like fashion, the instructions in memory are at unique addresses. When a particular task is needed, the address of the first instruction in the program is sent by the CPU to memory over the address bus. The address locates the instruction in memory, the CPU instructs the memory to read the instruction and it is sent over the data bus to the CPU. The CPU decodes the instruction and performs the directed operation. Each subsequent instruction in the program is addressed, recovered from memory, sent to the CPU and executed.

Address and Instruction/Data Bus

Addresses, over the address bus, are not only used to locate instructions in memory but are used to identify particular inputs or particular outputs. By addressing a particular input, the CPU has selected that input to supply input data; or addressing a particular output, the CPU will send data to that output to be transmitted to the next function. And there is another use of addresses. When an instruction calls for an arithmetic operation, (or other operations that require unique information), such as, ADD A and B, the data A and the data B must be supplied to the CPU before the operations can be performed. Data A and B and other data used for the program being executed are stored in another portion of memory, separate from the program. Data A and Data B are addressed over the address bus just like instructions and recovered and sent to the CPU. The instructions and the data are transmitted from memory to the CPU over the data bus; thus, this bus is usually called the instruction/data bus.

Timing and Control

All the CPU operations, all address, instruction, and data transfers, as shown in *Figure 6-1*, occur in a timed sequence determined by the timing and control signals derived from the CPU's clock. The clock is a circuit that outputs a series of repetitive pulses occurring at a set frequency or set frequencies. The clock pulses have fast rise and fall times so that circuits can be triggered on either edge to accurately time the operation of the circuits. The rise time is called the leading edge and the fall time the trailing edge of the pulses.

Clock signals must be very accurate. As a result, they are generated by phase-locked loops (PLLs), or for the greatest accuracy, by quartz crystal oscillators. Quartz crystals, of a particular cut and size, when excited with electricity, will oscillate at a very precise frequency. The clock signals precisely control the transfers, manipulations, and storage of information throughout the CPU and the accompanying total system.

Power Systems

Each digital processor has a complete unique power system. Sophisticated systems are required for the distribution of the supply voltages and the required currents, regulated to keep the variation of voltages to within tight limits, as the circuits switch rapidly from one state to another. Extensive use of bypass capacitors at critical junctions help to maintain voltages within limits as significant values of current are switched along the supply lines.

As the density of integrated circuits has increased, there is more need for heat sinks and cooling air distribution as the watts/in^2 dissipation increases. IC technology has led the way as circuit density increased within an IC to change the circuit type from bipolar to MOS (metal-oxide-semiconductor) to CMOS (complementary MOS) so that the power dissipation per circuit function has been reduced. As density further increased, the supply voltages for circuit operation have been reduced from 5V to 3V, and now 1.8V to again reduce the power dissipation per function. The tight regulation specifications still remain even with the reduction in the voltage values.

The CPU—
Program Counter

Figure 6-2 is a diagram of a generalized central processing unit (CPU). The main components are the program counter, the instruction register, the instruction decoder, the data address register, the arithmetic and logic unit (ALU), the timing and control circuits, and the permanent and temporary storage. As discussed previously, a digital code, called an instruction, organized in sequence into a program, is sent to the CPU to

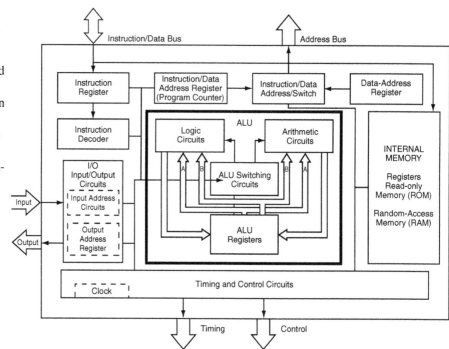

Figure 6-2: A generalized CPU

instruct it to execute a particular operation. The instruction came from a memory address contained in an instruction address register called the program counter. The program is stored in memory one address after another in sequence so the program counter holding the address can be incremented by one to step through the program instructions one step after the other. Thus, the name for the address register is the *program counter*. Each instruction address from the program counter addresses the next step in the program as the task proceeds.

Example 1. Program Counter

Using 4-bit addresses, show in a simple example how the program counter is incremented to sequence through a program to add 16 to 8.

Solution:

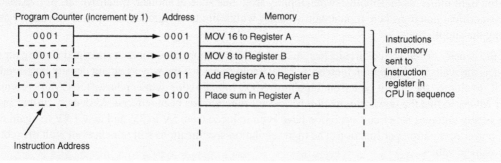

CPU—Instruction Register and Instruction Decoder

As the digital code representing the CPU instruction is retrieved from its memory location it is stored in a temporary storage register called the *instruction register*. Here it is recognized and decoded by the *instruction decoder* and directed to the appropriate circuits to execute the operation dictated by the instruction.

CPU—The Data Address Register

If the instruction requires that additional data be fetched from memory, then the next instructions will direct the CPU to place the address for the data in the *data address register*, send the address to memory to retrieve the data and store it in a temporary storage location in the CPU, either a register, or a RAM location. Through a multiplexing switch, the instruction and data address are sent to memory over the same bus, the data bus.

CPU—The Arithmetic Logic Unit (ALU)

The ALU provides the logical, computational, and decision-making capabilities of the CPU. Basic arithmetic operations, such as, addition, subtraction, multiplication, and division; basic logical decisions, as well as, greater than, less than, equal to, positive or negative are all performed by the ALU.

Registers for temporary storage of data brought from inputs or from memory are available in the ALU. The information in these registers is used by the CPU for completing the operation directed by the instruction addressed by the program counter. When the operation is completed the information is erased or replaced with new information to be used for executing the next instruction.

CPU—Internal Memory

There are internal memories contained within the CPU. They may be additional registers, read-only memory (ROM) or random-access memory (RAM). They store particular sets of instructions called subroutines, temporary data, and data routing information. The RAM is of the type that needs to be refreshed periodically. Some CPUs do not have any ROM or RAM, but usually have the additional registers. For these CPUs, the ROM or RAM is in the external memory shown in *Figure 6-1*.

Timing and Control

Each of the operations of the CPU is timed and controlled by circuits that operate at specific times. Many operations occur at the same time; others are sequenced so they operate after data is entered, or transmitted, or before another operation. The timing and control signals, generated from the master clock signals, not only time the CPU, but also are distributed throughout to time and control the complete system.

CPU—Input and Output (I/O)

Not all CPUs have the input and output selection circuits in the CPU; for some, these circuits are external as shown in *Figure 6-1*. *Figure 6-2* shows the I/O contained in the CPU. The input address registers determine the particular input that will receive data, and the output address registers determine the particular output used to couple out data to external destinations. If the CPU needs data, the CPU sends the address of the input to receive the data to the input address register and inputs the data from that input. The CPU inputs the data at a select time so that it is synchronized to the operation that is being executed. Likewise, after the CPU has executed an operation, the resultant data needs to be outputted to complete the task. The CPU sends the address of the output that is to couple out the resultant data to the output address register, and, synchronized by the clock, outputs the data.

Example 2. I/O Selection

Show with a simple example, using a 4-bit code, how a particular input is selected by the CPU.

Solution:

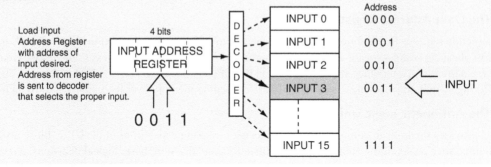

What is a Microprocessor?

When all the circuitry for the functions shown in *Figure 6-2* for a CPU are contained in an IC, the IC is known as a microprocessor. Attach to it the I/O functions, memory, and power supply, and one has a digital processor shown in *Figure 6-1*, or more commonly, a digital computer.

What is a Microcomputer?

When all the circuitry for a digital computer is contained on one integrated circuit, the unit is called a microcomputer. Even though there are self-contained memory and I/O circuits contained in a microcomputer, external circuits of the same type may be added, especially memory. As a result, there are many variations between microprocessors and microcomputers. Memory, I/O, signal conditioning, timing and control many times are added to adapt the particular IC to an application, or to a market requirement.

A particular type of microcomputer, now called a microcontroller unit (MCU), has been adapted to the industrial control market. A microcontroller unit from the MSP430 family manufactured by Texas Instuments will be used in *Chapter 7* to explain assembly-language programming and in *Chapter 10* to demonstrate the application of MCUs by providing the reader an opportunity for a hands-on project that can be built from contained instructions.

System Clarifications

System Buses

In *Figure 6-1* and *Figure 6-2* there are wide signal paths connecting the functional units in the diagrams. Each of these contains multiple wires connecting between units. Each is called a "bus" because it represents more than one wire making the interconnections between units. For example, if the memory in *Figure 6-1* has 65,536 different memory storage locations, then a binary address of 16 bits must be used to address each location. The address bus, as a result, is really 16 wires bundled together, each wire carrying a binary signal of 1 or 0 to make up the 16-bit word for the address.

The expansion of memory locations as bits are added to the address is shown in *Figure 6-3*. If the address is expanded to 24 bits, 16,777,216 memory locations can be addressed; if the address is expanded to 32 bits, 4,294,967,296 locations can be addressed. If each memory location has an 8-bit piece of binary information (called a byte), then 24 bits will locate 16 million bytes of information; more precisely, 16,777,216 bytes of memory, but shortened by industry use to a 16-Megabyte memory (16 MB). In like fashion, a 32-bit address will locate 4.2 billion bytes or is a 4.2-Gigabyte memory (4.2 GB).

Memory and data buses are the buses that must be the widest (to be able to handle the largest number of bits at a time) in order to carry the memory addresses and the instructions required by the system. Control

Address Bits	Locations	Address Bits	Locations	Address Bits	Locations	Address Bits	Locations
1	2	9	512	17	131,072	25	33, 554, 432
2	4	10	1024	18	262,144	26	67, 108, 864
3	8	11	2048	19	524,288	27	134, 217, 728
4	16	12	4096	20	1,048,576	28	268, 435, 456
5	32	13	8192	21	2,097,152	29	536, 870, 912
6	64	14	16,384	22	4,194,304	30	1, 073, 741, 824
7	128	15	32,768	23	8,388,608	31	2, 147, 483, 684
8	256	16	65,536	24	16,777,216	32	4, 294, 967, 296

Figure 6-3: Memory locations vs. address bits

buses and timing signal lines may have only a single line, but in most cases have multiple lines, but their buses hardly need to be as wide as the address and data buses.

Digital Information Nomenclature and Transfer

Binary strings of bits are identified in a number of ways. Long strings of bits are called *words*. Modern day digital computers use 16-, 32-, and 64-bit words. In *Figure 6-4a* a 16-bit word is shown. In any binary representations, the most significant bit (MSB) is on the left of the string, and the least significant bit (LSB) is on the right.

A group of 8 bits, as shown in *Figure 6-4a*, is called a "byte," and is a very common grouping used to identify memory capacity. A 1 MB (1 megabyte) memory has a storage capacity of one million locations with a byte (8 bits) at each location. Even though a memory may be organized differently, say two million locations with 4 bits per location, the capacity is still referred to as 1 MB. Years ago this 4-bit group was used extensively and called a "nibble."

A byte, or a number of bytes, is a common way of identifying other binary signals. A control signal may contain a certain number of bytes. A code may be made up of words that are each a byte, or a code may contain any number of bits. This will be further clarified in the section on *Digital Signal Representations*.

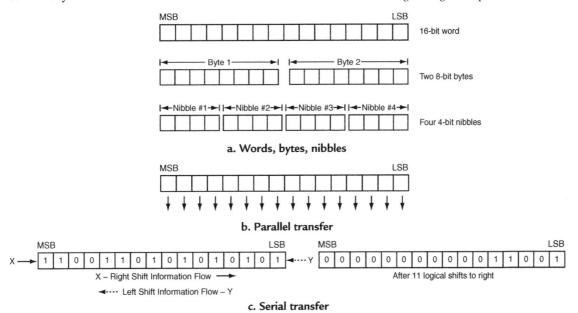

a. Words, bytes, nibbles

b. Parallel transfer

c. Serial transfer

Figure 6-4: Digital information nomenclature and transfer

Data Transfers

Within a digital computer, digital processor, digital system, or digital circuit, the binary bits that carry the information must be transferred from place to place to allow the system or circuit to perform its task. *Figures 6-4b and c* show the method of transfer. *Figure 6-4b* is a parallel transfer and *Figure 6-4c* is a serial transfer. This discussion centers on the signals within a digital processor, or within a self-contained digital system. Further discussion of the transfer of data between digital systems is contained in *Chapter 8*.

In *Figure 6-4b*, all the bits of binary information are transferred at the same time. If it is a 16-bit word as shown, all 16 bits are sent from one location to the other at the same time, in parallel. The highest speed digital processors use the parallel transfer so no time is lost in processing the binary information to act on it.

The serial transfer, shown in *Figure 6-4c*, takes longer in time to process the information. As shown, each bit of information is shifted in sequence to identify all the bits in the 16-bit word. Using the 16-bit word as an example, 16 clock-shifting pulses are required to identify all 16 bits. The shifting of the bits can either be in a right or left direction, as shown, and there are a number of different types of shifts—a logical, circulate, or arithmetic.

Logical Shifts

A right 11-step shift is shown in Figure 6-4c. As the bits are shifted right toward the LSB position, a detection circuit receives the LSB output and identifies the bit value as 1 or 0. The bits arrive serially, one bit after another, until all 16 bits of the word are identified. In a logical shift, bits of 0 values are inserted at the MSB position as the shifting occurs. For a left shift, the identifying circuit is at the MSB position rather than the LSB position, and the bits are inserted at the LSB position.

Arithmetic Shifts

Many times the instruction to the processor may only be for one shift because shifting a binary word to the right divides the binary value by 2. Likewise, shifting a binary word one bit position to the left, multiplies the binary value by 2. These types of shifts are particularly significant in arithmetic operations.

Example 3. **Arithmetic Shift Left for Multiplication**

Show an example, using an 8-bit word, to demonstrate how shifting a binary number one bit position to the left multiplies the binary value in the number by 2.

Solution:

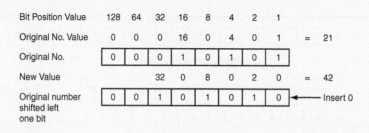

In a right circulate shift, the bit value in the LSB position is *circulated back and inserted at the MSB position. After 16 clock shifts,* the bits of the 16-bit word are shifted out and identified, and, after the shifting is complete, the same data is in the 16-bit word as before the shifting process began. Such shifts are very useful in arithmetic and logical shifts without destroying the original data present before the shifts.

Binary information can identify both positive and negative numbers. To do this, the MSB of the binary word is reserved to be a sign bit. If the bit is a 0, the binary number is positive; if the bit is a 1, the binary number is negative. During an arithmetic shift, the sign bit in the MSB position is maintained. Thus, when a shift occurs, the value in the MSB position is reinserted into the MSB position, so that it remains the same and the arithmetic value of the binary number is not lost. Examples for a 4-bit code are shown in *Figure 6-5*.

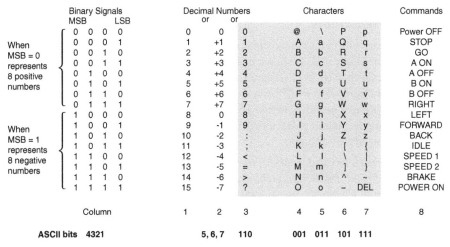

Binary Signals MSB LSB	Decimal Numbers or	or	Characters					Commands
0 0 0 0	0	0	0	@	\	P	p	Power OFF
0 0 0 1	1	+1	1	A	a	Q	q	STOP
0 0 1 0	2	+2	2	B	b	R	r	GO
0 0 1 1	3	+3	3	C	c	S	s	A ON
0 1 0 0	4	+4	4	D	d	T	t	A OFF
0 1 0 1	5	+5	5	E	e	U	u	B ON
0 1 1 0	6	+6	6	F	f	V	v	B OFF
0 1 1 1	7	+7	7	G	g	W	w	RIGHT
1 0 0 0	8	0	8	H	h	X	x	LEFT
1 0 0 1	9	-1	9	I	i	Y	y	FORWARD
1 0 1 0	10	-2	:	J	j	Z	z	BACK
1 0 1 1	11	-3	;	K	k	[	{	IDLE
1 1 0 0	12	-4	<	L	l	\			SPEED 1
1 1 0 1	13	-5	=	M	m	]	}	SPEED 2
1 1 1 0	14	-6	>	N	n	^	~	BRAKE
1 1 1 1	15	-7	?	O	o	-	DEL	POWER ON

When MSB = 0 represents 8 positive numbers (rows 0000–0111).
When MSB = 1 represents 8 negative numbers (rows 1000–1111).

Column	1	2	3	4	5	6	7	8
ASCII bits	4321	5, 6, 7	110	001	011	101	111	

Figure 6-5: Digital signals can represent numbers, letters, special characters, commands, and so forth.

Example 4. Arithmetic Shift for Recirculation

Show an example, using an 8-bit word, of how a right recirculate shift of the same number as the bits in the word reinserts the same word in a register after use of the word.

Solution:

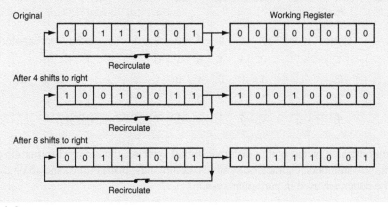

Parallel vs. Serial

One can see the parallel transfer of information is fastest because it takes significant time to shift out the bits for identification in a serial transfer. However, there is a significant tradeoff in hardware of increased circuitry, increased interconnections, increased power dissipation, and so forth. Serial operation calls for only one detection circuit at the LSB or MSB position to identify the bits. Parallel operation requires a circuit for each bit so the bits can be identified all at the same time. This multiplication of circuits, interconnections, more power occurs throughout the system.

The trade off then is one between speed of operation versus amount of hardware. But IC processing, device and circuit technology is having a tremendous impact on this tradeoff, as discussed in more detail in *Chapter 8*. The advances by ICs in density per chip, faster operating speeds and lower power operation and new circuit protocols are reducing the separation in this tradeoff and serial operation is gaining in use.

Digital Signal Representations

Figure 6-5 details that binary bits in digital information can commonly represent numbers, letters, characters and commands. A 4-bit binary code is shown that can represent 16 different entities. The 16 different entities can be the numbers from 0 to 15 (1st column); or they can be eight positive numbers from +0 to +7, and eight negative numbers from –0 to –7 (2nd column). As explained, the MSB of the code is used to tell whether the number is positive or negative. Or the 16 different codes can be used to identify the numbers from 0 to 9 and six special punctuation characters (3rd column). Or the 16 different codes could be used to identify 16 different commands (8th column).

In order to identify more characters and symbols, more bits must be added to the code. As an example, *The American Standard Code for Information Interchange* (ASCII), mentioned briefly in *Chapter 1* and contained in its complete form in *Chapter 8*, uses a 7-bit code. It identifies 52 upper and lower case alphabetic characters, 10 numbers from 0 to 9, 34 special data transfer and Teletype commands, and 32 other special characters for a total of 128.

Columns 4, 5, 6 and 7 of *Figure 6-5* are the 52 upper and lower case alphabetic characters and other special symbols that are identified in the ASCII code. Column 3, mentioned previously, is also used in the ASCII code. To fill out the 7-bit code, column 3 has bits 5, 6 and 7 at 110, and columns 4, 5, 6, and 7 have them at 001, 011, 101, and 111, respectively. As the combination of the 5, 6, and 7 bits change, the identities of the 16 codes change to new characters, numbers or symbols.

Example 5. ASCII Code

Identify what the given 7-bit codes represent using *Figure 6-5*.

Solution:

Code	Bit	7	6	5	4	3	2	1	Data Represented
1.		0	1	1	0	1	0	1	5
2.		1	0	0	1	0	1	0	J
3.		1	1	0	0	1	1	1	g
4.		0	1	1	1	1	1	1	?

What has been demonstrated is that within different digital systems, the binary information can represent many different things—numbers, characters, symbols, commands, instructions, and so on. System designers will define how the codes are used in particular systems.

Clock, Timing and Control Signals

As stated previously, a computer program is a series of steps that a digital processor must execute in sequence in order to accomplish a task dictated by the program. These steps in sequence occur at particular set times dictated by the timing and control signals. Within each step, instructions are dictating how electronic circuits are operating to perform the functions called for by the program. The instructions occur at specific times and the circuit operation occurs at specific times controlled by the timing and control signals.

Clock

The heart of the timing circuits is the clock. Its source is usually a crystal-controlled oscillator that generates signals at a very precise frequency. Its signal output is formed into rectangular pulses that have very fast rising and falling edges. Typical pulses are shown in *Figure 6-6a*. The rising and falling edges of the clock pulse provide precise times for controlling electronic circuit action. The clock may have just one series of pulses like phase 1(Φ_1), or it may have additional phases as shown in *Figure 6-6a*. The additional phases provide additional timing signals for the control of circuits. As shown in *Figure 6-6a*, some of the circuits controlled by the clock trigger on the rising edge of the clock pulse, while other circuits trigger on the falling edge of the pulse. Such alternatives in the triggering of circuits provide a wide selection and flexible means for timing the operation of electronic circuits.

Gated Latch

A specific example of how electronic circuits are timed is shown *in Figure 6-6b*. The electronic circuit shown is called a gated latch. It is used for temporary storage of digital data. The inputs to the gated latch are the binary signal D (either 1 or 0), and the clock. The outputs are Q and Q', which are complementary to each other—if Q = 1, Q' = 0 or vice versa. A signal that appears on D is only stored in the latch and appears on Q after it is "clocked in," i.e., the clock has appeared and has timed in the D signal. As shown in *Figure 6-6b*, Q only changes after D changes and a clock signal times the change into the latch.

The latch receives its name from the fact that it is a temporary storage electronic circuit that latches on to data and holds it. The gated latch means that data is gated in at a particular time.

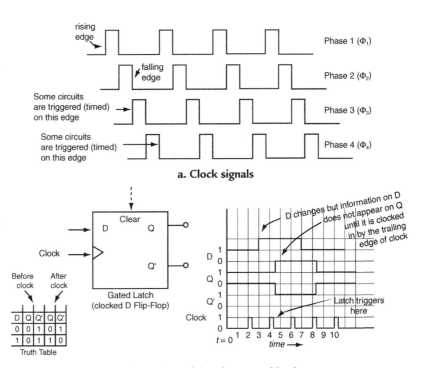

a. Clock signals

b. Timing of signals at gated latch

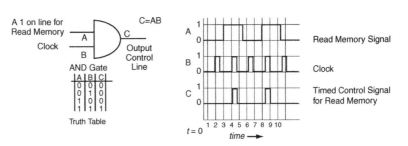

c. Timed control signal using AND gate

Figure 6-6: Clock signals for timing and control

A truth table, shown in *Figure 6-6b*, identifies the output Q and Q' values for each D input value. It identifies the state of the signals before and after the clock.

AND Gate Control

Another example of signal timing is shown in *Figure 6-6c*. Here a 2-input AND gate is used to time a control signal. The control signal required tells a memory to read information from memory. The address of the information has been received by the memory and decoded prior to the receipt of the control signal. An AND gate is used to provide the memory read signal at a precise time. As the truth table shows, both inputs to the AND gate must be a 1 for the output to be a 1. If both or one input is a 0, the output is a 0. By placing the memory read signal on the A input to the AND gate, when it is a 1, the memory is to be read. However, the control signal to actually tell the memory to read will not occur on the output of the AND gate until the clock signal is a 1. As a result, the memory is read at a precise time determined by the clock. The read signal on the input to the AND gate overlaps the clock signal in time, and can vary significantly in time position in relationship to the clock and still be timed correctly. The AND gate output, the memory read pulse in this case, turns out to be the same width as the clock pulse.

The fact that a clock may have different phases adds to the flexibility of the timing and control signals. For example, the clock used in *Figure 6-6b* might use Phase 2, while the clock used in *Figure 6-6c* might be Phase 4. This demonstrates the flexibility, mentioned previously, that a designer has to time the system circuits.

Interrupts

A signal that controls a digital processor at unexpected or random times is called an interrupt. It interrupts the digital processor from what it is doing and directs it to do something different, as indicated by the interrupt signal. A STOP signal terminates whatever the processor is doing. It usually occurs at random times depending on the need to shut down the processor. Or maybe the processor is following a program and input signals are required. When the inputs are available, the input circuits notify the digital processor that the inputs are present. This initiates an interrupt to the processor, which halts what it is doing and inputs the data. After the data is inputted, the processor continues from the place it was interrupted. The CPU keeps track of where the processor is when the interrupt occurred.

Similar action occurs at the outputs. The processor is required by the program to output data to an external unit. The processor addresses the I/O and selects an output. The output circuits send an interrupt to the CPU to signify that the output is ready. The interrupted processor switches to a routine to output the data. When the transfer to the output is complete, the processor returns to the program location directly after the location at which it was interrupted.

The application of a digital processor may be dictated by its response to an interrupt. Some processors respond very quickly to interrupts so that the overall performance to execute its program and complete a task is not affected. While other processors may be slow to respond to interrupts, and, therefore, if an application depends on many interrupts, the overall performance of the processor will be slowed a great deal. The ultimate speed at which the processor can accomplish the task is severely limited. Some digital processors only respond to an interrupt when they want to, not randomly or unexpectedly. Most modern digital processors respond quickly to interrupts that occur at random and unexpected times.

Status Bits

Digital processors operate using control signals derived from the condition of check bits called status bits. Status bits are stored in a register. A register is a chain of latches strung together to temporarily store a set number of bits; as an example, a 16-bit register stores 16 bits. Most registers store the number of bits in the word being used throughout the digital system. The status register is somewhat different. It holds a variety of different bits where the state of each bit is somewhat independent of the other bits in the register. Many of the bits are set independently and their value depends on the result of a particular processor operation. For example, what was the sign of a number as a result of an arithmetic operation—positive or negative? A status bit is set after the operation is executed to indicate the result. Was the result of an arithmetic operation greater or less than zero? A status bit is set to indicate the result. Was there a carry or a borrow when an arithmetic operation was performed? Is the number too large for the digital system to handle? The setting of status or condition bits after such operations, and the checking of the bits by the processor, contribute to the control of the operation of the digital processor as it executes its program.

Example 6. Status Register

The N bit of a status register is set when the result of an arithmetic operation is negative. Show an example of how this occurs.

Solution:

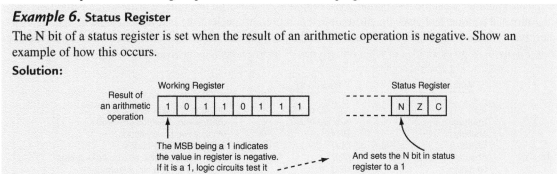

More About Software

Refer again to *Figure 6-4* for a short review. The digital information flowing through a digital processor flows as a given combination of bits—a 32-bit address code, a 16-bit instruction code, or an 8-bit character code. Circuits that identify and decode the digital information must identify the value of each bit (either a 1 or a 0) and act as a result of the value to decode the information. As stated previously, the program that the processor follows is a series of instructions in sequence. Each instruction has a given number of bits and a unique code for a particular instruction. The instructions come from memory to the processor over the data bus. Inside the processor the instructions are stored temporarily in the instruction register so that the instruction decoder circuits can decode them. The decoder evaluates the bits and identifies the action the processor must take to execute the instruction.

Humans write the computer programs. The instructions to the computer must be written in a language that humans understand; yet the instructions that the computer follows must be in digital codes that the computer understands. A conversion is required from the human language to the digital codes that the machine (processor) understands. The digital code that the machine understands is called *machine code*. A computer program written in machine code is called a *machine-language program*.

Machine-Language Programs

Humans can write programs in machine language. To do so, the programmer writes the program directly in the digital codes that the machine understands. No conversion is necessary. The machine can decode the instructions directly and execute them to accomplish the task required. However, the task is extremely difficult, tedious and time consuming, and if errors are made, and they will be regularly, it becomes an even more difficult and tedious task to find the errors and correct them.

Assembly-Language Programs

In order to make it easier to write the programs, the manufacturers of digital processors have designed their processor to respond to instructions that are closer to human language. These instructions are called *assembly-language instructions*. They are easier to understand than machine code but require the manufacturer to provide a program to convert the assembly-language instructions into machine code. Such a program is called an *assembler*. A computer is much more accurate in doing the conversion, and by processing an assembly language program for a particular processor using its assembler, all the instructions are converted very accurately into machine code for that processor.

Mnemonics

The operation or action that the assembly-language instruction causes the processor to perform is identified by an abbreviation called a *mnemonic*. The abbreviation used for the mnemonic gives a strong suggestion to the programmer what the instruction does. *Figure 6-7a* shows an example of arithmetic instructions and their directed actions, and gives the mnemonic that represents each of the instructions. The mnemonic is a short two or three letter symbol that identifies to the programmer the processor action caused by the instruction. *Figure 6-7b* gives an idea of what other types of instructions may be available in digital processors.

Arithmetic	Mnemonic	Action
Add	A or AD or ADD	Addition of two binary codes
Subtract	S or SU or SB	Subtraction of two binary codes
Multiply	MPY	Multiply two binary codes
Divide	DIV	Divide two binary codes
Absolute Value	ABS	Take absolute value of a binary number
Negation	NEG	Change sign of a binary number
Shift	ROL or ROR	Shift left or shift right
Increment	INC or INR	Add 1 to binary code
Decrement	DEC or DCR	Subtract 1 from binary code

a. Example of mnemonics for arithmetic instructions

Logical	Data Movement	Branch	Comparison
AND	Move	Unconditional	Less than
OR	Load	Conditional	Greater than
NOT	Store	Subroutine	Equal
XOR			

b. Examples of other processor instructions

Figure 6-7: Examples of digital processor instruction set

Operands

In an assembly-language instruction, the instruction itself describes the operation to be performed, but does not say what is to be operated on; therefore, *operands* (what is to be operated on) must be added to the instructions. For example, the instruction:

Mov A,B

The mnemonic MOV means that a move operation is to be performed and the operands are register A and register B. The contents of register A are to be moved to register B. Suppose that register B is the program counter; therefore, it contains the memory address of the next instruction of a program or subroutine. By loading register A with the address of the first instruction of a program, moving the contents of register A to register B a new program is started. Incrementing register B (subtracting one from its contents) with the instruction:

Inc B,

causes the processor to step to the next instruction. After the instruction is executed, the program loops back to the Inc B instruction and the processor steps to the next instruction. The processor steps through addresses of the instructions in sequence to execute the program.

Sophisticated Programming Languages

The writing of a computer program to perform a task consists of organizing the digital processor instructions into the correct sequence. It is a paper process that doesn't require the building of any hardware, but just understanding the processor's instructions and using them to manipulate existing hardware to perform the task required. Thus, programs are called *software*, and people that write programs are called software engineers or just programmers. It is the objective of programmers to write their programs in a language as close to human language as possible. They would also like to learn a particular programming language and not be restricted to using it only for one processor. They would like to apply their knowledge of the language to other processors solving other application problems. To satisfy this need, sophisticated programming languages have been developed.

Sophisticated programming languages are a step up and beyond assembly-language programming. They are, once learned, used for writing many different programs, using different processors. Such languages are referred to as high-level languages because they are somewhat general purpose because they are used to program different processors.

Whatever high-level language is used one thing is certain, the program must be converted to machine-language code. In earlier times this was a two-step process. First a program called a *compiler* converted the high-level language to assembly language. Then, an assembler was used to convert the program to machine code. Today most compilers convert the high-level language directly to machine code. In addition, many digital processors are members of a family of processors; the compiler for a particular processor usually handles the whole family of processors.

Software Summary

Figure 6-8 provides a summary of programming. A digital processor can be programmed directly in machine language, but it is very tedious and difficult to find errors. Or it can be programmed in assembly language, put through a specially designed program (an assembler) that converts the program to machine code. Or it can be programmed using a sophisticated high-level general-purpose language. The program must be put through a specially designed program (a compiler) that converts the high-level language instructions into machine code for the particular processor used. Fortran was an early high-level language. Today, "C", "C⁺", UNIX, JAVA are names of sophisticated languages for writing programs.

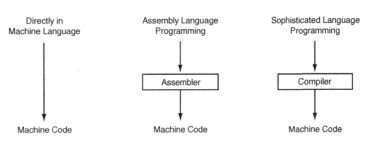

Figure 6-8: Programming computers

How Parts of a Processor Perform Their Functions

ALU—Arithmetic Logic Unit

The discussion now switches to how various parts of a processor perform their functions. The first of these is the arithmetic logic unit (ALU). An arithmetic function performed by the ALU is addition, shown in *Figure 6-9*. The central electronic circuit used for addition is an adder, shown *in Figure 6-9a*. The full-adder has three inputs—the two binary numbers to be added and a carry input. *Figure 6-9a* shows not only the

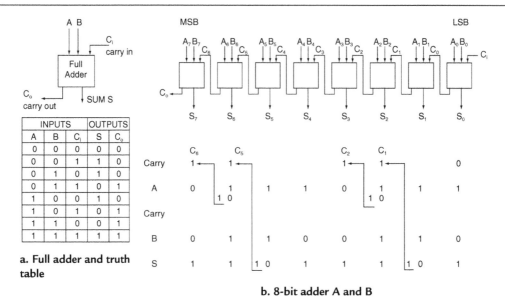

a. Full adder and truth table

b. 8-bit adder A and B

Figure 6-9: The addition function

full-adder block diagram, but also its truth table. A truth table, remember, catalogs the state of the outputs for all the states of the inputs. If A or B or C_i is a 1, the sum bit will be a 1. When A and C_i or B and C_i or A and B are a 1, the sum bit is a 0 and C_o will be a 1. When A and B and C_i are all 1s, the sum bit is a 1 and C_o is a 1.

Figure 6-9b shows an 8-bit adder and the addition of two 8-bit binary numbers A and B. Note how the C_o output of one stage of the adder becomes the C_i input to the next stage to the left. The example shows how the carry bit is generated and propagates to determine the sum bit at the next stage. The speed of operation of the adder is determined by how long it takes the carries to propagate through the adder. Using the adder multiple times, plus shifting, provides the multiplication function. Subtracting is performed by adding the one's (1's) complement of one of the binary numbers instead of the number itself, and multiple subtractions, plus shifting, results in a division function.

ALU—Logic Functions

Figure 6-10 shows three logic functions that are normally available in an ALU. Using A and B 4-bit binary numbers as examples, the logic operations are performed bit by bit giving the result C from LSB to MSB. A 1 appears as the result for the AND function only when A *and* B are a 1. A 1

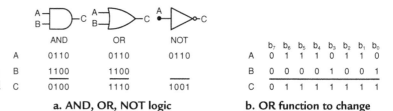

a. AND, OR, NOT logic

b. OR function to change bit value

Figure 6-10: Logic functions

appears as the result C when A *or* B *or both* are a 1 in the OR function. The complement of the input—a 1 if input is a 0, or a 0 if input is a 1—will appear as the result C for the NOT function. The electronic circuit that performs the NOT function is called an inverter.

An example of using the OR function to set particular bits in a binary number to a particular value is shown in *Figure 6-10b*. In the 8-bit binary number for A, 01110110, bits b_0 and b_3 and b_7 are 0. The program requires that bits b_0 and b_3 be set to a 1. By performing an OR function between A and B, where B is the binary number 00001001, the result C will have bits b_0 and b_3 set to a 1. The bits that were 1s in B will be set to a 1 in the result C.

Memory and Input/Output

Figure 6-11a shows the typical interface between a microprocessor and memory. This corresponds to what was shown in *Figure 6-1*, but details it just for memory. The address bus carries the binary code put out by the microprocessor for the address of information in

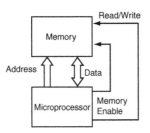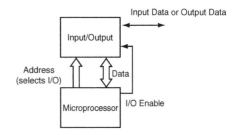

Figure 6-11: Data to and from memory and input/output

memory. The memory size determines the number of bits in the code. The data bus will either have data on it that is put there by the microprocessor to store in memory (write to memory), or it will have data or instructions that come from the address location (reading from memory) to the microprocessor. Whether the memory is being written to or read from is controlled by the read/write signal. In addition, whenever memory is to be used, whether writing or reading, an enable signal is sent to memory to activate it. The read/write and enable signals are timed control signals operating at precisely designed times.

Input/output or I/O circuits operate very similar to memory as shown in *Figure 6-11b*. The microprocessor sends out an I/O address on the address bus to specify which I/O is to be used. At the same time, a control signal tells the I/O that it wants to input data to the microprocessor over the data bus; or that it wants to output data that the microprocessor is placing on the data bus. As with memory, timed control signal enables the I/O circuits. They are not active until the enable signal arrives.

Addressing Modes

Program instructions tell a digital processor what to do, where to find the information it is to use with the instruction, and where to put the result after the instruction is executed. Addresses or addressing is needed to direct the processor to the correct location. *Addressing modes* are the means by which the instruction indicates the address. They are the designed ways that the instruction tells the processor how to locate the information it needs to use with the instruction. There are several common addressing modes for digital processors. Five different ones are shown in *Figures 6-12, 13, 14, 15* and *16*.

Immediate Addressing

Immediate addressing is diagrammed in *Figure 6-12*. The program counter contains a memory address that points to the operation code (op code) of the instruction—the operation the instruction wants the processor to perform. Following immediately after the op code, in the next memory location, is the data on which the instruction will operate. So if the instruction is addressed with immediate addressing, the code that describes the operation to be performed is in the memory location addressed by the contents of the program counter, and the data is in the next memory location. There is relatively little decoding. The instruction knows immediately where the data (operand) is located.

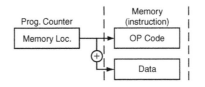

The op code is contained in the memory location pointed to by the PC followed, in the next memory location, by the data that is to be used.

Figure 6-12: Immediate addressing

Register Addressing

Figure 6-13 diagrams register addressing. Here the data is not contained in memory locations but in registers. The instruction contains the op code and specifies in which register the source data is located and, if need be, the register for the destination data.

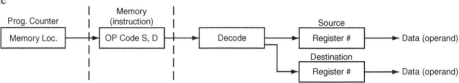

The source and destination data are contained in registers. The instruction contains the op code and specifies a register as a data location. In the case shown, data is contained in a register for both the source and destination.

Figure 6-13: Register addressing

Register Indirect Addressing

Figure 6-14 diagrams register indirect addressing. In register indirect addressing, instead of specific registers containing the data to be operated on as in *Figure 6-13*, now the specific registers contain the memory address of the data. Thus, loading different memory locations in registers causes the processor to operate on different data stored in memory.

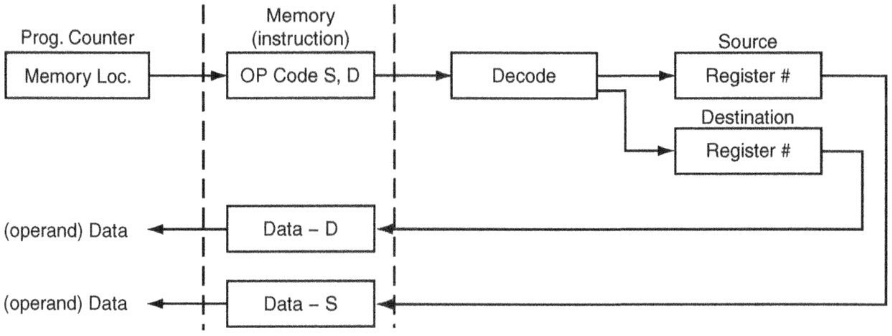

Instead of the register containing the data as in register addressing, the register contains the memory address of the data. Thus, by loading the register with different memory locations, different data is operated on by the instruction.

Figure 6-14: Register indirect addressing

Indexed Addressing

Figure 6-15 diagrams indexed addressing. The next memory location after the op code contains an index. The address of the data to be used is the sum of a value in a register and the value of the index. The instruction is used separately for the source and for the destination. Indexed addressing is used extensively for data that is grouped together in memory. The program can be modified quickly to select a different set of data by changing the index in the instruction.

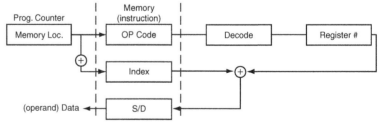

For index addressing, the address of the data is the sum of a value contained in a register and the value of the index. The selection of data addresses that appear in groups can be modified quickly by changing the index.

Figure 6-15: Indexed addressing

Example 7. Register Indirect Addressing

Show an example of register indirect addressing.

Solution:

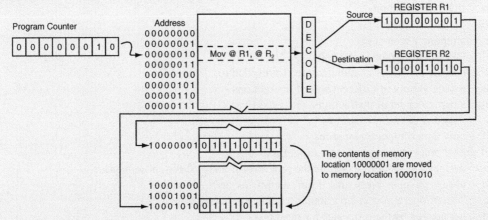

The program counter points to the memory address where the instruction MOV @R1,@R2 is located. R1 and R2 are register numbers and the @ sign indicates that the contents of the register is the address in memory where the information on which the instruction is to operate is located. The instruction, with MOV as the op code, says move the contents of the memory location whose address is the contents of R1 (the source) to the memory location whose address is the contents of R2 (the destination).

Direct or Symbolic Addressing

Figure 6-16 diagrams direct or symbolic addressing. In immediate addressing of *Figure 6-12*, the memory location following the op code contained the data to be operated on. In direct or symbolic addressing, the next location in memory after the op code is an address in memory that contains the data.

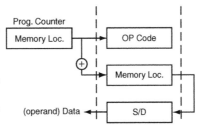

As with immediate addressing, the instruction op code is contained in the memory location, but now the next word does not contain the data for the instruction, but contains a memory address for the data.

Figure 6-16: Direct or symbolic addressing

Instruction sets for different processors use specific symbols and notations for their instructions and for their addressing modes. They usually are specific to the particular processor. In *Chapter 7*, there will be further discussion of the addressing modes used for the MSP430 family of microcontrollers.

Summary

This completes the discussion about the basic operation of a digital processor, some of its specific functions, and how the processor is made to do what is instructed by a program to perform a desired task. In the next chapter, the discussion centers on the details of programming the processor.

Chapter 6 Quiz

1. A digital processor, more commonly called a digital computer, has a unit that is the brain of the system called the:
   a. I/O—input/output.
   b. permanent memory.
   c. temporary memory.
   d. CPU (central processing unit).

2. Each digital processor is manufactured to respond to:
   a. a wide variety of different sets of instructions.
   b. a particular set of instructions.
   c. only one or two instructions.
   d. only input/output instructions.

3. A digital processor responds to a program that is:
   a. designed to randomly operate the processor in many different sequences.
   b. always changes every time it runs the processor.
   c. a set of operations in a particular sequence to accomplish a task.
   d. not needed by the processor for most tasks.

4. The instruction/data bus is used:
   a. to send addresses to locate instructions and data to be delivered to the CPU.
   b. to identify inputs and outputs to receive or output data for the CPU.
   c. to send timing information throughout the system.
   d. a and b only above.
   e. c only above.

5. Clock signals in the digital processor:
   a. precisely control the transfer, manipulation and storage of information throughout the processor.
   b. must be very accurate in time.
   c. are a series of repetitive pulses that have fast rise and fall times.
   d. all of above.
   e. a and c only above.

6. Power systems in digital processors:
   a. must have very accurate voltage regulators and good power dissipation control.
   b. require no precise voltage or current control.
   c. require little concern for power dissipation.
   d. operate with high voltage and high current.

7. The devices that have contributed most to low power dissipation in digital processors are:
   a. power transistors.
   b. bipolar logic transistors.
   c. CMOS—complementary-metal-oxide-semiconductor—integrated circuits.
   d. a mix of bipolar and MOS devices.

8. The data bus carries to memory a digital code representing:
   a. the instruction address and the data address.
   b. only the instruction address.
   c. only the data address.
   d. none of the above.

9. The arithmetic logic unit (ALU) in the CPU:
    a. provides the I/O capabilities.
    b. provides the clock capabilities.
    c. provides the storage capabilities.
    d. provides the logical, computational, and decision making capabilities.
10. Read-only memory (ROM), random-access memory (RAM) and registers are:
    a. logical circuits contained in the CPU.
    b. types of memory that are or maybe contained in a CPU.
    c. data transmission circuits contained in a CPU.
    d. I/O circuits contained in a CPU.
11. A microcontroller unit (MCU):
    a. is an industrial control computer made up from individual ICs.
    b. is the smallest possible microcomputer.
    c. is a microcomputer IC that is adapted to the industrial control market.
    d. is a computer made up of individual ICs, but designed for low-power use.
12. The MSB (most significant bit) of a word is:
    a. is the second bit in the code representing the word.
    b. is the left-most bit in the code representing the word.
    c. is the right-most bit in the code representing the word.
    d. is the middle bit in the code representing the word.
13. In a parallel data transfer:
    a. all bits arrive at a point at the same time.
    b. all bits do not arrive at a point at the same time.
    c. all bits are delayed one bit at a time.
    d. all bits arrive at a point one after another.
14. In a serial data transfer:
    a. all bits arrive at a point at the same time.
    b. all bits are collected, delayed, and then arrive at the same time.
    c. all bits are delayed, then arrive at a point at the same time.
    d. all bits arrive at a point one after another in sequence.
15. The ASCII code can identify:
    a. numbers only.
    b. letters only, not special characters.
    c. numbers, letters, special characters, commands.
    d. commands only.
16. Clock signals inside a digital processor:
    a. may trigger electronic circuits only on the falling edge.
    b. may trigger electronic circuits into action on either the rising or falling edge of the clock pulse.
    c. may trigger electronic circuits only on the rising edge.
    d. don't trigger electronic circuits on the rising or falling edges.
17. An interrupt signal to a digital processor:
    a. speeds up the operation of a digital processor.
    b. controls a digital processor at unexpected or random times.
    c. acts just like any other digital processor control signal.
    d. none of the above.

18. A mnemonic is a:
    a. short two or three letter symbol that represents a program instruction.
    b. random set of letter symbols that varies continuously.
    c. long set of letter symbols that is an instruction in itself.
    d. symbol that has no relationship to assembly-language programming.
19. Programs written in high-level languages:
    a. can be written for different processors using the same language.
    b. must be converted to machine code to run the processor.
    c. use a compiler to convert the high-level language to machine code.
    d. all of above.
    e. c only above.
20. Addressing modes for a digital processor:
    a. are always immediate addressing.
    b. are the designed ways the instruction tells the processor what to do.
    c. are the means by which the instruction indicates the action to be taken by the processor.
    d. c only above.
    e. b and c only above.

Answers: 1.d, 2.b, 3.c, 4.d, 5.d, 6.a, 7.c, 8.a, 9.d, 10.b, 11.c, 12.b, 13.a, 14.d, 15.c, 16.b, 17.b, 18.a, 19.d, 20.e.

Examples of Assembly-Language Programming

Introduction

Many times the easiest way to understand how to do something is to work with examples. That is the subject of this chapter. By looking at small subprograms that have been written to accomplish specific tasks, the reader will be introduced to assembly-language programming. The objective is to provide a base of understanding of how an assembly-language program is formulated so that programs can be deciphered, at least to obtain a "feel" for what the program is trying to accomplish. In no way will this chapter be a thorough coverage of assembly language, its format, its detail, its uniqueness, but, hopefully, by taking small segments of programs and discussing them, line by line, enough information will be transmitted to accomplish the basic understanding desired.

A Processor for the Examples

In order to be specific about the programs discussed and the tasks, a Texas Instruments MSP430 Family microcontroller has been chosen to use for the programming examples because it is readily available, well-supported with documentation and applications information, and has relatively inexpensive evaluation tools. The family of microcontrollers is designed specifically for industrial control, instrumentation, and measurement tasks with low-power, extended battery-life applications as prime design objectives. These specifications are not necessarily important to its choice for this chapter. Rather, the easy-to-understand architecture, instruction set, and family structure contributed significantly to the selection.

About the MSP430 Family

In Texas Instruments' words, "The MSP430 devices constitute a family of ultra low-power, 16-bit RISC microcontrollers with an advanced architecture and rich peripheral set. The architecture uses advanced timing and design features, as well as a highly orthogonal structure to deliver a processor that is both powerful and flexible." The architecture is called "von Neumann" since all program, data memory and peripherals share a common bus structure. RISC means reduced instruction set computer, and defines a specific design approach for the microcontroller. There are only 27 core instructions, which, through the technique of combining core instructions—called emulation—is expanded into a set of 51 instructions. The core instructions are built into hardware, while the emulated instructions are formed by the assembler (the program that interprets the assembly-language mnemonics and produces machine code).

Family Block Diagram

A MSP430 Family system block diagram is shown in *Figure 7-1*. Note the 16-bit memory address bus (MAB), the 16-bit memory data

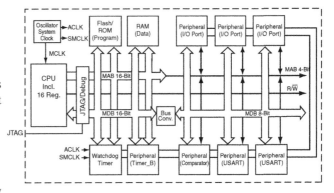

Figure 7-1: MSP430 family block diagram
Courtesy of Texas Instruments Incorporated.

bus (MDB), and the bus conversion for the I/O, USART and comparator. In *Chapter 10*, the MSP430F1232, part of a family of MSP430F12XX devices, will be used in an application. How the MSP430F12XX devices vary in the family is shown in *Table 7-1*.

MSP430F12XX Devices of the Family

430 Device	Main	Memory Flash	RAM	I/O(8)	BOR	WDT	TA	C	USART	ADC
F122	4kB	256B	256B	3		X	X	X	1	slope
F123	8kB	256B	256B	3		X	X	X	1	slope
F1222	4kB	256B	256B	3	X	X	X		1	SAR 10
F1232	8kB	256B	256B	3	X	X	X		1	SAR 10

Table 7-1: Devices of the MSP430F12XX family

The MSP430F12XX devices have program memory that is Flash memory. The devices are identified with a F in the device number as shown in *Table 7-1*. The Flash memory, which is made up of a large main memory and a smaller information memory, provides in-system programmability that permits flexible code changes, and, for remote systems that are battery operated, field upgrades. Flash memory is electronically erasable programmable ROM (EEPROM), and is programmable and erased by applying a voltage. The MSP430F12XX devices vary in program memory size from 4 kB to 8 kB, and all have the same size RAM. They have three 8-bit I/Os, a watchdog timer (WDT), and 16-bit PWM timer (TA), a USART communication interface, and ADCs. Some have no comparators (C), some have brownout reset (BOR), and the ADC varies from slope to SARs. They are packaged in 28-pin packages. The brownout reset is a function that resets the microcontroller when the power supply voltage reaches a critical low value. When the power supply voltage is re-established, the microcontroller starts again from the RESET condition.

MSP430 Family Characteristics

The MSP430F1XXX family, which extends through the F13x, F14x, F15x, and F16x devices, includes devices with more USARTs and timers, hardware multipliers, 12-bit ADCs, an I²C communications bus, and SVSs—supply voltage supervisors. These devices are in 64-pin packages.

Another family group, the MSP430F4XX devices, extends the family into 64-pin and 80-pin packages. The devices have up to 60 kB of program memory and 2 kB of RAM, and most have 12-bit ADCs. All have LCD drivers—from 96 to 160 segments.

A segment of the family is based on ROM programming, the MSP430C or P3XX devices. They have similar LCD drivers to the F4XX devices, but do not have Flash memory. There are devices with 32 kB of program memory and 1 kB of RAM, but the most exotic have 6-channel, 14-bit ADCs that are packaged in 64-pin packages. Other devices are in 100-pin packages and have 32 kB of program memory, 1 kB of RAM, an 8-bit interval timer, a 16-bit timer A, a USART, and a hardware multiply. Such a variety of devices allow the designer of control systems a wide choice of design options.

The CPU

The CPU for the family is the same. As mentioned previously, it is a 16-bit RISC CPU. It consists of a 16-bit ALU, 16 registers and instruction control logic. The register arrangement is shown in *Figure 7-2a*. Note the common memory address bus (MAB) and memory data bus (MDB). Four of the registers are for special purposes: program counter, stack pointer, status register and constant generator. The rest are general-purpose registers. The constant generator supplies instruction constants, and is not used for storage. The sixteen fully-addressable, single-cycle 16-bit registers and orthogonal architecture provides versatility and simplicity in system applications.

Program Memory and Data Memory

A map of memory available for the MSP430 family is shown in *Figure 7-2b*. There are 64KB (65,536) of addressable memory spaces divided over the address spaces from 0 to hexadecimal 0FFFFh (1111 1111 1111 1111 in straight binary). The special-function registers and peripheral module addresses are from 0 to 01FFh. Recall that an h after the address notation means it is in hexadecimal format and that 01FFh is really a 16-bit word with bits of 0000000111111111. In hexadecimal notation, when the hexadecimal address starts with the MSB of A,B,C,D,E or F, a zero is placed in front of the hexadecimal value to make sure the address is identified correctly, for example, 0BE14h.

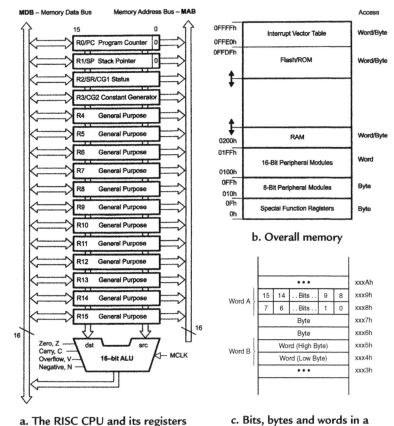

a. The RISC CPU and its registers

b. Overall memory

c. Bits, bytes and words in a byte-organized memory

Figure 7-2: CPU, registers and memory map
Courtesy of Texas Instruments Incorporated

The memory addresses (memory space) from 0200h to 0FFFFh are shared by data and program code memory. The space from 0FFE0h to 0FFFFh is reserved for a table of interrupt vectors in Flash/ROM (Flash for F devices) and more Flash/ROM is devoted to program, branch control tables and data tables below the address 0FFDFh. The remaining addresses are used for Flash/ROM and RAM (random access memory) and are used for program and data storage.

Words of data, which occupy 16 bits or 2 bytes, are only located at even addresses, while bytes can be located at odd or even addresses. If a data word is located at an even address, the low byte is at the even address and the high byte is at the next odd address. The typical arrangement is shown in *Figure 7-2c*. Word A shows the actual bits of the high and low bytes, while word B is just identified by the position of the "high byte" and the "low byte."

Note also that if a peripheral module is a 16-bit module, its address will be between 0100h and 01FFh. If it is an 8-bit module, its address will be between 010h and 0FFh. The addresses from 0 to 0Fh are reserved for special-function registers, SFRs. The functions served by the various portions of memory are shown in *Figure 7-2b*, and shows that some of the functions are only accessible with 8-bit (byte) or 16-bit (word) instructions, while others are accessible with either 8-bit or 16-bit instructions.

Instructions are fetched from program memory with 16-bit addresses, while data memory can be addressed either using 16-bit or 8-bit instructions. Program code can either be in Flash/ROM or RAM because the Flash/ROM and RAM are connected via the same two buses: the memory address bus (MAB) and the memory data bus (MDB). In addition to program code, data can be placed in the Flash/ROM section of the memory map, a significant advantage for data tables.

Peripherals

The variation of peripherals is one of the major advantages of the MSP430 family. A general overview of the peripheral variations were pointed out in the family discussion, but more specific variations are shown in *Figure 7-1*. Shown are variations of the available I/O ports, as well as a comparator and a USART (**U**niversal **S**ynchronous/**A**synchronous **R**eceiver/**T**ransmitter). Within the family, also available are different ADCs, different timers, and even a hardware multiplier. Most of the peripherals operate in byte format, and modules with 8-bit data buses are connected by bus-conversion circuitry to the 16-bit CPU. Most of the peripherals use a 5-bit memory address bus.

Operation Control and Operating Modes

The contents of the special-function registers, mentioned previously, control the operation of the different MSP430 functions. The bits contained in the register(s) select system operation, enable interrupts, provide information about the status of interrupt flags (caution signals that tell a program whether it can continue or not) and define the operating modes of the peripherals.

Mode	Status Register Bits				CPU	Clock Functions			
	SCG0	SCG1	OSCOFF	CPUOFF		MCLK	SMCLK	ACLK	DCO
AM	0	0	0	0	ON	ON[1]	ON[1]	ON[1]	ON[1]
LPM0	0	0	0	1	OFF	OFF	ON	ON	ON[2]
LPM1	1	0	0	1	OFF	OFF	ON	ON	OFF
LPM2	0	1	0	1	OFF	OFF	OFF	ON	OFF
LPM3	1	1	0	1	OFF	OFF	OFF	ON	OFF
LPM4	1	1	1	1	OFF	OFF	OFF	OFF	OFF

Notes:
1. Various modules are active as required.
2. If DCO is used as clock source.

Figure 7-3: Operating modes of MSP430 family

Because the microcontroller that is used for the example digital processor has been designed to operate at low power, and many of its applications are battery powered, there are a number of operating modes specially directed to saving power consumption. Six operating modes, AM through LPM4, are shown in *Figure 7-3*. AM is the active mode where the CPU is powered as well as all other modules that are designated to be active by the program. Modes LPM0 to LPM4 are so-called low-power modes with successively less power dissipated. If the operating mode is one of the LPM modes, anytime the CPU is required by the program, it must be called into the active mode by the program. To simplify the operation for the examples in this chapter, the only modes used will be the active mode and the LPM3 mode.

Watchdog Timer

There is another component within the MSP430 microcontroller, the watchdog timer that is particularly associated with remote low-power operation. It is shown in *Figure 7-1*. It is called a watchdog timer because its primary function is to perform a controlled system restart after a software problem occurs. This is for system protection in case an application is in a remote battery-operated location and some glitch causes a software failure. After a set time interval, a system reset is generated and the program is restarted. What is important is that if the system is operating properly and the watchdog timer is active, the program must reset the watchdog timer before its time interval expires, otherwise the system will be reset. If the watchdog timer function is not necessary, the timer can be used as an interval timer. Such use is in one of the program examples.

System Reset

To make sure a system application always starts the same way, a reset of the system is initiated by the turn-on of power, called a power-on reset (POR). There is also another reset, called power-up clear (PUC), that is for resetting if the watchdog timer has expired, or there is some system violation. Reset is considered a system interrupt.

Interrupts

In *Chapter 6*, an interrupt was described as a signal that interrupts the digital signal processor from what it is doing and directs it to do something different as indicated by the interrupt signal. It may control the digital processor at unexpected or random times.

One of the most common types of interrupts is from one of the peripheral modules, such as an I/O unit. The processor has had to wait on an input until it is available. Now it is available and signals the processor with an interrupt signal, and the processor accepts the input. If an-

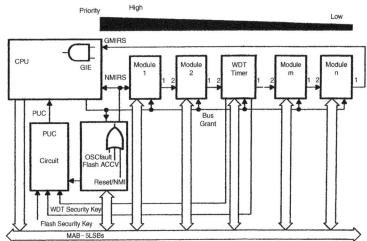

Figure 7-4: MSP430 interrupt priority scheme
Courtesy of Texas Instruments Incorporated

other interrupt were to occur simultaneously, the MSP430, as shown in *Figure 7-4*, has an interrupt priority scheme. The peripheral modules that are nearer the CPU in the connection chain have the higher priority in case two signals were to appear at the processor at the same time. While the interrupt occurs, all other interrupts are blocked by default. For specific devices the modules included have specific hardware positions in the chain. Each device's interrupts are described in an interrupt vector table in the data sheet for the device.

Oscillators and Clock Generators

Included in the microcontroller is a built-in oscillator that uses only an external crystal. The common oscillator uses a watch crystal and oscillates at 32,768 Hz, but using a higher-frequency crystal, it can oscillate at frequencies from 1MHz to 8MHz. In addition, there is a digitally-controlled oscillator that is digitally tuned. Such flexibility makes it easy to select a particular clock operating frequency.

The MSP430 basic clock system is shown in *Figure 7-5*. For the MSP430F12XX microcontroller used for this chapter, the LFXT1 oscillator is the low/high frequency crystal oscillator mentioned above. The DCO oscillator is a RC-type oscillator and is digitally controlled to adjust the frequency. Other family devices have a second crystal oscillator, XT2, that can oscillate at frequencies from 450kHz to 8MHz.

The main system clock, MCLK, can use either LFXT1 or DCO as its source controlled by the state of the selection bits SELM. By software commands setting the state of the DIVM bits, the source for MCLK can be divided by 1, 2, 4, and 8. The state of the DCOR bit, which chooses either an internal or external resistor, defines the fundamental frequency of the DCO. Then the state of the RSEL bits selects one of eight nominal frequency ranges defined in the specific device data sheet. The three DCO bits divide the DCO range selected by the RSEL bits into eight frequency steps approximately 10% apart. Because the DCO is a RC-type oscillator, its frequency varies with temperature, voltage and from device to device. The five MOD bits set the conditions to adjust and stabilize the DCO frequency.

The action of the three RSEL bits and the three DCO bits to set the DCO frequency after the fundamental frequency is set is shown in *Figure 7-5b*. The three RSEL bits, based on their binary value, select one of eight mominal frequency ranges for the DCO. The ranges are defined for a specific device in the device's data sheet. The three DCO bits, based on their binary value, divide the DCO range selected by the RSEL bits into eight frequency steps, approximately 10% apart. Thus, setting the binary value of the RSEL and DCO bits will result in a DCOCLK frequency for the system. The typical ranges and steps are shown in *Figure 7-5b*.

The auxiliary clock, ACLK, uses LFXT1 as its source, and divides LFXT1 down by 1, 2, 4, and 8 based on the state of the DIVA bits.

The subsystem clock, SMCLK, uses either XT2CLK or DCO as its source, again divided by 1, 2, 4, or 8 based on the state of the DIVS bits. However, when XT2CLK is not present, as is the case for the MSP430x11xx and x12xx devices, an internal connection is made in the MSP430 that connects LFXT1CLK in its place.

The choice of which clock system to use is based upon the application. Systems requiring very precise timing with little variation allowed will use the high-frequency crystal oscillators as sources. Systems with very nominal speed and accuracy for the timing and require very low power will use the DCO.

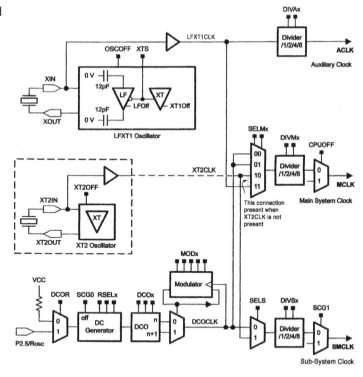

Note: XT2 Oscillator

The XT2 Oscillator is not present on MSP430x11xx or MSP430x12xx devices. The LFXT1CLK is used in place of XT2CLK.

a. Clock system block diagram

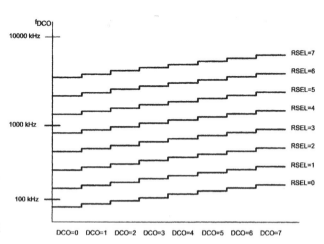

b. Typical DCOx range and RSELx steps

Figure 7-5: MSP430 basic clock system
Courtesy of Texas Instruments Incorporated

Timers

Timers are digital counters that use a clock at a set frequency as the source to establish time intervals by counting a certain number of input pulses. Thus, specific time periods can be established either by the number of pulses counted, or by changing the frequency of the pulses.

The timers in the MSP430 family are 16-bit counters that are extremely versatile. Their sources can be programmed to be any one of those shown in *Figure 7-5*. Some of the counters can be programmed to be 8-, 10-, or 12-bit counters. Each timer has capture/compare register blocks that sense when the counter has reached a particular count (capture) and compare the count to a set target. An output signal from the capture/compare block can be used as an interrupt or as an external signal. These timers are particularly useful to keep track of elapsed time, to set time intervals within which specific action occurs or is to occur, and to produce resets, alerts or warnings.

Addressing Modes

Addressing modes were discussed in general in *Chapter 6*. Now the specific modes used in the MSP430 family will be discussed—the format, the symbols used, and a description of the modes. The seven addressing modes are shown in *Figure 7-6*; note the column **As/Ad**. **As** are bits in an instruction that define the addressing mode used for the source, and **Ad** are bits in an instruction that define the addressing mode used for the destination. In *Figure 7-6*, addressing modes 1, 2, 3 and 4 have bits in the As and Ad column; therefore, they can be used to address both the source and the destination. Modes 5, 6 and 7 can be used for the source only. Here is a short discussion of each addressing mode:

As/Ad	Addressing Mode	Syntax	Description
1. 00/0	Register mode	Rn	Register contents are operand
2. 01/1	Indexed mode	X(Rn)	(Rn + X) points to the operand
			X is stored in the next word
3. 01/1	Symbolic mode	ADDR	(PC + X) points to the operand
			X is stored in the next word. Indexed mode X(PC) is used.
4. 01/1	Absolute mode	&ADDR	The word following the instruction contains the absolute address.
			X is stored in the next word. Indexed mode X(SR) is used.
5. 10/–	Indirect register mode	@Rn	Rn is used as a pointer to the operand.
6. 11/–	Indirect autoincrement	@Rn+	Rn is used as a pointer to the operand. Rn is incremented afterwards by 1 for .B instructions and by 2 for .W instructions.
7. 11/–	Immediate mode	#N	The word following the instruction contains the immediate constant N. Indirect autoincrement mode @PC+ is used.

Figure 7-6: Addressing modes

1. Register Mode—The symbol is Rn

If register mode addressing is used, the content of the register is the operand. For example, the instruction Mov R1,R2 means that register addressing is used for both the source, register R1, and the destination, register R2. The contents of R1 are moved to R2. R2 is changed but R1 remains the same. Register mode can be used either for the source or the destination or both.

2. Indexed Mode—The symbol is X(Rn)

The X is an index that is added to the contents of Rn to form an address that is either the source of or the destination for the operand. For example, for the instruction Mov 2(R1),4(R2). The operand at the source address (R1 + 2) is moved to the destination address (R2 + 4). The X index is stored in the next word after the instruction; the source in the first word and the destination in the second word. The contents of R1 and R2 are not affected.

3. Symbolic Mode—A symbol name such as ADDR

A symbolic name is given to the address of the operand, either the source or the destination or both. For example, the instruction Mov ADDR,END says to move the contents at the source address ADDR to the destination address END. The symbol ADDR and END are assigned digital words that are substituted by the assembler to make up the proper address.

4. Absolute Mode (&ADDR)

The & symbol is added in front of the operand, &ADDR. The & symbol indicates that the absolute operand address is contained in the word following the instruction. Absolute mode can be used for both the source and the destination. For example, the instruction Mov &ADDR,&END says move the contents of the source address ADDR to the destination address END. However, no calculations are involved as for symbolic mode. The absolute address for both the source and destination are in the words following the instruction, the source in the first word, the destination in the second word.

5. Indirect Register Mode (@Rn)

The @ symbol is added in front of a register number, @Rn. This is an addressing mode that is valid only for the source. It indicates that the *contents of the source* are to be used *as the address* of the operand. For example, the instruction Mov @R1,0(R2) says to move the contents at the source address, the contents of R1, to the destination address. Since indirect register mode cannot be used for the destination, the substitute for the destination operand is 0(R2), which means the destination address is the contents of R2. R1 and R2 are not modified.

6. Indirect Autoincrement (@Rn+)

Besides the @ symbol added in front of a register number a plus sign (+) is added after the register, @Rn+. This is the same addressing mode as for the indirect register mode except the source register content is incremented by one for a byte operation and by two for a word operation after the instruction is completed.

7. Immediate Mode (#N)

The # symbol is added in front of the operand, usually a constant number, #N. The # symbol, states that the number indicated, which is contained in the word following the instruction, is the source operand. The immediate mode can only be used for source addressing. For example, the instruction Mov #9, ADDR says that the constant 9 is to be moved to the destination ADDR (symbolic addressing). When executed, the program counter points to the word following the instruction and moves its contents (the number 9) to the destination ADDR.

More on MSP430 Control

It will be important to the understanding of assembly-language programming to look further how the MSP430 microcontroller is controlled. One of the principal features of its design is the use of registers to implement the control. The state of a particular bit or particular bits in a register determines the operating condition or action of a particular function inside the MSP430.

The Status Register

The status register, SR, shown in *Figure 7-7*, is a prime example. It is register R2 of the sixteen 16-bit registers in the CPU shown in *Figure 7-2a*. The status register, R2, has nine active bits; the remaining seven are available for future expansion. The LSB is the zero bit; the eight bit is the MSB. Each of the nine bits has a specific control over the CPU, or its state dictates that a particular action has occurred. For example, the four bit is labeled "CPUOFF." If the four bit is set (to a 1), the CPU will be off. Program execution stops,

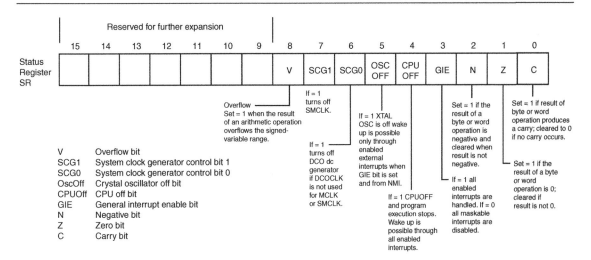

Figure 7-7: Status register R2

but the RAM, the port registers and any enabled peripherals stay active. The CPU is awakened when any enabled interrupt occurs.

The five bit, labeled "OSCOFF," if set (to a 1), the crystal oscillator enters the off mode. The DCO remains ON so the CPU can be running. The RAM contents, the ports, and the registers are maintained. Wake up is possible only through enabled external interrupts.

The three bit is the general-interrupt-enable bit, GIE. If set, all enabled maskable interrupts are handled; if reset (to a 0), all maskable interrupts are disabled, GIE is cleared by interrupts and set by a return from interrupt, RETI instruction, as well as other appropriate instructions. The six and seven bit, labeled SCG0 and SCG1, respectively, determine, through their bit combination, which clock is active. If SCG0 is set (to a 1), the DCO dc generator is turned off; however, this only happens if the DCO is not being used as a source for MCLK or SMCLK. If SCG1 is set, SMCLK is turned off. It must be noted, as discussed in *Figure 7-3*, that the bits OSCOFF, CPUOFF, SCG0 and SCG1 work together to define an operating mode, not independently to provide various control. The eight bit, labeled V, is an overflow bit. It is set when the result of an arithmetic operation overflows the signed-variable range.

The zero, one and two bits are labeled C, Z, and N, respectively. The C or carry bit is set when a byte or word operation called for in an instruction produces a carry. It is cleared if no carry occurs. The Z or zero bit is set if the result of a byte or word operation is zero; if the result is not zero, it is cleared (set to a 0). The negative bit, N, is set if the result of a byte or word operation is negative, and is cleared when the result is not negative. Instructions in the program will test the C, Z, or N bits and the CPU will respond as directed by the program instructions. Operations as a result of an instruction, or the instruction itself, can set the bits so that the CPU is controlled accordingly.

Basic Clock System Control Registers

The basic clock system is set up (configured) by using three control registers, the DCOCTL (the digitally-controlled oscillator control register), and the two basic clock system control registers, BCSCTL1 and BCSCTL2. In addition, SCG1, SCG0, OSC0FF and CPUOFF bits in the status register control the operating mode as described. The DCOCTL register and a brief description of its bits and what they control is shown in *Figure 7-8*. The code represented by the state of the DCO bits defines one of eight frequency steps within the DCO frequency range set by the RSEL bits in the BCSCTL1 control register. This was explained previously (*Figure 7-5*). The state of the five MOD bits set a modulation constant used to adjust the DCO frequency. At power up, the power-up control signal (PUC) loads the DCOCTL register with 060h to set the initial DCO frequency.

The two basic clock system control registers, BCSCTL1 and BCSCTL2, are shown in *Figure 7-9*, along with a brief description of the control affected by the bits of each register. BCSCTL1 controls basic clock system 1 and BCSCTL2 basic clock system 2. Referring to *Figure 7-5* and BCSCTL1 in *Figure 7-9*,

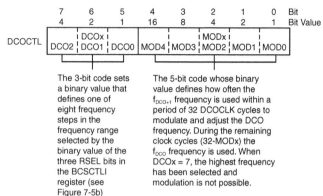

Figure 7-8: The digitally-controlled oscillator (DCO) control register

the XTS bit determines if the LFXT1 oscillator will operate with a low-frequency or high-frequency crystal to produce the LFXT1 clock source. The states of the DIVA bits determine if clock source LFXT1 is going to be divided by 1, 2, 4 or 8 to produce the clock ACLK. The RSEL bits 0, 1, and 2 determine the nominal frequency range of the DCO as previously discussed for *Figure 7-5b*.

Referring to *Figure 7-5* and BCSCTL2 in *Figure 7-9*, the SELM bit states determine if DCO, XT2 or LFXT1 are going to be the source for the MCLK clock. The DIVM bit states determine if the clock source is going to be divided by 1, 2, 4 or 8 to produce MCLK. Likewise, the 3 bit, the SELS bit, state determines if DCOCLK, XT2CLK or LFXT1CLK will be the source for the SMCLK clock. The DIVS bit states determine if the source to SMCLK will be divided by 1, 2, 4 or 8. The DCOR bit controls whether current is going to be supplied to the DCO from an internal or external resistor to control oscillations. The complete clock system for the MSP430 can be set up initially using instructions to the CPU to set the bits of the DCOCTL, BCSCTL1 and BCSCTL2 registers.

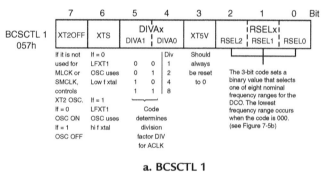

Figure 7-9: Basic clock system control registers

Watchdog Timer

The WDTCTL register controls the watchdog timer. It is shown in *Figure 7-10*, and a description of the control that each bit applies in a particular state is included. When the watchdog timer function is active, the WDTTMSEL bit must be 0 to be in the watchdog mode and the WDTHOLD bit must be 0; if WDTHOLD is set, the counting stops.

If the watchdog timer is active, software should periodically reset the watchdog timer by writing a 1 to the WDTCNTCL clear bit to prevent the timer interval from expiring and restarting the system. Setting WDTCNTCL (to a 1) restarts the counter, WDTCNT, at 0000h. The WDTSSEL bit selects the clock source for WDTCNT, when = 0 the source is SMCLK; when = 1 the source is ACLK. The state

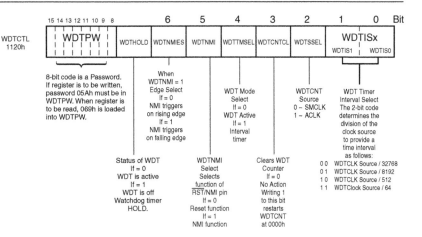

Figure 7-10: Watch dog timer control register

of the WDTIS bits determines the time interval of the clock, either with SMCLK or ACLK as the source. The code for the time interval is shown in *Figure 7-10*.

The five bit, WDTNMI, controls whether a pin, RST/NMI, is a reset input or a nonmaskable interrupt input (NMI). When the state of WDTNMI is a 0, the RST/NMI input is a level-sensitive reset input, and when the state is a 1, it is an edge-sensitive nonmaskable input. When WDTNMI is set to 1, the six bit, WDTNMIES, controls whether the input triggers on the rising (WDTNMIES = 0) or falling edge (WDTNMIES = 1) of the input signal.

Timer_A Control Register

There is a 16-bit, general-purpose timer in the MSP430F1232 device used for this chapter, called Timer_A. Its control register, TACTL, is shown in *Figure 7-11*, with a description of the control each bit applies in a particular state.

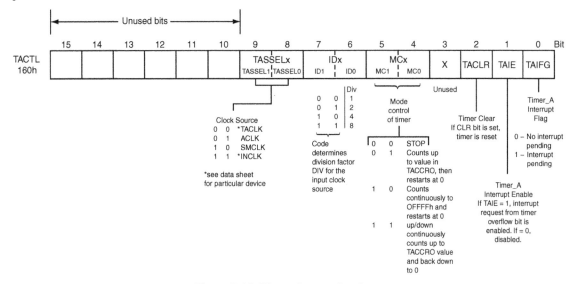

Figure 7-11: Timer_A control register

The TASSEL bit states determines the clock source to be used for Timer_A, either internal clocks ACLK or SMCLK or an external source TACLK. After clock selection, the state of the ID bits control whether the clock source is passed directly to Timer_A, or whether it is divided by 2, 4 or 8. The state of the MC bits set the mode of the timer as shown. The modes of the timer are further clarified in *Table 7-2*.

MC1	MC0	Mode
0	0	Stop
0	1	Up (from 0 to value TACCRO)
1	0	Continuous (from 0 to 0FFFFh, then restart at 0)
1	1	Up/Down (counts up to TACCRO value, then back to 0)

Table 7-2: Mode of Timer_A

To clear the counter, the two bit, TACLR, is set to a 1. The remaining one bit and zero bit control the response to an interrupt generated when Timer_A reaches a specific value. The interrupt sets a flag, TAIFG, and the TAIE bit enables the interrupt if it is set to a 1 or disables the interrupt if it is reset to a 0.

Input/Output Control

Previous discussion stated that the I/O ports in the MSP430 could be programmed to be inputs or outputs. For the family device used for this chapter there are three I/O ports, 1, 2 and 3. *Figure 7-12* shows the registers that can be programmed to configure the external pins of the MSP430. There are PxIN input registers; there are PxOUT output registers; there are PxDIR direction registers and there are PxSEL function-select registers.

All I/O ports are initially inputs when the machine powers up. If the zero bit of P1DIR is set to a 1, then the external Pin1.0 will be an output; if its state is a 0, the external Pin1.0 will be an input. Any input signal from pin P1.0, when programmed as an input, will be stored in the zero bit of the P1IN register. When pin P1.0 is programmed as an output by P1DIR, the P1OUT register zero bit is output to P1.0. Correspondingly, the other bits of P1IN and P1OUT will either receive data into their register or output data from their register based on the programming in the P1DIR register.

External pins can be used by other modules rather than I/O ports 1, 2 and 3. The PxSEL bits of the function-select register controls the selection. When the PxSEL bit for a particular pin is set to a 1, that pin will be used by a module other than port 1 or 2 or 3. The data sheet for the particular device, with its package pin layout, will indicate if pins have multiple function capability. When the multiple capability is available, the PxSEL must be configured to select the proper pin function.

To summarize, the PxDIR register bits are set (= 1) to dictate if the external pins of the I/O ports are to be outputs. The initial condition is that all I/O ports are inputs. Any input signal will be placed in the PxIN register(s). Any output pins will receive signals from the PxOUT register(s). If a particular external pin is not to be an I/O output or input from Port 1, 2 or 3, then the bits of the PxSEL register(s) are set to select the function that is to be on the pin. The functions available for the pin are called out on the data sheet for a particular device.

Further Thoughts

Some further thoughts and concepts need to be examined before an actual assembly-language program is explained. The first of these is symbolic notation.

Symbolic Notation

Recall that the actual 1s and 0s (machine language) that direct the circuits inside of a digital processor for a particular program must be formulated or coded for each instruction. This has been mentioned previously but it bears repeating. The language used for our programming is assembly language. To convert assembly-language programming to machine language an assembler, a computer program that converts the mnemonic instructions used in assembly-language programs into the 1s and 0s of machine code, is required. The assembler has been developed to recognize symbolic representations such as the mnemonics used for the assembly-language instructions; such as symbolic names used to identify register bits, system commands, system names, or system signals. When the assembler sees the respective symbolic name it has been programmed to insert a specific binary number that represents the symbol. Unique reference lists have been developed for the assembler of a specific device or family of devices that assign the binary numbers to symbolic names used for the devices. As the assembler reads the assembly-language program and encounters a symbolic name, it inserts the respective binary number and "assembles" the machine code for the program.

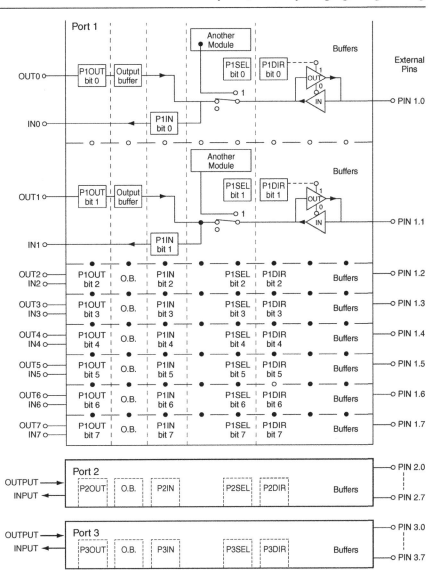

Figure 7-12: I/O Ports 1, 2 and 3

Table 7-3 is an example portion taken from a "Standard Register and Bit Definitions for the Texas Instruments MSP430 Microcontroller Family" reference list contained in the Appendix.

* STATUS REGISTER BITS *

#define C	(0x0001)
#define Z	(0x0002)
#define N	(0x0004)
#define V	(0x0100)
#define GIE	(0x0008)
#define CPUoff	(0x0010)
#define OSCoff	(0x0020)
#define SCG0	(0x0040)
#define SCG1	(0x0080)

/*Low Power Modes coded with bits 4 – 7 in SR*/

if ndef_IAR_Systems_ICC/Begin #defines for assembler/

#define LPM0	(CPUoff)
#define LPM1	(SCG0 + CPUoff)
#define LPM2	(SCG1 + CPUoff)
#define LPM3	(SCG1 + SCG0 + CPUoff)
#define LPM4	(SCG1 + SCG0 + OSCoff + CPUoff)

Table 7-3: Reference list for assembler

Notice, first of all, that the symbolic names in *Table 7-3* are the same ones used to identify the bits in the status register, and the reference list is defining a binary number associated with the symbolic name. For example, the binary number for GIE is the 16-bit hexadecimal number 0008h, which is 0000 0000 0000 1000 in straight binary. If this number is loaded into the status register, it sets the GIE bit. Thus, if the program wants the GIE bit set, an instruction can use GIE as the source operand and SR as the destination, and the assembler knows, because of the reference list, to load the hexadecimal number 0008h into SR which sets GIE. Similarly, the binary number assigned in the reference list will set the bit with the symbolic name in one of the control registers if that symbolic name is used in the appropriate instruction.

Combining symbolic names and their associated binary numbers will set multiple bits in the control registers. This occurs in defining the bits to be set for the MSP430 low-power modes shown in *Table 7-3*. For example, to place the MSP430 system in the LPM3 mode, the SR bits SCG1, SCG0 and CPUoff must be set. One operand of (SCG1 + SCG0 + CPUoff) can be specified and the assembler will combine the binary numbers specified in the reference list and insert them in the SR as shown in *Figure 7-13*. The operand calls out the symbolic names and the respective bit corresponding to the name is set, even when multiple names are used in the operand.

Format and Symbols

A final thought before discussing the actual programming. The format for the lines of code is as follows, shown with an example instruction:

```
Label          Instruction         Operands            Comment
ADCLoop        bis.b               #CLK,&P2OUT         ;Clock high
```

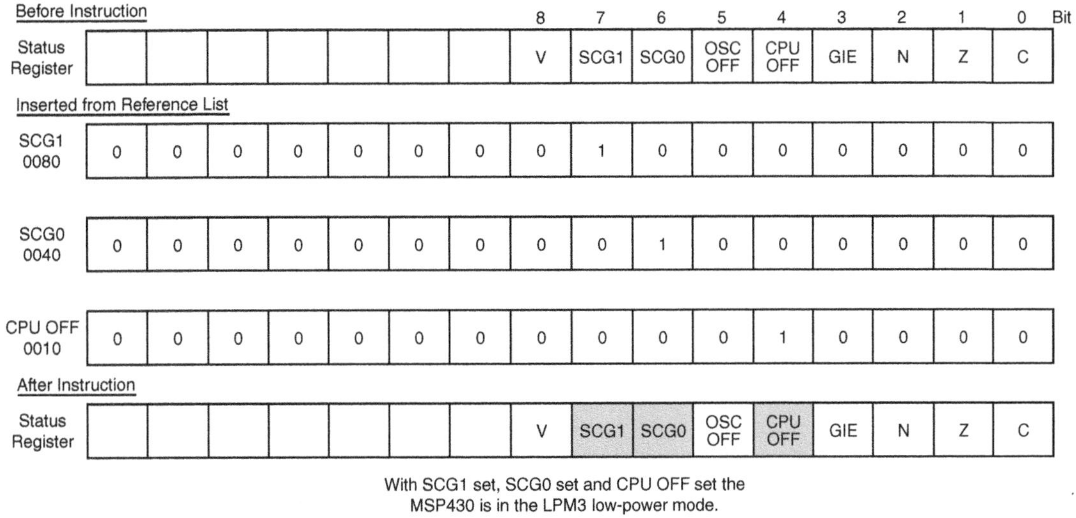

With SCG1 set, SCG0 set and CPU OFF set the
MSP430 is in the LPM3 low-power mode.

Figure 7-13: Substitution for symbolic names in status register

Labels

Labels identify particular positions in the program. They are used extensively to identify the beginning of a program subroutine. When a program needs a particular subroutine, the program can do a subroutine jump to the particular label associated with the subroutine.

Instructions

The actual instruction appears in the instruction column. In the instruction, bis.b, "set bits in destination", the .b means it is a byte instruction dealing only with eight bits, the lower byte, of a 16-bit word. When the instruction is a word instruction where all 16 bits are involved, there need be nothing or .w can be used.

Operands

An operand is the part of the instruction which will operated on by the instruction. Operands are the portion of an instruction designated by an op code to be the quantity to be operated on by the instruction. They appear in the operand column with the source always listed first separated from the destination by a comma. The source may have a symbol in front of it, and the same for the destination. The symbols will correspond to the syntax column in *Figure 7-6* identifying the addressing mode used. In the code line example shown above, the source CLK has a # sign in front of it, and the destination, register P2OUT, has an & in front of it. The # sign indicates immediate addressing for the source, and the & means absolute addressing is used for the destination.

Hexadecimal Numbers

Hexadecimal numbers will have special identification in most cases. As discussed previously, any hexadecimal number that starts with A through F will have a zero in front of it to make sure the number is identified correctly. A small h is included in the number to identify it as hexadecimal, otherwise the assembler assumes the number is a decimal number. Or, as in the portion of the reference list shown in *Table 7-3*, the format 0x0000 may also be used for a hexadecimal number.

Comments

The comments column contains hints to someone reading the program what the original programmer had in mind when the line of code was written—what the line of code should accomplish. Many times the comment column is also a refresher to the original programmer. A semicolon must precede all comments. In the explanations that follow of assembly-language programming, no time will be spent on the comments. The reader may use these for extra understanding of the program.

Programming Examples

Introduction

In order to explain how to develop a program using assembly language, several subprograms that perform different tasks will be explained in detail to help grasp the concept of programming, learn some of the programming details and get familiar with the format necessary for assembly-language programming. Obviously, sophisticated programmers use high-level languages, but assembly-language programming is used here because it offers an opportunity to grasp the fundamentals of programming so that higher-level language programming can be implemented with less difficulty. It offers fundamental concepts that aid in the understanding of programming in the higher-level languages.

Subprogram No. 1

General Description

The program that will be described is a portion of a total program using a TLV0831 ADC that interfaces to a MSP430F12X microcontroller. The total program includes sampling an analog input voltage, converting it to a digital code, shifting the data into the MSP430, and transmitting the data to a personal computer (PC). The subprogram that is described here is the portion of the total program that deals with initiating the digital conversion and shifting the data into a temporary storage register in the MSP430. Essentially, this subprogram implements a shift register using software.

The block diagram of the system and the interconnections are shown in *Figure 7-14a*. A timing diagram of the events as they occur is shown in *Figure 7-14b*.

Here is a brief description of the application:

1. The TLV0831 is an 8-bit ADC. It samples its analog input, converts the signal to digital and stores the 8-bit digital output in an output register.

2. The TLV0831 data is then shifted out of the output register by the MSP430 into the ADCData register in the MSP430.

3. The TLV0831 data is coupled out on output DO. DO is connected to pin P2.3 of the MSP430, which is programmed to be an input.

4. Pins P2.0 and P2.1 are programmed to be outputs from the MSP430. P2.0 provides a chip select signal to activate the TLV0831, while P2.1 provides clock pulses to the TVL0831.

I/O Port 2, one of the three available, is used for this application. The content of register P2DIR is set to determine which I/O pins are to be inputs and which are to be outputs. The P2IN register inputs and captures the input data from any pins that are inputs, while the P2OUT register outputs the respective data onto the pins that are outputs. The P2.3 input is coupled to a register in the MSP430 called by the symbolic name of ADCData. The 8 bits from the output data register of the TLV0831 are shifted out serially onto DO and end up in this register. It takes 9 shifts to do this—one for a start bit and the remainder for the eight bits of data. The signal on P2.1 acts as the clock for the TLV0831 to shift out the bits onto DO. The start timing is controlled by the signal on P2.0 of the MSP430, which is connected to CS of the TLV0831. A logic low on CS activates the TLV0831 and initiates the A-to-D conversion. The MSP430 is operated in the LPM3 low-power mode. The watchdog timer is used as an interval timer, set at 64 ms, and when it times out it generates an interrupt to wake up the system and initiate a conversion.

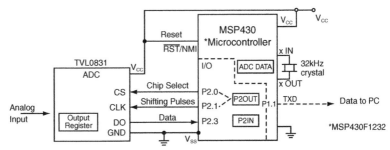

a. Block diagram showing interconnections

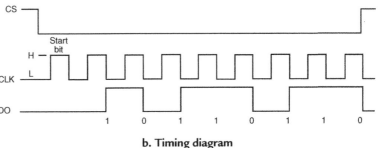

b. Timing diagram

The Initial Conditions

The subprogram No. 1 assembly-language program is shown in *Figure 7-15*. Normally, the A, B, C, and D notations are not present; they have been added to aid in the discussion of the program. The reference list that was discussed previously for the MSP430 applies to this program. It is contained in the Appendix. The A and B portions of this program are the same type of reference list, but they are specific only to this program. The assembler takes the "#define" and the "equ" and substitutes the numbers defined for the symbolic name. Software engineers call it "syntaxic substitution"—substituting numbers for words (the symbolic notations) in the program.

The destination operand is shifted left one position as shown. The carry bit (C) is shifted into the LSB and the MSB is shifted into the carry bit (C).

c. Rotate left through carry

Figure 7-14: Systems application implemented by Subprogram No. 1
Courtesy of Texas Instruments Incorporated

Section A and B

Section A defines the specific registers that are going to be used, R6(BitCnt) to count bits, R5(RxTxData) to receive and transmit data to the PC, and R11(ADCData) the register to store the data received from the ADC. Section B continues the same type definitions with the "equ" notation. Here the programmer has assigned specific hexadecimal numbers complementing the reference list but specific to this program. The subprogram will use ADCData, TXD, CS, CLK and DO. The remaining substitutions are used by the portion of the total program that transmits data to the PC and that calibrates the DCO. That portion is not included for the sake of brevity.

```
A.      ; Dedicated CPU registers used
    (1) #define    BitCnt   R6
    (2) #define    RXTXData R5
    (3) #define    ADCData  R11
        ;
B.      ; User definitions, 9600 Baud HW/SW UART, MCLK = 37.5x32768 = 1228800
    (1) Bitime     equ      0128                  ; 104 us
    (2) Delta      equ      150                   ; Delta = (target DCO)/(32768/4)
    (3) TXD        equ      002h                  ; TXD on P1.1
    (4) CS         equ      001h                  ; P2.0 Chip Select
    (5) CLK        equ      002h                  ; P2.1 Clock
    (6) DO         equ      008h                  ; P2.3 Data Out
    (7) LF         equ      0ah                   ; ASCII Line Feed
    (8) CR         equ      0dh                   ; ASCII Carriage Return
        ;

        Label   Instruction Operands             Comment
        ;-----------------------------------------------------------------------
C.              ORG      0F000h                   ; Program Start
        -----------------------------------------------------------------------
    1.  Reset      mov      #0300h,SP             ; Initialize F12x stackpointer
    2.             call     #Init_Sys             ; Initialize system

        ;-----------------------------------------------------------------------
D.      Init_Sys; Subroutine sets up Modules and Control Registers
        -----------------------------------------------------------------------
        Label   Instruction Operands             Comment
    16. StopWDT    mov      #WDTPW+WDTHOLD,&WDTCTL ; Stop Watchdog Timer
    17. SetupBC    mov.b    #DIVA1+RSEL2+RSEL0,&BCSCTL1  ; ACLK/4   RSEL=5
    18. SetupP1_2  bis.b    #TXD,&P1SEL           ; P1.1/TA0 for TXD function
    19.            bis.b    #TXD,&P1DIR           ; TXD output on P1
    20. SetupP2    bis.b    #CS,&P2OUT            ; CS, Set
    21.            bis.b    #CS+CLK,&P2DIR        ; CS and Clk Output direction
    22. SetupTA    mov      #TASSEL1+TACLR,&TACTL ; SMCLK, clear timer
    23. SetupC0    mov      #OUT,&CCTL0           ; TXD Idle as Mark
    24.            call     #Delay                ; Time for crystal to stabilize

    25.            bis      #MC1,&TACTL           ; Start timer in Continous Mode
    26.            call     #Set_DCO              ; Set DCO to target frequency
    27. SetupWDT   mov      #WDT_ADLY_16,&WDTCTL  ; WDT 16ms*4 Interval Timer
    28.            bis.b    #WDTIE,&IE1           ; Enable WDT Interrupt
    29.            eint                           ; General Interrupt Enable
    30.            ret                            ; Return from subroutine
                                                 ;

E.      Label   Instruction Operands             Comment
    3.  Mainloop   bis      #LPM3,SR              ; Enter LPM3
                                                 ;
    4.  Meas_ADC;  Shift    TVL0831 data into ADCData, R15 used as counter
    5.             bic.b    #CS,&P2OUT            ; Chip Select low
    6.             mov      #09,R15               ; 9 bits *1 start* + 8 data
    7.  ADC_Loop   bis.b    #CLK,&P2OUT           ; Clock high
    8.             bic.b    #CLK,&P2OUT           ; Clock low
    9.             bit.b    #DO,&P2IN             ; DO -> C (carry)
    10.            rlc.b    ADCData               ; C -> ADCData

    11.            dec      R15                   ; All shifted in?
    12.            jnz      ADC_Loop              ; If not --> ADC_Loop
    13.            bis.b    #CS,&P2OUT            ; Chip Select high

    14.            call     #TX_ADC_2PC           ; ADC result --> PC
    15.            jmp      Mainloop              ; Repeat
                                                 ;
```

Figure 7-15: Subprogram No. 1—an assembly-language program—a software shift register
Program courtesy of M.E. Buccini and Texas Instruments Incorporated

Section C

Section C begins with the following line:

```
Label       Instruction     Operands        Comment
            ORG             0F000h
```

The "ORG" instruction, called an assembler directive, tells the assembler where in memory to put the start of the program. For this program, it starts at the hexadecimal location F000 (1111 0000 0000 0000 in straight binary).

1. The first line of code is:

```
Label       Instruction     Operands
RESET       mov             #0300h,SP
```

RESET is a label to identify a location to which the program goes when the system power is turned on. The instruction "move source to destination" means to move the source, the hexadecimal number 0300h, to the destination SP, the symbolic name for the stack pointer. The assembler knows that SP means register R1 in the CPU, as shown in *Figure 7-2a*, and loads 0300 into R1. Recall that the stack pointer stores the return address from a subroutine call so that the program can proceed after it finishes a subroutine. The source is addressed with immediate addressing and the destination with symbolic addressing.

2. The second line of code is:

```
Label       Instruction     Operands
            call            #Init_Sys
```

The instruction "call" is a subroutine call. It directs the program, with immediate addressing, to a subroutine at a memory location identified as "Init_Sys."

Section D

Section D is the subroutine "Init_Sys." It is the portion of the program that sets the initial conditions of the system by setting bits in the control registers that were discussed previously. The subroutine starts at line 16.

16. The sixteenth line of code is:

```
Label       Instruction     Operands
StopWDT     mov             #WDTPW + WDTHOLD,&WDTCTL
```

The line of code is labeled "StopWDT" as a clue of what is happening. The instruction "mov" means to move the source WDTPW + WDTHOLD to the destination WDTCTL. Immediate addressing (# sign) is used for the source and absolute addressing (& sign) for the destination. WDTCTL is the watchdog timer control register shown in *Figure 7-10*. The symbolic names WDTPW and WDTHOLD load, from the reference list in the Appendix, hexadecimal numbers that correspond to the symbolic names. 05A00h is a password that allows the instruction to write to the WDTCTL, and WDTHOLD sets the HOLD bit to a 1. This holds/stops the watchdog timer.

17. The seventeenth line of code is:

```
Label       Instruction     Operands
SetupBC     mov.b           #DIVA1 + RSEL2 + RSEL0,&BCSCTL1
```

The line of code is labeled "SetupBC" for setup basic clock. The instruction "move source to destination, byte mode" means to move the lower byte of the source into the destination, BCSCTL1, the basic clock system control register. Immediate addressing is used for the source and absolute addressing for the destination. The symbolic names in the source, DIVA1, RSEL2 AND RSEL0, when moved to the BC-SCTL1, set the respective bits to a 1. Referring to *Figure 7-9*, setting DIVA1 divides the source for ACLK, LFTX1 by 4, and setting RSEL2 (value = 4) and RSEL0 (value = 1) means RSEL=5, or the fifth resistor combination to set the nominal frequency of the DCO.

18. The eighteenth line of code is:

```
Label           Instruction        Operands
SetupP1_2       bis.b              #TXD,&P1SEL
```

The line of code sets up Port 1, thus, labeled "SetupP1_2." The instruction "set bits in destination, byte mode", means that the binary number associated with the symbolic name TXD, which is 002h according to Section B, is used to set the control register P1SEL. 002h, or 0000 0010 in binary, sets the one bit in P1SEL. Setting the one bit in P1SEL means that I/O pin P1.1 will be used by another function other than Port 1; in this case, for the TXD function to transmit data to the PC.

19. The nineteenth line of code is:

```
Label           Instruction        Operands
                bis.b              #TXD,&P1DIR
```

The instructions "set bits in destination, byte mode" again uses the hex number assigned to the symbol TXD (002h) to set the one bit in the direction control register, P1DIR, shown in *Figure 7-12*. Bit one when set means that pin P1.1 will be an output.

20. The twentieth line of code is:

```
Label           Instruction        Operands
SetupP2         bis.b              #CS,P2OUT
```

The line of code is labeled "SetupP2" to indicate it is setting up Port 2. The instruction "set bits in destination, byte mode" means the hex number assigned to symbol CS (001h per Section B) is used to set the output register P2OUT. The zero bit of P2OUT is set, and thus, in the high state.

21. The twenty-first line of code:

```
Label           Instruction        Operands
                bis.b              #CS + CLK,&P2DIR
```

The instruction "set bits in destination, byte mode" means now the source is CS + CLK; therefore, both hex numbers assigned to the symbols CS and CLK will set bits in the destination, the P2DIR direction register. This sets I/O pin P2.0 and pin P2.1 as outputs. Since the zero bit of P2OUT is in the high state, pin P2.0 will be in the high state. P2.0 is the chip select line for the TLV0831. Since it is high, the TLV0831 is not active.

22. The twenty-second line of code is:

```
Label           Instruction        Operands
SetupTA         mov                #TASSEL1 + TACLR,&TACTL
```

Timer_A is being setup by the line of code labeled "SetupTA." The instruction "move source to destination" means to move the hex numbers associated with the symbolic names TASSEL1 and TACLR to the Timer_A control register, TACTL, shown in *Figure 7-11*. Setting TASSEL1 selects the SMCLK clock as the Timer_A source and setting the CLR bit resets Timer_A.

23. The twenty-third line of code is:

```
Label           Instruction        Operands
SetupCO         mov                #OUT,&CCTL0
```

The label "SetupCO" explains that the OUT bit in a capture/compare control register is being set. The instruction "move source to destination" is setting the OUT bit of the capture/compare control register, CCTLO. Effectively, the TXD bit state is output onto pin P1.1 with this instruction. Since the P1OUT is a 1, TXD will be a 1 or a MARK in transmit language. A 0 is defined as a SPACE.

24. The twenty-fourth line of code is:

```
Label        Instruction        Operands
             call               #Delay
```

The instruction "call" means the program is calling a subroutine labeled "Delay." This subroutine, not shown in our subprogram, provides a time delay with software. The crystal oscillator used as the source for the clocks needs time to stabilize. The instruction calls the subroutine, which when executed, provides the time delay needed for the oscillator to stabilize.

25. The twenty-fifth line of code is:

```
Label        Instruction        Operands
             bis                #MC1,&TACTL
```

The instruction "set bits in destination" sets the MC1 bit in the TACTL control register, shown in *Figure 7-11,* by inserting the assigned hex number. With MC1 = 1 (and MC0 = 0), Timer A is set into the continuous mode and starts counting from 0 to 0FFFFh. When it gets to 0FFFFh it restarts from 0.

26. The twenty-sixth line of code is:

```
Label        Instruction        Operands
             call               #Set_DCO
```

The instruction "call" this time is calling the subroutine "Set_DCO" which is not shown in our subprogram. It is a subroutine that calibrates the high-speed, digitally-controlled oscillator (DCO). For this program, the DCO is calibrated to 1,228,800 Hz (cycles per second), and configured to be the source for the main system clock, MCLK, and subsystem clock, SMCLK.

27. The twenty-seventh line of code is:

```
Label        Instruction        Operands
SetupWDT     mov                #WDT_ADLY_16,&WDTCTL
```

Labeled "SetupWDT" to explain that the watchdog timer is being set up, the instruction "move source to destination" moves the hex number assigned to the source WDT_ADLY_16 by the reference list to the destination, the watchdog timer control register, WDTCTL, shown in *Figure 7-10*. The assigned hex number 05A1E provides the 5A that is required when the WDTCTL is being written to, and sets bits WDTTMSEL, WDTCNTCL, WDTSSEL and WDTIS1. Setting WDTTMSEL makes the WDT an interval timer; setting CNCTL clears the WDTCNT counter and restarts it at zero; setting WDTSSEL selects ACLK for the counter source; and setting WDTIS1 chooses a 512 division factor for the time interval which sets the time interval between pulses to be 62.5 ms (milliseconds).

28. The twenty-eighth line of code is:

```
Label        Instruction        Operands
             bis.b              #WDTIE,&IE1
```

The instruction "set bits in destination, byte mode" takes the source hex number assigned to the symbolic name WDTIE (01h) and places it in the interrupt enable register, IE1. It sets the zero bit, which enables the watchdog timer interrupt. As a result, because this signal is active when the watchdog timer is in the interval timer mode, the watchdog timer interrupt is enabled.

29. The twenty-ninth line of code is:

```
Label          Instruction        Operands
               eint
```

The instruction "enable (general) interrupts" sets the GIE bit in the status register shown in *Figure 7-7* and says "all interrupts are enabled." This allows the interrupt generated when the WDT interval timer times out to interrupt the system, wake it up from the LPM3 mode and be active.

30. The thirtieth line of code is:

```
Label          Instruction        Operands
               ret
```

The instruction "return from subroutine" tells the program to return to the code address following the subroutine call, in this program to line 3. The program has completed all initial conditions and now returns to do its main operations.

Section E—Main Application

3. The third line of code is:

```
Label          Instruction        Operands
Mainloop       bis                #LPM3,SR
```

The label "Mainloop" identifies this line of code as the start of the main portion of the program. The instruction "set bits in destination" takes the hex number assigned to the source, symbolic name LPM3, by the reference list, and sets bits in the destination, SR. As shown in *Figure 7-3*, the hex number for LPM3 sets the bits SCG1, SCG0 and CPUOFF in the status register. This sets the system in the LPM3 low-power mode. Recall that in LPM3, the CPU is inactive but peripherals and the ACLK clock are active, and, in this application, that the WDT interval timer awakens the system.

4. The fourth line of code is:

```
Label          Comment
Meas_ADC       ;Shift TLV0831 data into ADCData, R15 used as counter
```

The comment for the label "Meas_ADC" identifies that part of the program that initiates the ADC measurement, and, after the data is present, shifts the data into the register ADCData. The register R15 will be used to count off the number of shifts. Its contents determine the number of shifts.

5. The fifth line of code is:

```
Label          Instruction        Operands
               bic.b              #CS,&P2OUT
```

The instruction "clear bits in destination, byte mode" means that the hex number assigned to CS (001h) in Section B will clear bits in the destination, P2OUT, the output register. Immediate addressing is used for the source, absolute addressing for the destination. The zero bit of P2OUT is cleared and, as a result, pin P2.0, is a 0, or low. P2.0 is the CS signal to the TLV0831. Since it is low, it activates the TLV0831 and starts the ADC conversion.

6. The sixth line of code is:

```
Label          Instruction        Operands
               mov                #09,R15
```

The instruction "mov" means to move the source to the destination. There is immediate addressing for the source; register addressing for the destination. As a result, the number 9 is inserted as the contents of (moved to) register 15 to determine how many bits are going to be shifted onto DO. As discussed earlier, a start bit is required for outputting serial data. Since the data is eight bits, the number 9 is loaded into R15 with the "mov" instruction. If the ADC were converting to a larger number of bits than eight, then R15 would have to be loaded with a correspondingly larger number.

7.   The seventh line of code is:

```
Label          Instruction        Operands
ADC_Loop       bis.b              #CLK,&P2OUT
```

"ADC_Loop" is a subroutine label. The program will continue to a decision point and then loop back to this label. The instruction "set bits in destination, byte mode" means that the source hex number assigned to "CLK" in Section B (002h) will be used to set bits in the lower byte of the destination, register P2OUT, the output register for Port 2. Thus, the one bit of P2OUT will be set, and pin P2.1 will have a 1 or high output. P2.1 is connected to CLK of the TLV0831; therefore, CLK is high.

8.   The eighth line of code is:

```
Label          Instruction        Operands
               bic.b              #CLK,&P2OUT
```

The instruction "clear bits in destination, byte mode" means that the same source hex number assigned to "CLK" (002h) will be used to clear bits in the lower byte of the destination, the output register P2OUT. This line of code clears bits rather than set them as in line 7. As a result, the one bit of P2OUT is cleared to a 0, and pin P2.1 will have a 0, or a low, on it. Thus, CLK for the TLV0831 is now low. A low on CLK shifts the data onto the output DO, onto the pin P2.3 of the MSP430 and into register P2IN. The shifting of data occurs when the CLK line of the TLV0831 goes low as shown in the timing diagram of *Figure 7-14b*.

9.   The ninth line of code:

```
Label          Instruction        Operands
               bit.b              #DO,&P2IN
```

The instruction "test bits in destination, byte mode" means that the source hex number assigned to "DO" in Section B (008h) will be used to designate that the eight bit of the destination P2IN will be tested. And the result of the operation will affect the carry bit of the status register in the MSP430. Only the status register bits are affected. If the eight bit of P2IN is a 0, carry will be a 0; if the eight bit is a 1, carry will be a 1.

10.   The tenth line of code is:

```
Label          Instruction        Operands
               rlc.b              ADCData
```

The instruction "rotate left through carry" means that the contents of the ADCData register is rotated left one position and the carry bit of the status register is shifted into the LSB and the MSB is shifted into the carry bit. Symbolic addressing is used. *Figure 7-14c* illustrates the result of the rlc.b instruction. The carry bit from the previous instruction becomes the carrier of the data. When the carry bit is a 0, the ADCData register bit is a 0; when the carry bit is a 1, the ADCData register bit is a 1. The ADCData register becomes the temporary storage for the output data from the TLV0831 until all data is transferred. After all the data is collected, the ADCData register can be operated on by the MSP430 CPU.

11.   The eleventh line of code is:

```
Label          Instruction        Operands
               dec                R15
```

The instruction "decrement destination" means to subtract one from the contents of register R15. Register addressing is used. Register R15 has the number 9 in it. Nine minus one means the content of R15 is now 8.

At the same time, the register contents are tested and status bits are set in the status register. The Z bit is the one noted in the next instruction, so it is the bit of interest. Here is the rule for the test of the Z bit:

```
Status Bit      Rule
Z               Set if destination register contains 1, reset otherwise
```

If destination register R15 is other than zero, then the Z bit is set to a 1.

12. The twelfth line of code is:

```
Label           Instruction        Operands
                jnz                ADC_Loop
```

The instruction "jump if not zero" tests the status register Z bit. If Z is not 0, the program jumps to the line in the program that has the label ADC_Loop, which is line 7. Symbolic addressing is used. The program again runs through line 7, 8, 9, 10 and 11. This is called a subroutine jump, and the subroutine loop being lines 7, 8, 9, 10 and 11.

When the program returns to line 7, it again sets the TLV0831 CLK high. Line 8 then sets this same CLK low to shift the second bit to the output DO and pin P2.3 of the MSP430. The result is rotated into ADC-Data and R15 is decremented. The program again tests the Z bit and finds it is not zero and jumps back to line 7. The program continues in the loop rotating each bit in and subtracting 1 from R15 until the content of R15 is zero (9 counts).

When the status bit Z is zero as a result of R15 being zero, the program now does not jump but continues to line 13.

13. The thirteenth line of code is:

```
Label           Instruction        Operands
                bis.b              #CS,&P2OUT
```

The instruction "set bits in destination, byte mode" means that the hex number assigned to CS (001h) will be used to set bits in the lower byte of the output register P2OUT. Thus, the zero bit of P2OUT will be set and pin P2.0 will have a high output. P2.0 is the chip select for the TLV0831, and with it high, the TLV0831 is deactivated.

14. The fourteenth line of code is:

```
Label           Instruction        Operands
                Call               #TX_ADC_2PC
```

The instruction "call" tells the program to go to the label TX_ADC_2PC that is a subroutine in the program that transmits data from the register ADCData to a personal computer using a UART. The program will go through the subroutine TX_ADC_2PC, which is left out to keep the discussion brief. When it is finished, it returns to the program step after the subroutine call, step 15.

15. The fifteenth line of code is:

```
Label           Instruction        Operands
                jmp                Mainloop
```

The instruction "jmp" is called an unconditional jump instruction. It is addressed with symbolic addressing. The program jumps to Mainloop, which is the label on line 3 that is the start of Section E, the measuring portion of subprogram No. 1. Thus, the program is ready to start another measuring cycle by initiating a conversion by the ADC.

Subprogram No. 2

General Description

All systems need a clock to synchronize timing of events occurring as the system operates. This subprogram sets up the MSP430 clock MCLK to use LFXT1 as its source. LFXT1 is operated in the high-frequency crystal oscillator mode using a crystal between 1 MHz and 8 MHz. As mentioned in Subprogram No. 1, the crystal oscillator requires a certain time to stabilize; therefore, the program is setup to test the crystal oscillator, and only after it is stable, will it use LFXT1 as the source for MCLK. MCLK drives a software loop that takes exactly 10 clock cycles; therefore, it produces a clock signal that divides MCLK by 10.

One other feature of the MSP430 is that there is a "fail safe" mechanism built into the clock system. Since the crystal oscillator needs time to stabilize, there is a default mode which uses the DCO as a clock source until the crystal oscillator is up and running properly. Even though the DCO is not as accurate as the crystal timing, the DCO keeps the system timed and operating properly from the start.

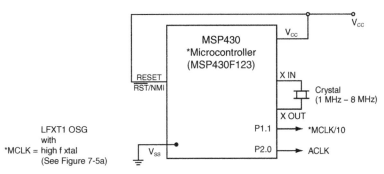

a. Block diagram showing pin connections

The block diagram for the application is shown in *Figure 7-16a* and the timing diagram in *Figure 7-16b*. Pin P1.1 of I/O Port 1 is used as an output for the MCLK divided by 10 signal, and pin P2.0 of I/O Port 2 is used for an external clock, ACLK. Note that the Pin P1.1 output is an asymmetrical waveform.

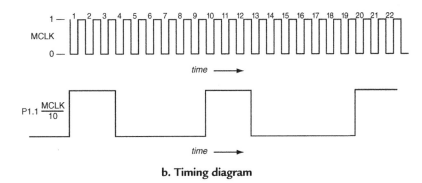

b. Timing diagram

Figure 7-16: Subprogram No. 2 application—outputting clocks

Section A—Initial Conditions

The subprogram is shown in *Figure 17*. It is understood that for this subprogram the same reference list that was used for Subprogram No. 1 is used again. Other very specific reference lists could be used here as in Subprogam No.1, but are not necessary. Any reference list is to be used by the assembler to insert specific hexadecimal numbers assigned to particular symbolic names.

Section A begins with the following line of code:

```
Label          Instruction      Operands
               ORG              0F000h
```

The assembler directive "ORG" tells the assembler to put the start of the program in memory location 0F000h. The same location used for Subprogram No. 1.

```
; *********************************************************************
  Label      Instruction  Operands                   Comment
; ---------------------------------------------------------------------
A.             ORG        0F000h                     ; Program Start
; ---------------------------------------------------------------------
 1.  RESET     mov.w      #300h,SP                   ; Initialize stackpointer
 2.  StopWDT   mov.w      #WDTPW+WDTHOLD,&WDTCTL      ; Stop WDT
 3.  SetupBC   bis.b      #XTS,&BCSCTL1              ; LFXT1 = HF XTAL
 4.  SetupOsc  bic.b      #OFIFG,&IFG1               ; Clear OSC fault flag
 5.            mov.w      #0FFh,R15                  ; R15 = Delay
 6.  SetupOsc1 dec.w      R15                        ; Additional delay to ensure start
 7.            jnz        SetupOsc1                  ;
 8.            bit.b      #OFIFG,&IFG1               ; OSC fault flag set?
 9.            jnz        SetupOsc                   ; OSC Fault, clear flag again
10.            bis.b      #SELM1+SELM0,&BCSCTL2      ; MCLK = LFXT1
                                                    ;
11.            bis.b      #001h,&P2DIR              ; P2.0 = output direction
12.            bis.b      #001h,&P2SEL              ; P2.0 = ACLK function
13.            bis.b      #002h,&P1DIR              ; P1.1 = output direction
                                                    ;
14. Mainloop   bis.b      #002h,&P1OUT              ; P1.1 = 1
15.            bic.b      #002h,&P1OUT              ; P1.1 = 0
16.            jmp        Mainloop                   ; Repeat
```

a. Asymmetrical waveform for output clock with MCLK/10 frequency

```
14. Mainloop   xor.b      #002h,&P1OUT              ; P1.1 = Toggle
15.            jmp        Mainloop                   ; Repeat
```

b. Symmetrical waveform for output clock with MCLK/12 frequency

Figure 7-17: Subprogram No. 2—assembly-language program—outputting clocks

1. The first line of code is:

```
Label        Instruction        Operands
RESET        mov.w              #300h,SP
```

When power is turned on, the program goes to the line of code labeled RESET for its instruction "mov.w". The instruction "move source to destination" loads the number 0300h into the stack pointer. This initializes the stack pointer. Note the .w notation has been used to identify the instruction as a word instruction. Immediate addressing is used for the source, and symbolic addressing for the destination. The reader should now be familiar with these notations so reference to them will be discontinued unless pertinent to the discussion.

2. The second line of code is:

```
Label        Instruction        Operands
StopWDT      mov.w              #WDTPW + WDTHOLD,&WDTCTL
```

The label "StopWDT" explains the instruction is stopping the watchdog timer. The instruction "mov.w" moves the hexadecimal numbers assigned to the symbolic names of the source, WDTPW and WDTHOLD, to the watchdog timer control register WDTCTL to set the respective bits. The password WDTPW is 5A00h to write to the WDTCTL and set WDTHOLD. This holds or stops the watchdog timer, and thus, it will not interrupt the system.

3. The third line of code is:

```
Label        Instruction        Operands
SetupBC      bis.b              #XTS,&BCSCTL1
```

Setup basic clock is what the label "SetupBC" means. The instruction "bis.b" means "set bits in destination, byte mode" and the binary number associated with the symbolic name of the source, XTS, will set that bit in the basic clock control register BCSCTL1 shown in Figure 7-9a. With XTS set, the LFTXT1 clock will operate with a high-frequency crystal oscillator as the source.

4. The fourth line of code is:

```
Label          Instruction       Operands
SetupOsc       bic.b             #OFIFG,&IFG1
```

As indicated by the label "SetupOsc", the instruction is used to setup the crystal oscillator used for the clock. The instruction "clear bits in destination, byte mode" means that the bit in the hex number associated with the source OFIFG will be used to clear a flag in the destination register IFG1. IFG1 is an interrupt flag register. The bit OFIFG is an interrupt flag for the crystal oscillator. If the crystal oscillator is not "up and running" the flag is set. Recall that the crystal oscillator needs a certain time delay before it is operating properly. When the OFIFG flag is not set, the oscillator is running properly. This instruction clears the flag so it is in the correct condition.

5. The fifth line of code is:

```
Label          Instruction       Operands
               mov.w             #0FFh,R15
```

The instruction "move source to destination, word mode" means that the hex number 0FFh will be loaded into register R15. R15 is going to be used as a counter whose content determines the time delay that is setup to allow the crystal oscillator to stabilize.

6. The sixth line of code is:

```
Label          Instruction       Operands
SetupOsc1      dec.w             R15
```

"SetupOsc1" is a label identifying a subroutine loop that is associated with the crystal oscillator delay that is required. The instruction "decrement destination" subtracts one from the contents of R15.

7. The seventh line of code is:

```
Label          Instruction       Operands
               jnz               SetupOsc1
```

The instruction "jump if not zero" tests the Z (zero) bit in the status register. If the result of the operation in line 6 is not zero, Z will be zero, and the program will jump to the subroutine label "SetupOsc1" which is line 6. The program will stay in this subroutine loop until Register 15 contents are decremented to zero. This produces a time delay of the time that is required to cycle through the loop until R15 = 0. The time delay is determined by the value loaded into R15 in line 5.

When R15 = 0, then the Z bit will be set and the program does not jump to line 6 but continues to line 8.

8. The eighth line of code is:

```
Label          Instruction       Operands
               bit.b             #OFIFG,&IFG1
```

The instruction "test bits in destination, byte mode" means that the source bit, the oscillator fault interrupt flag, OFIFG, in the destination interrupt flag register IFG1, will be tested. If the crystal oscillator is not completely stable, the flag will be set.

9. The ninth line of code is:

```
Label          Instruction       Operands
               jnz               SetupOsc
```

The instruction "jump if not zero" again tests the Z bit. If the result of the operation in line 8 is not zero, i.e. the flag is set, Z = 0 and the program will jump to the subroutine label "SetupOsc" which is line 4. Thus, the program returns to line 4 where it clears the oscillator fault interrupt flag bit, OFIFG, in register IFG1 and reloads R15 for an additional delay time. The loop of line 6 and 7 decrements R15 until the delay is complete. The fault flag OFIFG is tested again to see if it is set by line 8. If the oscillator is stable, OFIFG will not be set, the result will be zero and the program does not jump back on line 9, but continues to line 10. If the flag is set, then the oscillator is still not stable, and another pass through line 4, 5, 6, 7, 8 and 9 adds additional delay.

10. The tenth line of code is:

```
Label         Instruction      Operands
              bis.b            #SELM1 + SELM0,&BCSCTL2
```

The instruction "set bits in destination, byte mode" will set the bits SELM1 and SELM0 of the source, in the destination register BCSCTL2 as a result of assigned hex numbers from the reference list. Referring to *Figure 7-9b*, with SELM1 and SELM0 both equal to 1, the source LFXT1 is selected for the MCLK clock. What has happened is the program has assured that the high-frequency crystal oscillator is up and running and stable before it is used as a source for the main system clock, MCLK.

11. The eleventh line of code is:

```
Label         Instruction      Operands
              bis.b            #001h,&P2DIR
```

The instruction "set bits in destination, byte mode" loads the source 001h into the destination Port 2 direction register, P2DIR. This sets the zero bit in P2DIR and makes pin P2.0 an output.

12. The twelfth line of code is:

```
Label         Instruction      Operands
              bis.b            #001h,&P2SEL
```

The instruction "set bits in destination, byte mode" loads the source, again 001h into the special function register P2SEL and sets the zero bit. This means that the pin P2.0 is an output for an external clock ACLK, rather than the Port 2 output register P2OUT.

13. The thirteenth line of code is:

```
Label         Instruction      Operands
              bis.b            #002h,&P1DIR
```

The instruction "set bits in destination, byte mode" means that the source 002h is loaded into the destination, the direction register, P1DIR, to set the one bit. This sets pin P1.1 as an output.

Section B—Mainloop

14. The fourteenth line of code is:

```
Label         Instruction      Operands
Mainloop      bis.b            #002h,&P1OUT
```

The label "Mainloop" indicates this is the start of a subroutine. The instruction "set bits in destination, byte mode" loads 002h into the destination, the output register P1OUT, and sets pin P1.1. This means that P1.1 is in the high state.

15. The fifteenth line of code is:

```
Label         Instruction      Operands
              bic.b            #002h,&P1OUT
```

The instruction "clear bits in destination, byte mode" clears the one bit, identified by the source 002, in the destination P1OUT output register. Thus, pin P1.1 is cleared to zero, a low state.

16. The sixteenth line of code is:

```
Label          Instruction        Operands
               jmp                 Mainloop
```

The instruction "jump unconditionally" directs the program to jump to the line of code labeled "Mainloop", line 14. Thus, the program remains in the loop and cycles from line 14 to line 15 to line 16 and back to line 14 resulting in a square wave clock output on pin P1.1 as shown in *Figure 7-16b*. The clock driving the CPU is MCLK. It takes four clock cycles for the program to execute line 14, four clock cycles for executing line 15, and two cycles to execute line 16; thus producing an asymmetrical square wave clock output on P1.1 that is one-tenth the frequency of MCLK. This relationship is shown in the timing diagram of *Figure 7-16b*. Thus, two clocks result from the subprogram, one on P1.1 which is one-tenth the frequency of the high-frequency crystal oscillator, LFXT1, and the other an external clock, ACLK, on P2.0.

Section B—Mainloop Modification

Because the program in *Figure 7-17a* produces an asymmetrical waveform it may not be as desirable as a symmetrical wave; therefore, the main loop instructions can be modified to produce a symmetrical wave. Steps 14 and 15 can be modified as shown in *Figure 7-17b*, and step 16 is ommitted.

With steps 14 and 15 modified, the program proceeds from step 14 as follows:

14. The fourteenth line of code is:

```
Label          Instruction        Operands
Mainloop       xor.b              #002,&P1OUT
```

The label "Mainloop" is the same as previously and indicates this is the start of a subroutine. The instruction "Exclusive OR of source with destination, byte mode" does an exclusive OR logic operation with the source 002h and the output register P1OUT and places the result in the destination, the P1OUT register. Since pin P1.1 is the 1 bit of the P1OUT register, the result of the exclusive OR will appear on pin P1.1. In the first execution of line 14, if the 1 bit is 0, the XOR will toggle the state of the 1 bit—it will be a 1 and pin 1.1 will be a 1. In the next pass, the bit will be toggled to a 0.

15. The fifteenth line of code is:

```
Label          Instruction        Operands
               jmp                 Mainloop
```

The instruction "jump unconditionally" directs the program to jump to the line of code labeled "Mainloop", line 14. Thus, the program remains in the loop and cycles from step 14 to step 15 and back again. The clock, as previously, driving the CPU is MCLK. As a result, for this modification, it takes four cycles to execute line 14 and two cycles to execute line 15. P1.1 will now have a symmetrical square wave output with a frequency equal to MCLK/12.

Subprogram No. 3

General Description

Here is a program that outputs a visual signal when an input voltage is at or greater than a particular value. The block diagram is shown in *Figure 7-18a*. This time a TLC549 ADC is used. It is an 8-bit analog-to-digital converter that converts the input analog voltage into an 8-bit code that is shifted into the MSP430 microcontroller register R11 labeled ADCData. There is an LED (light-emitting diode) on I/O output pin P1.0.

When the input voltage is equal or greater than $+0.5V_{CC}$, then the value of the contents of R11 will be equal to or greater than $+0.5V_{CC}$, and the LED will be lit. For any input voltage less than $+0.5V_{CC}$, the LED will not light; therefore, one can start at $V_{IN} = 0$ and adjust the input voltage toward V_{CC}. When the input voltage is at $+0.5V_{CC}$ the LED will glow to indicate $+0.5V_{CC}$ had been reached.

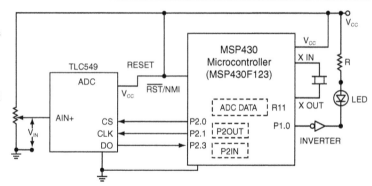

a. Block diagram showing interconnections

The assembly-language program is shown in *Figure 7-19*. Again as with Subprograms No. 1 and No. 2, "Section A," "Section B," "Section C," "Section D" and "Section E" have been added to aid in describing the program.

Sections A and B

Section A and B are specific to this subprogram. Section A again is clarifying that register R11 is identified with a label "ADCData" and that register R12 is identified with a label "Counter." They are two of the 16-bit working registers as shown in *Figure 7-2a*.

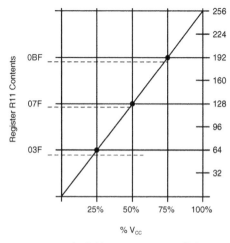

b. R11 content versus % V_{CC}

Section B is again an addition to the standard reference list in the Appendix for the MSP430. In the section,

Figure 7-18: System application implemented by Subprogram No. 3 energizing an output when input is greater than $+0.5V_{CC}$

the specific hexadecimal numbers shown are assigned to symbolic names that the assembler substitutes into the program when the symbolic names are used in the program.

Section C—Initial Conditions

The program starts in memory at address 0F000h established by the ORG instruction.

```
       Label         Instruction        Operands
                     ORG                0F000h
```

The assembler begins this program at the same location in memory (0F000h) used for Subprograms No. 1 and No.2. Then initial conditions are setup for the program with steps 1 through 5. Bits in the control registers discussed in *Figures 7-7 to 7-11* will be set to control the initial conditions.

1. The first line of code is:

```
       Label         Instruction        Operands
       RESET         mov.w              #0300h, SP
```

The type of addressing mode should now be fairly well understood so reference to the addressing modes will be omitted to simplify the discussion. The label "RESET" identifies where the program starts when

```
   ;
A. #define         ADCData  R11
   #define         Counter  R12
B. CS              equ      001h              ; P2.0 - Chip Select
   CLK             equ      002h              ; P2.1 - Clock
   DO              equ      008h              ; P2.3 - Data Out
   ;--------------------------------------------------------------------
C.                 ORG      0F000h            ; Program Start
   ;--------------------------------------------------------------------
   1.  RESET       mov.w    #300h,SP          ; Initialize 'x112x stack
   2.  StopWDT     mov.w    #WDTPW+WDTHOLD,&WDTCTL ; Stop Watchdog Timer
   3.  SetupP2     mov.b    #CS,&P2OUT        ; /CS set, - P2.x reset
   4.              bis.b    #CS+CLK,&P2DIR    ; /CS and CLK outputs
   5.  SetupP1     bis.b    #001h,&P1DIR      ; P1.0 output

D. 6.  Mainloop    call     #Meas_549         ; Call subroutine
   18.             bic.b    #01h,&P1OUT       ; P1.0 = 0
   19.             cmp.w    #07Fh,ADCData     ; ADCData > 0.5Vcc?
   20.             jlo      Mainloop          ; Again
   21.             bis.b    #01h,&P1OUT       ; P1.0 = 1
   22.             jmp      Mainloop          ; Again
                                             ;
   ;--------------------------------------------------------------------
E. Meas_549;       Subroutine to read TLC549, data is shifted into ADCData
   ;               (R11), Counter (R12) is used as a bit counter.
   ;--------------------------------------------------------------------
   7.              mov.w    #8,Counter        ; 8 data bits
   8.              clr.w    ADCData           ; Clear data buffer
   9.              bic.b    #CS,&P2OUT        ; /CS reset, enable ADC
   10. ADC_Loop    bit.b    #DO,&P2IN         ; (4) DO -> C (carry)
   11.             rlc.w    ADCData           ; (1) C -> ADCData
   12.             bis.b    #CLK,&P2OUT       ; (4) Clock high
   13.             bic.b    #CLK,&P2OUT       ; (4) Clock low
   14.             dec.w    Counter           ; (1) All bits shifted in?
   15.             jnz      ADC_Loop          ; (2) If not --> ADC_Loop
   16.             bis.b    #CS,&P2OUT        ; /CS set, disable ADC
   17.             ret                        ; Return from subroutine
                                             ;
```

Figure 7-19: Subprogram No. 3—An assembly-language program energizing an output when input is greater than +0.5V$_{cc}$

power is turned on, or a reset is performed. The instruction "move source to destination" moves the source 0300h to the stack pointer, the special function register identified by the symbolic name SP. When a program completes a subroutine it will return to the address on the stack.

2. The second line of code is:

```
Label           Instruction        Operands
StopWDT         mov.w              #WDTPW + WDTHOLD,&WDTCTL
```

The word instruction "move source to destination" sets bit in the destination, the watchdog timer control register WDTCTL, so that the watchdog timer is put on hold. The symbolic name WDTPW (5A00h) allows writing to WDTCTL and the symbolic name WDTHOLD sets that bit in WDTCTL to hold the watchdog timer and stop it from interrupting the system.

3. The third line of code is:

```
Label           Instruction        Operands
SetupP2         mov.b              #CS,&P2OUT
```

The label "SetupP2" identifies the instruction as one to setup the I/O P2. The "move source to destination, byte mode" takes 001h assigned to the source CS and moves it to the P2OUT register to set the zero bit of P2OUT or pin P2.0 to a 1. P2.0 is the chip select line to the TLC549.

4. The fourth line of code is:

```
Label            Instruction          Operands
                 bis.b                #CS + CLK,&P2DIR
```

The instruction "set bits in destination, byte mode" means that bits in the P2DIR control register will be set to control whether the pins of the P2 I/O will be outputs per *Figure 7-12*. If the pins are not set they will be inputs. 001h of CS sets the zero bit of P2DIR and 002h of CLK sets the one bit of P2DIR; therefore, pin P2.0 and pin P2.1 are outputs from the MSP430. Since P2.0 is a 1, or high, and is the chip select line for the TLC549, the TLC549 is inactive.

5. The fifth line of code is:

```
Label            Instruction          Operands
SetupP1          bis.b                #001h,&P1DIR
```

This line of code is going to setup I/O P1 as indicated by the label "Setup P1". The instruction "set bits in destination, byte mode" with the source, 001h, sets the zero bit of the direction control register P1DIR so that pin P1.0 is an output.

Section D—Main Application

6. The sixth line of code is:

```
Label            Instruction          Operands
Mainloop         call                 #Meas_549
```

The label "Mainloop" identifies the location in the program as the start of the main part of the program—the part of the program that measures the input to the TLC549 ADC. The "call" instruction tells the program to jump to the subroutine labeled "Meas_549." The "Meas_549 subroutine starts with the seventh line of code.

7. The seventh line of code is:

```
Label            Instruction          Operands
                 mov.w                #8,Counter
```

The instruction "move source to destination, word mode" means that the source, the hex number 8 will be moved to register 12 which has been assigned the symbolic name "Counter." It will be used to count the eight bits of the data output of the TLC549 ADC.

8. The eighth line of code is:

```
Label            Instruction          Operands
                 clr.w                ADCData
```

The instruction "clear destination, word mode" means that register R11 identified with the symbolic name "ADCData" will be cleared to zero.

9. The ninth line of code is:

```
Label            Instruction          Operands
                 bic.b                #CS,&P2OUT
```

The instruction "clear bits in destination, byte mode" means that the hex number assigned to CS (001h) will be used to clear the zero bit of the destination, the output register P2OUT; therefore, pin P2.0 will be reset to 0 or a low. Since P2.0 is the chip select line of the TLC549, this activates the TLC549 to measure its input analog voltage and convert it to an 8-bit digital code representing the value of the input voltage.

10. The tenth line of code is:

```
Label          Instruction          Operands
ADC_loop       bit.b                #DO,&P2IN
```

The label "ADC_loop" identifies the line of code as the start of a subroutine loop. The instruction "test bits in destination, byte mode" means that the source hex number assigned to "DO" in section B (008h) will be used to designate that the eight bit of the destination P2IN will be tested. The result of the operation will affect the carry bit of the status register in the MSP430. Only the status register bits are affected. If the eight bit of P2IN is a 0, carry will be a 0; if the eight bit is a 1, carry will be a 1.

11. The eleventh line of code is:

```
Label          Instruction          Operands
               rlc.w                ADCData
```

The instruction "rotate left through carry" means that the ADCData register is rotated left one position and the carry bit of the status register is shifted into the LSB and the MSB is shifted into the carry bit. Refer to the diagram in *Figure 7-14c*. The carry bit from the previous instruction becomes the carrier of the data. When the carry bit is a 0, the ADCData register bit is a 0; when the carry bit is a 1, the ADCData register bit is a 1. The ADCData register becomes the temporary storage for the data as the eight bits of data are shifted into the register.

12. The twelfth line of code is:

```
Label          Instruction          Operands
               bis.b                #CLK,&P2OUT
```

The instruction "set bits in destination, byte mode" means the hex number assigned to the source CLK (002h) will be used to set the one bit of the P2OUT register so that pin P2.1 will be at a high level. P2.1 is tied to the CLK input of the TLC549.

13. The thirteenth line of code is:

```
Label          Instruction          Operands
               bic.b                #CLK,&P2OUT
```

The instruction "clear bits in destination" means that the same bit in the P2OUT register as in the previous instruction is now cleared back to 0, or a low level. The pin P2.1, being the clock for the TLC549, means that when the clock goes low the next bit from the ADC data is shifted out on the DO line of the TLC549.

14. The fourteenth line of code is:

```
Label          Instruction          Operands
               dec.w                Counter
```

The instruction "decrement destination" means to subtract one from the contents of register R12, the register identified by the symbolic name "Counter." Since this is the first pass through the loop, R12 will now have a contents equal to seven, since the register was originally loaded with the value eight.

15. The fifteenth line of code is:

```
Label          Instruction          Operands
               jnz                  ADC_loop
```

The instruction "jump if not zero" tests the status register Z bit which will be a 1 or 0 based on the result of the instruction in line 14. If the result of line 14 is not zero, Z will be 0, and the program jumps to the line in the program that has the label ADC_loop, which is line 10. When the result of line 14 is zero, Z will be 1, and the program will not jump, but continue on to the next instruction. The program will continue in the loop from line 10 to line 15 until the contents of R12, the counter register, reach zero. When the contents have the value of zero, it means that the eight data bits have been shifted out onto DO. When the contents of R12 is zero, the program does not jump, but continues to line 16.

16. The sixteenth line of code is:

```
Label           Instruction       Operands
                bis.b             #CS,&P2OUT
```

The instruction "set bits in destination, byte mode" means the hex number 001h assigned to CS is used to set the zero bit of the P2OUT register so that pin P2.0 is set to a high level. Since P2.0 is the chip select of the TLC549, the TLC549 is disabled and its conversion ceases.

17. The seventeenth line of code is:

```
Label           Instruction       Operands
                ret
```

The instruction "return from subroutine" means the program picks up the return address from the stack pointer which is the address of the next line of code after the subroutine call. As a result, the program returns to line 18, the next instruction after line 6.

18. The eighteenth line of code is:

```
Label           Instruction       Operands
                bic.b             #01h,&P1OUT
```

The instruction "clear bits in destination, byte mode" means that the zero bit of the P1OUT register designated by the source 01h will be cleared; therefore, pin P1.0 will be cleared to a zero, or low level.

19. The nineteenth line of code is:

```
Label           Instruction       Operands
                cmp.w             #07Fh,ADCData
```

The instruction "compare source and destination" means that ADCData is compared to the hex number 07Fh, and the bits in the status register are set accordingly.

20. The twentieth line of code is:

```
Label           Instruction       Operands
                jlo               Mainloop
```

The instruction "jump if lower" means that the result of the operation in line 19 governs what happens in this instruction. If ADCData register contents are lower than 07Fh, then the program jumps to "Mainloop", another subroutine Meas_549 is called and another ADC conversion is accomplished as the program goes through the subroutine from line 7 through line 17. This continues again if ADCData contents are still lower than 07FH (which is 127 of a total of 256 of the full-scale content of ADCData. The value 127 is less than $0.5V_{CC}$, where V_{CC} is represented by the full-scale value of 256.

When the ADCData register contents are greater than 07Fh, then the program does not jump back to "Mainloop" but continues on to line 21.

21. The twenty-first line of code is:

```
Label           Instruction       Operands
                bis.b             #01h,&P1OUT
```

The instruction "set bits in destination, byte mode" means that the zero bit of the P1OUT register designated by the source 01h will be set to a one, or a high level. As a result, pin P1.0 will be set to a 1. This high level on P1.0 will light the LED that is connected to P1.0.

22. The twenty-second line of code is:

```
Label           Instruction         Operands
                jmp                 Mainloop
```

The instruction "jump" tells the program to jump unconditionally to the line of code labeled "Mainloop" or line 6. Line 6 calls the subroutine "Meas_549" and the whole measuring process begins again.

Variation of Threshold

The threshold voltage at which the system turns on the LED can be adjusted based on the binary number used for the comparison in the instruction of line 19. The relationship of the contents of R11, the register labeled as ADCData, to the percentage of V_{CC} is shown in *Figure 7-18b*. For +0.5V_{CC}, the binary number used in the comparison instruction was 07Fh, or one less than the binary number of 08F representing exactly +0.5V_{CC}. In like fashion, the binary number used for line 19 is 03Fh for +0.25V_{CC} and 0BFh +0.75V_{CC}. Other binary numbers per *Figure 7-18b* would adjust the trigger threshold to a selected percentage level of V_{CC}.

Summary

In this chapter the reader is exposed to the techniques used to program in assembly language. The Texas Instruments MSP430 microcontroller was chosen as the digital processor to use to explain assembly-language programming. Using its specific instruction set, the basics of writing an assembly-language program were discussed. Three assembly-language programs were discussed in detail to help the reader understand the concepts of assembly-language programming. With an assembly-language program, an assembler—a specific software program written to convert the assembly-language program into machine code—must be used before the program can be applied in a system. The next chapter will deal with the techniques of data transmission.

Chapter 7 Quiz

1. A RISC microcontroller is:
    a. a reduced, minimized component CPU.
    b. a much more complicated CPU design.
    c. based on a reduced-instruction-set CPU.
    d. a CPU with reduced peripherals around it.
2. A von Neumann architecture:
    a. is rectangular and triangular in nature.
    b. has a separate bus for program memory and data memory.
    c. has a separate bus just for peripherals.
    d. has program, data memory and peripherals all sharing a common bus structure.
3. A peripheral module in the MSP430 family can be:
    a. either a 16-bit or an 8-bit module.
    b. can only be a 16-bit module.
    c. can only be an 8-bit module.
    d. a module with only 5-bits.
4. The peripherals in the MSP430 family:
    a. use 16-bits exclusively for addressing.
    b. use both 8-bit and 16-bit addresses.
    c. use 8-bits exclusively for addressing.
    d. use 12-bits exclusively for addressing.
5. The operating mode of the MSP430 microcontroller is:
    a. determined by the I/O input number one.
    b. determined by the state of the CPU.
    c. determined by four control bits in the status register.
    d. all of above.
6. Interrupts control the digital processor:
    a. at specific well defined times.
    b. at unexpected or random times.
    c. at the same time every time.
    d. at regular predetermined repeating times.
7. Timers are used in a MSP430 system:
    a. to keep track of elapsed time.
    b. to set time intervals within which specific actions occur or are to occur.
    c. to produce resets, alerts or warnings.
    d. none of above.
    e. all of above.
8. When a source or destination in a MSP430 instruction have the form &ADDR, the addressing mode is:
    a. symbolic mode.
    b. register mode.
    c. absolute mode.
    d. indexed mode.
9. The MSP430 status register, R2:
    a. has nine active bits.
    b. has bits whose state dictates that a particular action has occurred.

    c.   is one of sixteen 16-bit registers in the CPU.

    d.   all of above.

    c.   none of above.

10. The MSP430 status register bit:

    a.   N is set when the result of a byte or word operation is negative.

    b.   Z is set when the result of a byte or word operation is zero.

    c.   C is set when the result of a byte or word operation produces a carry.

    d.   all of the above.

    e.   none of the above.

11. The MSP430 clock system control registers are:

    a.   registers R4, R5 and R6.

    b.   BCSCTL1, BCSCTL2 and DCOCTL.

    c.   registers R7, R8 and R9.

    d.   registers R13, R14 and R15.

12. If the XTS bit in the BCSCTL1 control register is set to a 1:

    a.   The LFXT1 oscillator in the clock system can operate with a high-frequency crystal.

    b.   the LFXT1 oscillator is OFF.

    c.   the LFXT1 oscillator in the clock system can operate with a low-frequency crystal.

    d.   it is a "don't care" condition for the LFXT1 oscillator.

13. When the SELS bit in the BCSCTL2 control register is reset to 0:

    a.   the DCOCLK is OFF.

    b.   the SMCLK is divided by 8.

    c.   the source for the SMCLK clock is LFXT1 oscillator.

    d.   the source for the SMCLK clock is DCOCLK.

14. In the MSP430, the watchdog timer control bit WDTTMSEL:

    a.   is set to 1 so that the watchdog timer is an interval timer.

    b.   is reset to 0 to have the watchdog timer inactive.

    c.   is not a factor in the operation of the watchdog timer.

    d.   is the bit that restarts the watchdog timer.

15. The WDTCTL control register must have:

    a.   all its high-byte bits at 0.

    b.   a 069h password automatically inserted in the high byte when WDTCTL is read.

    c.   a password of 05Ah in the high byte if the instruction is to write to WDTCL.

    d.   all its high-byte bits at 1.

    e.   only b and c above.

    f.   only a above.

16. In the MSP430 system all I/O ports:

    a.   are initially outputs when the system powers up.

    b.   remain constant as the applications program proceeds.

    c.   vary with each step of the program.

    d.   are initially inputs when the system powers up.

17. To set an external pin of an I/O port to be an output:

    a.   the associated bit of PxDIR direction register must be set to 1.

    b.   the associated bit of PxSEL function-select register must be set to a 1.

    c.   the associated bit of the PxIN register must be set to a 1.

    d.   the associated bit of the PxOUT register must be set to a 1.

18. When an external pin on the MSP430 I/O port is programmed to be an input:
    a. the direction register bit associated with the pin is reset to a 0.
    b. the PxIN register bit associated with the pin is set to whatever the input data dictates.
    c. the PxOUT register bit associated with the pin is inactive.
    d. all of above.
    e. a and c only above.
    f. none of above.

19. When an assembler program for the MSP430 sees a symbolic name:
    a. it has been programmed to interrupt the processor.
    b. it has been programmed to reset the system.
    c. it has been programmed to insert a specific binary number that represents the symbol.
    d. it has been programmed to disregard the symbolic name.

20. Symbolic name reference lists prepared for the MSP430 family:
    a. are used exclusively for the I/O bits.
    b. are used extensively for setting up initial conditions for the system.
    c. are used sparingly in assembly-language programming.
    d. are used to develop special symbols unrelated to actual register bits.

21. Labels:
    a. identify particular positions in a program.
    b. bear no relationship to the program.
    c. are only used at the end of a program.
    d. are not very useful in programming microcontrollers.

22. The .b in an instruction means:
    a. it is dealing with a 16-bit word.
    b. the instruction is part of a subroutine.
    c. the instruction is to be used later in the program.
    d. it is a byte instruction dealing only with the 8 bits in the lower byte of a word.

23. Operands are:
    a. special types of AND logic circuits.
    b. the portion of the instruction that identifies what quantities will be operated on using the instruction.
    c. special amplifiers used in signal conditioning a signal.
    d. the first component in an instruction line.

24. Hexadecimal numbers:
    a. use bit positions that are entirely different than binary codes.
    b. cannot be manipulated easily in binary systems.
    c. use numbers from 0 to 9 and letters from A to F to identify the 16 possible codes when using a 4-bit code.
    d. use no special notations to identify them in programs.

25. Assembly-language programming:
    a. helps to grasp the concept of programming.
    b. helps to learn programming details.
    c. helps to get familiar with programming format.
    d. all of above.

    e.   b and c only above.

26. In assembly-language programming for the MSP430:
    a.   "syntaxic substitution" is the technique of substituting numbers for words in a program.
    b.   the program instructions are converted to machine code by an assembler.
    c.   specific registers used for given tasks may be defined in a reference list.
    d.   the numbers used in the "syntaxic substitution" are defined in a reference list.
    e.   all of above.
    f.   c and d only above.

27. In assembly-language programming for the MSP430, a label:
    a.   has many uses but one important one is to identify a subroutine.
    b.   only provides reference to a particular action in a program.
    c.   has little meaning in a program.
    d.   is the most prominent way to set initial conditions.

28. In an assembly-language program for the MSP430:
    a.   a .w after an instruction means a decimal instruction.
    b.   a .w after an instruction means a hexadecimal instruction.
    c.   a .w after an instruction means to branch to another location.
    d.   a .w after an instruction means it is an operation using a word (two bytes).

29. In assembly-language programming for the MSP430:
    a.   a # sign before an operand means it is register-mode addressing.
    b.   a # sign before an operand means it is immediate addressing.
    c.   a # sign before an operand means it is absolute-mode addressing.
    d.   a # sign before an operand means it is symbolic-mode addressing.

30. In MSP430 programming using assembly language:
    a.   a reference list is very important to syntaxic substitution.
    b.   the programming depends totally on syntaxic substitution.
    c.   a reference list is not important to syntaxic substitution.
    d.   all syntaxic substitution reference lists are constant for any application.

Answers: 1.c, 2.d, 3.a, 4.b, 5.c, 6.b, 7.e, 8.c, 9.d, 10.d, 11.b, 12.a, 13.d, 14.a, 15.e, 16.d, 17.a, 18.d, 19.c, 20.b, 21.a, 22.d, 23.b, 24.c, 25.d, 26.e, 27.a, 28.d, 29.b, 30.a.

Data Communications

Introduction

A typical requirement of systems described in this book is that digital information must be transported from one location to another, from one piece of digital equipment to another. The two locations may be very close to each other, or they may be separated by a great distance. In this chapter, data communication systems will be discussed and several techniques used to transmit and receive digital data will be examined.

The Data Transmission System

Figure 8-1 shows a typical digital data communications system. Any digital communication must have a transmitter, receiver and a transmission medium. The transmitter prepares the digital information for transmission, the receiver detects and presents the digital information in original form, and the transmission medium transports the information, hopefully without modifying it or producing errors. The transmission medium may be twisted pair wire, wires in cables, fiber optic cable or wireless transmissions.

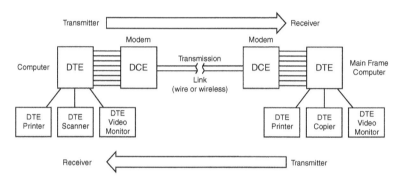

Figure 8-1: Data communication system

DTE and DCE

In *Figure 8-1*, a data terminal equipment, DTE, is coupled to a piece of data communications equipment, a DCE. The most common DCE is a modem that converts the digital data into signals that match the requirements of the transmission medium. A very common arrangement is a modem that couples to a telephone line. The DTE in this common case is a computer. In fact, the DCE (modem) is contained right in the computer, and the DTE and DCE combination becomes the transmitter for this data communications system.

At the receiving end, another DCE (again, another modem) receives the data from the transmission medium, decodes it and presents it to a DTE for transformation, manipulation, modification and/or display. As shown in *Figure 8-1*, each combination of DTE and DCE can either be a transmitter or a receiver depending on the direction of transfer of data.

Also shown in *Figure 8-1* is the fact that a DTE can be a computer, a printer, or a video monitor, and that beside the DTE to DCE and DCE to DTE data communication, there is and can be data transfers from a DTE to a DTE.

Parallel and Serial Transmission

There are two main methods of communicating digital data from one place to another, either parallel transfer or serial transfer. *Figure 8-2* shows the difference between the two. Parallel transfer is shown in *Figure*

8-2a. Here there are as many separate signal lines as there are bits of data in the digital signal. If the bits in the data change, all bits change at the same time. In other words, there is an information front that moves together on the lines, and when a change in the data is made all lines change at the same time.

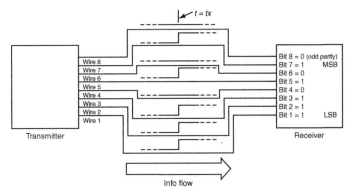

a. Parallel transfer of information for ASCII "W" (odd parity)

Contrast this to serial transfer of data shown in *Figure 8-2b.* Here there is only one line, and if one would sit on the line, the digital bits representing the information would pass by one bit after the other—in series, thus, the name serial data. Additional data must be added to the digital data to make sure it is recognized. A start bit must be added to tell when the data starts, and a stop bit is added to tell when the data stops. A parity bit, which will be

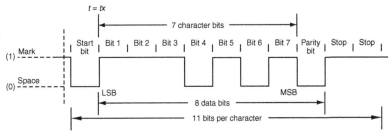

b. Serial transfer of information for ASCII "W" (odd parity)

Figure 8-2: Parallel and serial data transmission

discussed later, is added to aid in correcting errors.

Major Differences

The major differences between parallel and serial transfer are shown in *Figure 8-3*. As noted, serial transfers require only one line, while parallel transfers require a line for each bit of the multiple-bit character being transferred. If a moment of time is picked, t_x in *Figure 8-2a*, each

	PARALLEL	SERIAL
Lines Required	One line/bit	Single line
Bit Sequence	On all lines at same time	One bit following another
Speed	Faster	Slower
Transmission line length	Usually a short distance	Both long and short distances
Cost	More expensive	Less expensive
Critical Characteristic	Time relationship of bits	Needs start, stop bit

Figure 8-3: Comparison of parallel and serial communications

line will have a bit value corresponding to the digital code for the character being transferred, 1110101 for *Figure 8-2a*. While in a serial transfer, as shown in *Figure 8-2b*, the bit values will come one after another at t = t_x. First, the start bit, then the character bits, a parity bit and then the stop bit. Thus, the speed of transferring of data for serial communications is slower than for parallel. In parallel communications, all the bits arrive at the same time, while in serial communications, one must wait until all bits arrive.

It is very difficult in parallel communications to keep the time relation between bits the same for each line as the distance of the transmission increases; therefore, the connecting cables are usually short—a computer to a printer, or one computer to another, or a computer to a video monitor. There are parallel communications that occur over long distances that use what is called a packet technique and over special transmission lines or on microwave links. These are discussed briefly, and then explained a bit further for the USB protocol, but are really beyond the scope of this book. More detailed texts are required to explain it fully. Serial communications, on the other hand, using the latest technology, normally occurs over very long distances.

The equipment required for serial communications is less expensive since only one line is of concern rather than multiple lines in parallel communications. If 10-bit characters are being transmitted in parallel, the equipment multiplies by at least 10 times over what it is for serial communications. In parallel communications, one of the critical characteristics is the time relationship of signals on the line. In serial communications, additional information—a start and stop bit—must be added to be able to recognize and detect the information.

Example 1. Shift Right from Register

Show the bit storage in the lower byte of Register R5 for an ASCII capital N and the waveform generated as the ASCII code is shifted right out of the register. Odd parity is to be used. The ASCII code is shown in *Figure 8-5.*

Solution:

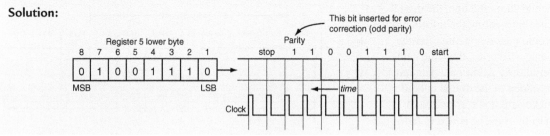

Protocols

"Protocol is the name given to hardware and software rules and procedures for making sure that any transmission errors are detected." [1] Data communications must follow certain rules and procedures as noted by the above quote whether it be the hardware used, the electrical signal levels, the signal timing, or the software used. *Figure 8-4* shows one of the earliest protocols, the RS-232 interface. It was used, and is still used today, to connect together all types of data communications equipment.

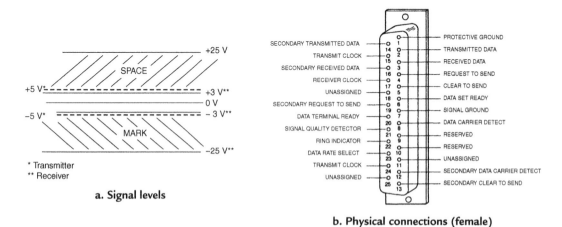

Figure 8-4: RS-232 protocol

---

[1] *Understanding Data Communications*, G.E. Friend, et al. ©1984, Texas Instruments Incorporated.

In *Figure 8-4a*, the electrical characteristics of an RS-232 signal are shown. The two binary levels are identified as "Mark" and "Space." In the RS-232 protocol, the receiver recognizes any positive signal from +3V to +25V as a space, and any negative signal from –3V to –25V as a mark. The transmitter, on the other hand, by specification, produces a space signal level between +5V and +25V, and a mark signal level between –5V and –25V. The mechanical connector and its associated pin connections are shown in *Figure 8-4b*.

Now look again at *Figure 8-2b*. Here the mark is identified as the 1 level and the space the 0 level. In modern day electronics, due to the influence of integrated circuits, and due to T^2L and CMOS logic circuitry, the 1 or mark is a high level from +2.4V to +5V, and the 0 or space level is a low level of +0.4V to 0V. When the RS-232 protocol was set, the maximum transfer speed was 20,000 bits per second, and the modem speeds were no higher than 9600 baud per second. USB, which will be discussed later, is a much more recent protocol for serial data communications and can transfer data at 4 million bits per second.

High-Speed Data Transmissions

As indicated, parallel data communications are limited by the length of the parallel wire cables; therefore, different techniques are used for such communications. Microwave, fiber optics, satellites are used for the transmission medium. The digital data is grouped into frames and packets to allow the data to be transmitted at millions of bits per second. Error detection and correction bits are added to the format so that the data can be communicated efficiently and without error at great speeds. This is possible because of the very wide bandwidth provided by microwave, fiber optic and satellite transmission links. Even though digital data communications require more bandwidth than analog signals, the very wide bandwidth is sufficient and available to allow high-speed digital data transmissions at ever increasing speeds.

Serial Data Communications Advances

The most common data communications today are serial communications. Even though the bits of a character flow in series one after another, the advances in technology, especially in the speed at which digital ICs can process digital information, have advanced so that the transfer speeds have kept up with the industry. As a result, the emphasis for the rest of this chapter will be on serial data communications. The discussion starts with a return to *Figure 8-2b*.

A Return to the Format

The two levels, Mark and Space, will be examined further. These terms come from the telegraph era. A pen attached to the armature of the sounder in a telegraph system would make a **mark** on paper moving under the pen as the armature was activated with the incoming signal. With no activation of the armature, the paper would just **space**. As the names for the two levels continued to be used, the mark was the state, of two available, in which there was a current. Space was identified as the state of no current. Still further use gave the idling state the name of "Mark" even though current was flowing. In this book, corresponding to accepted IC logic level, a "Mark" is the high level or a 1, and a "Space" is the low level or a 0. The RS-232 levels discussed before are really negative logic levels with a "Mark" being the most negative voltage level or a 1, and the least negative voltage level (the most positive level) a "Space" or a 0.

Start, Data and Stop Bits

As shown in *Figure 8-2b*, a start bit identifies the start of the data transfer. It is generated by changing the level from a 1 to a 0. Following the start bit are the bits used to determine the data. Seven bits are used in this example because a common binary code used for text data transfer is the ASCII (American Standard Code for Information Interchange) code shown in *Figure 8-5*. The ASCII code for "W" was used in *Figure 8-2*.

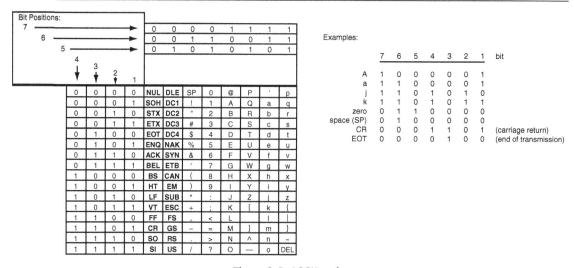

Figure 8-5: ASCII code

Following the seven bits of data, there is a parity bit and one or two stop bits. In case the data is eight bits, there would be a parity bit and only one stop bit. The stop bit is a continuous 1 level or idle condition.

Parity Bit

The parity bit is a bit of information added to the original data to allow for error detection. The bit is added by the transmitter to make the sum of all 1 bits in the character transmission either odd or even. The error detection method is called odd parity if the sum of the 1 bits is made odd; it is called even parity if the sum of the 1 bits is made even. *Figure 8-6* shows examples of how the transmitter adds the bits to make odd and even parity.

Bit	7-bit ASCII Code							Parity Bit (odd)	Parity Bit (even)
	1	2	3	4	5	6	7		
B	0	1	0	0	0	0	1	1	0
Q	1	0	0	0	1	0	1	0	1
3	1	1	0	0	1	1	0	1	0
z	0	1	0	1	1	1	1	0	1

These bits added by transmitter

Figure 8-6: Odd and even parity

At the receiver, circuits count the number of 1 bits in the character that is transferred. The system has been set up previously to operate either with odd or even parity. Suppose the system is operating using odd parity. If the counters always count odd numbers of 1s as the characters are transmitted, the receiver processes the data as correct. If, however, the 1 count turns up even, the receiver flags the information as incorrect and probably asks for it to be retransmitted. Even parity calls for the receiver to count an even number of 1s, and the data will be processed as correct as long as the count remains even. The receiver only flags the data as incorrect when the count is odd.

Example 2. Odd and Even Parity

What will the odd and even parity bit be for the digital codes given?

Solution:

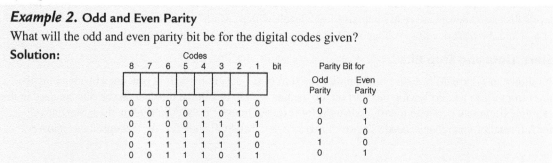

8	7	6	5	4	3	2	1	bit	Odd Parity	Even Parity
0	0	0	0	1	0	1	0		1	0
0	0	1	0	1	0	1	0		0	1
0	1	0	0	1	1	1	1		0	1
0	0	1	1	1	0	1	0		1	0
0	1	1	1	1	1	1	0		1	0
0	0	1	1	1	0	1	1		0	1

Baud Rate

In *Figure 8-2b* seven bits are used for the ASCII character and four bits are added—a start, parity and two stop bits. The total bits per character is eleven; therefore, the number of baud is 11. Suppose the rate of transmission is 10 characters per second. The baud rate will be characters per second × total bits per character, or, in this example, $10 \times 11 = 110$ baud per second. Modern telephone modems operate commonly at 56,000 baud per second.

Example 3. Baud Rate

What is the baud rate of an 8-bit data word with a start, parity and one stop bit when the transmission rate is 500 characters per second?

Solution:

No. of bits in serial word = $8 + 1 + 1 + 1 = 11$ characters
Transmission rate = 500/sec
Baud rate = $500 \times 11 = 5500$ baud/sec

Shift Registers

The shift register was discussed previously in *Chapter 6*. It is a main component of a serial communication system, and data can be manipulated in a number of ways, as shown in *Figure 8-7*, in order to arrive as serial data. In *Chapter 7*, the method shown in *Figure 8-7f*, rotate data left, was used to transfer data to the data register in the microcontroller, and is the same as the circulate example, in this case, left, discussed previously in *Chapter 6*.

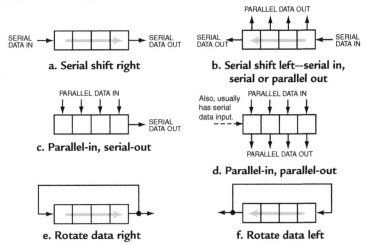

a. Serial shift right

b. Serial shift left—serial in, serial or parallel out

c. Parallel-in, serial-out

d. Parallel-in, parallel-out

e. Rotate data right

f. Rotate data left

Figure 8-7: Various types of shift registers

Courtesy of Master Publishing, Inc.

Example 4. Parallel In—Serial Out Shifting

Show the contents of register R10 for each clock cycle as a 4-bit word is transferred in a parallel transfer and stored in R10. A logical shift right is then made to examine the bits, one by one. The 4-bit code loaded in R10 is 0110.

Solution:

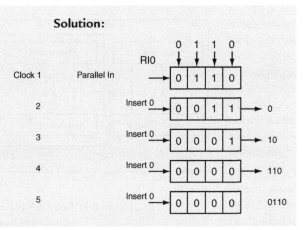

USART Serial Communications

A universal synchronous/asynchronous receiver/transmitter called a USART is a DCE used extensively for serial communications. There are two protocols used—one for synchronous transmit/receive and the other for asynchronous. In the asynchronous mode, the serial bit stream is at a programmed transmission rate determined by an internal clock in the transmitter. In the synchronous mode, the transmission rate is provided by a common clock, either in the transmitter or the receiver.

A simplified block diagram of a USART is shown in *Figure 8-8a*, and the format for the data, a typical serial format, is shown in *Figure 8-8b*. The block diagram shows an output, TXD, for the transmitted data, and an input, RXD, for the received data. Most USARTs can transmit and receive at the same time. If they cannot do the dual function, there is a R/W (read/write) control line that determines the mode of operation. The USART has a sync signal to set whether the operating mode is synchronous or asynchronous, and some additional control signals. The USART is in the synchronous mode when the sync signal is a 1.

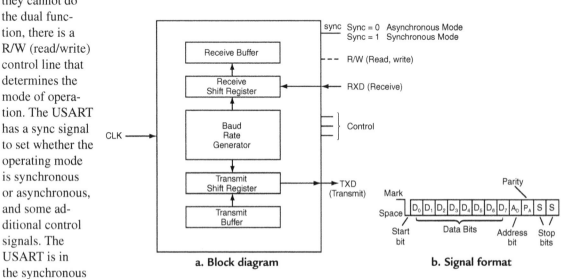

a. Block diagram

b. Signal format

Figure 8-8: Simplified USART

Synchronous Serial Communications

For synchronous serial communications there is a master unit and a slave unit. Since there is a common clock, the master generates the clock and the slave depends on this clock for its timing. The data format is still as shown in *Figure 8-8b*.

Figure 8-9 is a block diagram of two USARTs communicating

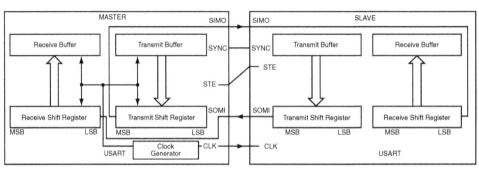

SYNC = 1 for synchronous operation

Figure 8-9: Two USARTs communicating in synchronous mode

Courtesy of Texas Instruments Incorporated

with each other in the synchronous mode. The left unit is the master, which supplies the clock, and the right unit is the slave. The master transmits data at the clock rate. The slave uses the clock to shift information in and out. The STE signal, controlled by the master, enables the slave to transmit data as well as receive data. The master and slave send and receive data at the same time. Data is shifted out of the transmit shift register on one clock edge and shifted in to the receive shift register on the opposite edge. The timing is shown in *Figure 8-10*.

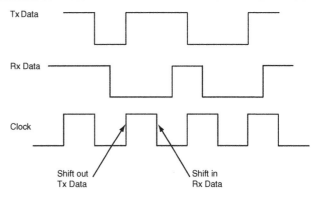

Figure 8-10: Shifting out Tx data and shifting in Rx data

The master output of the transmit shift register is coupled through the slave-in, master-out (SIMO) line to the slave receive shift register, while the slave out of the transmit shift register is coupled through the slave-out, master-in (SOMI) line to the master receive shift register. The data moves at a synchronized rate determined by the clock supplied by the master. The right unit could just as well be the master and the left the slave, and the operation is the same. The baud rate is programmed into and controlled by a baud-rate generator that is derived from the clock in the master.

Asynchronous Serial Communications

Asynchronous serial communications between two USARTs is shown *in Figure 8-11*. There again is a master and a slave, and the data format is the same as *Figure 8-8b*, but the frames of data do not

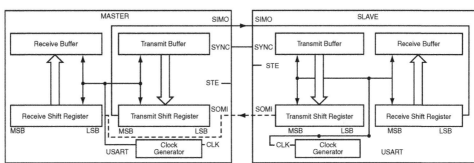

Figure 8-11: Two USARTs communicating in asynchronous mode

Courtesy of Texas Instruments Incorporated

always arrive in regular periods. There may be significant random idle periods between frames (greater than 10 bit times) as shown in *Figure 8-12*. There is no physical interconnection of clock signals from master to slave. The programmed master clock sets the transmission asynchronous serial communications rate.

As shown in *Figure 8-11*, the master is the transmitter and the slave is the receiver. When the first signals are received, the receiver adjusts its clock to match the clock rate of the received signal and uses this clock

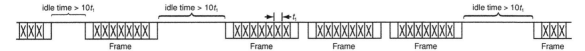

Figure 8-12: Asynchronous serial communication

149

to shift in the received data. Transmission in the asynchronous mode is only one way. In order for the slave to transmit to the master, the roles of the slave and master must be reversed. The slave becomes the master, that originates the clock, and the master becomes the slave. No interconnections need change, but control signals must change. Depending on the direction that the data is to flow, the roles of the master and slave reverse as the flow of data reverses.

The UART Function with Software.

Subprogram No. 1 of *Chapter 7* essentially implemented a shift register using software. Expanding on the technique used there, the UART function can be implemented with software. It is not covered here, but this would be a challenging project for a team of students that come in contact with this book.

Technology Advances

Two advances in technology will be cited to demonstrate new techniques that have been developed to increase the transfer rate of digital information using serial communications. The first is the Inter-IC serial bus.

I²C Bus

A serial communications proprietary protocol that was developed by Philips Semiconductor[2], is the I²C bus. It was developed principally for inter-IC control, thus the name I²C. All ICs that are I²C-bus compatible have on-chip interfaces that communicate directly with other I²C-bus compatible devices. Serial, 8-bit, bi-directional data transfers can be made in three modes:

1. Standard—100 kbits/sec
2. Fast—400 kbits/sec
3. High-Speed—3.4 Mbits/sec

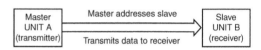

a. Unit A sends information to Unit B

The two-line bus has a serial data line (SDA) and a serial clock line (SCL). It is a synchronous system and requires a clock. The unit that initiates the data transfer is the master. It also is the unit that generates the clock, and initiates, permits, and terminates the transfer. If the master wants to communicate with another unit, it sends the address of that unit on the data line. The unit that is addressed is called the slave. The master and the slave can be either a transmitter or a receiver. Examples are shown in *Figure 8-13*. *Figure 8-13a* shows a master transmitting to a slave receiver; and *Figure 8-13b* shows the slave transmitting back to the master, now used as a receiver. The bus design allows multiple masters and slaves on the bus.

b. Unit A receives information from Unit B

Figure 8-13: I²C master and slave can be transmitter or receiver

I²C Protocol

Electrical Connections

Figure 8-14a shows the interconnection of devices inside units connected to the bus. Essentially, the SDA line and the SCL line are held in the high level by pull-up resistors until control transistors are activated to pull the line low. It is a large wired AND connection with open collector (bipolar) or open drain (MOS) connections from the devices to the lines. As SDA is activated by data and SCL by clock pulses, the lines are pulled low by the active devices. A low level is defined as a maximum of 0.3 V_{dd}, and a high level as a minimum of 0.7 V_{dd}. V_{dd} is typically the T²L logic level of 5V.

---

[2] I²C-Bus Specification, V2.1, Philips Semiconductor.

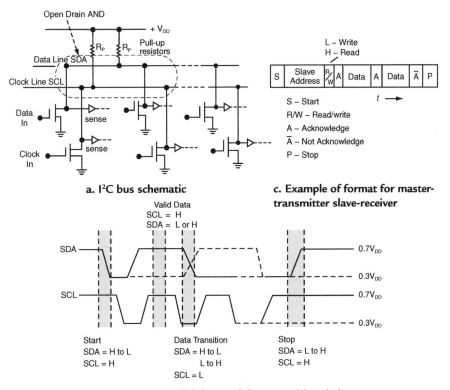

a. I²C bus schematic

c. Example of format for master-transmitter slave-receiver

b. Start, stop, valid data and data transition timing

Figure 8-14: I²C protocol

Signal Timing

Figure 8-14b shows the necessary timing of information on the bus. To generate the necessary start bit, SDA must be pulled from high to low while the SCL line is high. Data on SDA is valid only while SCL is high, and data cannot change (without error) unless SCL is low. A stop bit is generated when SDA goes from low to high when SCL is high. Thus, the start bit, data bit and stop bit requirement of the serial format is satisfied.

Example 5. I²C Data

Determine the data bits in the I²C waveform shown. See *Figure 8-14b* for signal protocol.

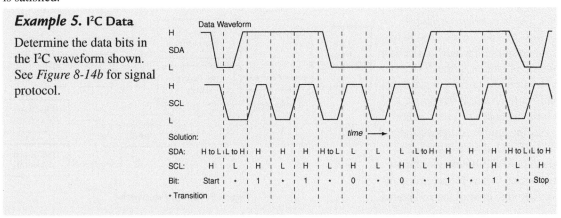

Format

An example of the serial format for a master-transmitter to a slave-receiver is shown in *Figure 8-14c*. The format starts with a start bit, then the address from the master to identify the slave, then a low on the R/W bit, and finally the data. Data continues to be sent by the transmitter as acknowledgement bits (A) are placed on the bus by the receiver. When no acknowledgement is received, the master-transmitter sends a stop bit.

USB

Another still more advanced serial data communications protocol is the universal serial bus (USB). It is being used extensively to communicate data from DTE to DTE, from DTE to DCE and from DCE to DTE. Using USB, serial data can be transferred at three different rates. Using USB low speed, the transfer rate is 1.5 million bits/sec; using USB full speed, the transfer rate increases up to 12 million bits/sec; and using USB high speed, the transfer rate is up to 480 million bits/sec. The discussions in this chapter center on USB low speed and full speed. The reader is left to investigate the specifications for USB high speed.

The connecting cable used is shown in *Figure 8-15*. It is a 4-wire system, using a twisted pair for D+ and D– data lines, and power lines of V_{BUS} and GND. It uses a unique feature of differential detection of data on the D+ and D– lines.

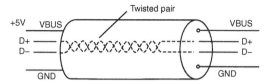

Figure 8-15: 4-wire USB cable

USB Network

A typical USB network is shown in *Figure 8-16*. It consists of a host, which contains a host controller, and separate USB devices. These devices, as shown, can either be a function or a hub. A function is a USB device that is able to transmit or receive data or control information over the bus. It contains information about its capabilities and the resources that it needs. Examples of functions are mouse controllers, light pens, keyboards, printers, scanners, and so forth.

Hubs are USB devices that expand the USB bus interconnections. They allow the attachment of multiple USB devices. The host, as shown in *Figure 8-16*, can be connected to a function or a hub, and that hub can be connected to other hubs or other functions. In addition, there is overriding software that manages the bus. USB permits the host to configure a hub and monitor and control its ports.

The host is responsible for knowing when devices are connected or disconnected from the bus, for managing the data flow between USB devices, and for the status of the bus. The host assigns a unique address to a device attached to it. It determines if the new device is a hub or a function. If the device is a function, the host recognizes this and configures it. If the device is a hub, the host's software establishes the unique addresses and end points for all devices attached to the hub. All USB devices support a common means for accessing information to control the end points.

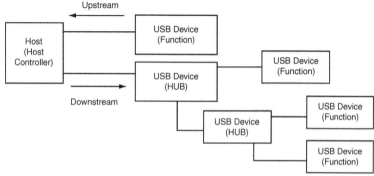

Figure 8-16: USB network

USB Electrical Connections

Figure 8-17 shows the USB electrical interconnections of the bus. The host controller is required to have a root hub that contains a transceiver. All hubs, including the root hub, are required to support both full-speed and low-speed data transfers. Functions may just support low speed.

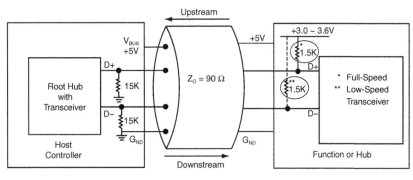

Figure 8-17: USB interconnections

Transmission from the host is called downstream; transmission to the host is called upstream. At the host, root hub, and any external hub, the D+ and D– lines at downstream ports each have a 15 kΩ pull-down resistor to ground. On a port feeding upstream from a device or hub, a 1.5 kΩ pull-up resistor is connected between the D+ line and a voltage supply from +3.0V to +3.6V. If it is a low-speed device, the 1.5 kΩ resistor is connected from the D– line to the voltage source. An external hub is a special case that has both 1.5 kΩ resistors on up-stream ports and 15 kΩ pull-down resistors on the downstream ports. The impedance of the USB cable is 90 Ω. V_{BUS} is nominally +5V at the source. The host supplies power to USB devices directly connected to it. A hub supplies power to its connected devices; however, some connected devices have internal power sources.

Bus Transceivers

The details of the transceivers are shown in *Figure 8-18.* Note that there are differential receivers for the data lines, and also single-ended receivers, one for each data line. The single-ended receivers are used for control purposes. There are output buffers that drive the data lines when transmitting data. Each of the output buffers have an enable input because the buffers must have a 3-state high-impedance output when not enabled. This means the buffers, when not enabled, are

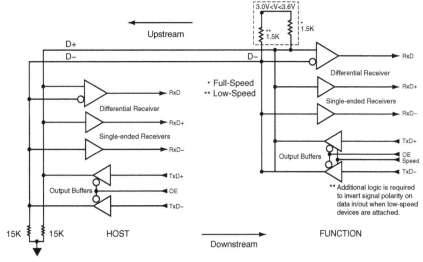

Figure 8-18: A possible USB system showing receivers at host and function

no load on the bus. The output buffer for the function transceiver has another input to control whether it is operating at low speed or full speed.

When the host is transmitting, its output buffers are enabled and drive the data lines differentially; the function output buffers are disabled. When the function is transmitting, its output buffers are enabled and drive the data lines differentially; the host output buffers are disabled. When not transmitting the output buffers are disabled. At low speed, two changes occur—the 1.5 kΩ resistor's termination is to the D– line instead of the D+ line, and the logic levels are reversed.

Data Line Waveforms

The differential signals plotted against time are shown in *Figure 8-19*. The data signals swing between 2.8V ($V_{OH(min)}$) and 0.3V ($V_{OL(max)}$). When D+ is greater than 2.8V and D– is less than 0.3V, the differential logic state is a 1; when D+ is less than 0.3V and D– is greater than 2.8V, the differential logic state is a 0. The point where the waveforms cross is called

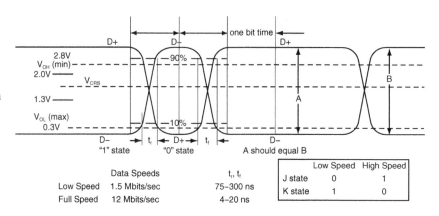

Figure 8-19: Data line switching waveforms

V_{CRS}, the voltage crossover point. It must be between 1.3V and 2.0V. The amplitude A should be approximately the same for each transition. The logic states are also called the J and K states. The J and K states are inverted, as shown in *Figure 8-19*, for low speed and full speed operation. This is the reason for the extra logic inverters in the function called for in *Figure 8-18*. The rise and fall times of the waveforms must be 75–300 nS for low-speed operation and 4–20 nS for full-speed operation.

USB Signal Protocol

The USB is a polled bus. The host controller initiates all data transfers. All bus transactions involve the transmission of three packets diagramed in *Figure 8-20*. Each transmission begins when the host controller, on a scheduled basis, sends a "token packet" describing the type and direction of the transmissions, the USB device address, and an end-point number. The USB device that is addressed selects itself by decoding the appropriate address fields.

Example 6. USB Host-to-Function Addressing

A USB network has the host transmitting downstream to four functions with addresses as shown. The host transmits an address 0010. What happens?

In a given transaction, data is transferred either from the host to a device, or from a device to a host. The direction of transfer is specified in the token packet. After the direction is set, the source sends data packets, or else indicates it has no data to transfer. When a transfer is received, the destination, in general, responds with a

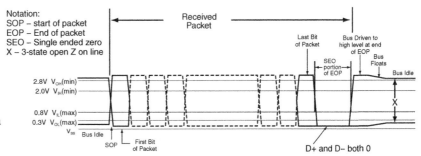

Figure 8-20: Example of a possible packet and data format

"handshake" packet that indicates the transfer was successful. Each packet is preceded by sync signals, called a sync field, and includes a control portion, a data portion and an error-correction field. The error correction is for single and double bit errors. The sync field, which is really a clock signal, is transmitted and encoded along with the differential data, and allows the receiver(s) to synchronize their bit recovery clocks. After the last bit of the packet, as shown in *Figure 8-20*, both D+ and D−, as single-ended signals, are driven to 0.

The data line is then driven to the high level to indicate an end to the packet. The output buffers are then driven to their high-impedance state so that the data bus floats. There are limitations on the capacitance loading and propagation delays so that signal reflections can be controlled.

Data Transfers

There are three types of information transfers: sync, control and data. The sync, control and data transfers between the host and a USB device can be one-way or two-way. The sync transfers synchronize the receivers. The control transfers configure devices when they are first connected or when there is any change in the device status. Data transfers are of three types: bulk, interrupt and real time. Bulk data transfers are for large amounts of data between a host and a printer or scanner that requires data accuracy. Interrupt transfers, as from a mouse for a computer, is for data that may be presented at random times. Real-time data transfers, such as voice transmissions, that need to occur in real time and have no error correction, but must continue to be transmitted, are called isochronous transfers.

Data Encoding

The transfers are encoded using a "nonreturn-to-zero" encoding as detailed in *Figure 8-21*. If the data is 0, the encoding changes; if it is a 1, it does not change. This encoding, as shown in *Figure 8-22*, produces a clock type sync field at the beginning of a packet. This is the signal that synchronizes the receiver of the packet. The rules for

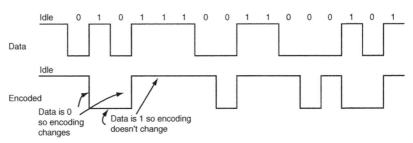

Figure 8-21: Nonreturn-to-zero encoding

the sync pulse preceding the packet are that seven 0s and a 1 are to be transferred as the sync field. When encoded it produces the sync field shown in *Figure 8-22*.

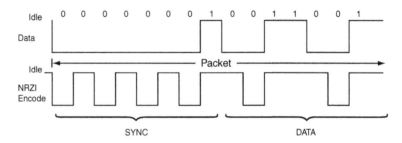

Figure 8-22: Sync signals prefixed to packet

Example 7. NRZ Encoding

For the data waveform shown, what would the encoded waveform be when using nonreturn-to-zero (NRZ) encoding? The rules for encoding are: if data = 0, encoding waveform changes; if data = 1 encoded waveform does not change.

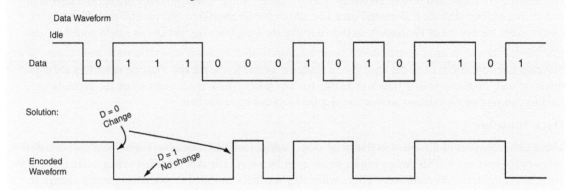

Summary

Data communications using parallel and serial techniques has been the subject of this chapter. The protocols, data format and techniques showed that parallel communications are the fastest but most expensive. Advances in technology have caused the less expensive serial communications to keep pace with the requirements of the hardware. New transmissions using I²C and USB show how these protocols are keeping pace. In the next chapter, power systems and their control will be discussed.

Chapter 8 Quiz

1. A digital data communication system is:
   a. made up of a transmitter and a transmission medium.
   b. made up of a transmitter, receiver and a transmission medium.
   c. made up of a receiver and a transmission medium.
   d. made up of just the transmission medium.

2. In digital communications, a parallel transfer means that:
   a. digital data is transferred at a faster speed.
   b. if bits in digital data change, they all change at the same time.
   c. it is the most economical way of transferring data.
   d. each bit of the digital data has a separate data line.
   e. c only above is correct.
   f. a, b and d above are correct.

3. In digital communications, a serial transfer means that:
   a. it is the fastest way of transferring data.
   b. the transfer of data is just on one line.
   c. the digital data arrives one bit after another in sequence.
   d. it cost less to transfer data than using parallel transfers.
   e. a only above is correct.
   f. b, c and d above are correct.

4. A communications protocol is:
   a. a set of communication system schematics.
   b. a set of rules that applies only to digital data communications hardware.
   c. the hardware and software rules and procedures for making sure that transmission errors are detected.
   d. a set of rules that applies only to digital data communications software.

5. Serial data communications is the prime digital data transfer method because:
   a. advances in IC technology have kept up with the bit transfer rates required by industry.
   b. it needs more equipment (hardware) than parallel transfers.
   c. it is much more expensive than parallel transfers.
   d. it is faster than parallel transfers.

6. In serial digital data communications, what must be added to the data?
   a. Only start and stop bits.
   b. Just parity bit.
   c. Start, stop and parity bit for error detection and correction.
   d. Some error-detection and error-correction scheme.

7. Odd parity:
   a. is when an additional bit is added to the data to make the sum of the 1 bits odd.
   b. means it is a strange set of rules.
   c. is when an additional bit is added to the data to make the sum of the 1 bits even.
   d. improves the baud rate.

8. A USART is:
   a. a universal synchronous/asynchronous receiver/transmitter DCE.
   b. a DCE used extensively for serial data communications.
   c. a DCE that can only be used for synchronous transmissions.

    d.   a DCE that can only be used for asynchronous transmissions.

    e.   only c above.

    f.   only a and b above.

9. When two USARTs communicate synchronously:

    a.   there is no common clock.

    b.   the clock is generated externally.

    c.   there are two independent clocks, one in transmitter and one in the receiver.

    d.   there is a common clock either in the transmitter or the receiver.

10. When two USARTs communicate asynchronously:

    a.   there is a common clock either in the transmitter or the receiver.

    b.   there is no physical interconnection of clock signals.

    c.   the clock is generated externally.

    d.   the frames of data arrive in regular periods.

11. I^2C-Bus means:

    a.   a combination of amplifier integrated circuits.

    b.   a current squared times a capacitor value.

    c.   a serial communications bus protocol developed for inter-IC control.

    d.   a bus used for parallel data transfers.

12. I^2C is:

    a.   a 2-line bus synchronous system that requires a clock.

    b.   a 2-line bus that has a master and a slave.

    c.   a 2-line bus where the transmitter or receiver can be the master or slave.

    d.   a 2-line bus where the master initiates the data transfer and generates a clock.

    e.   a 2-line bus where the unit that is addressed is the slave.

    f.   b, c and d only above.

    g.   all of a, b, c, d and e above.

13. The universal serial bus (USB) is:

    a.   a 2-wire system with a twisted pair for the data lines.

    b.   a system that has all hubs in its network.

    c.   a 4-wire system with differential signal detection of data on the D+ and D– data lines.

    d.   a system that operates using +2.5V.

14. The four wires of the USB system are:

    a.   D+, D–, V_{BUS} and GND.

    b.   GND, V_{DD}, V_{OH} and V_{OL}.

    c.   GND, V_{CC}, $V_{OH(min)}$ and $V_{OL(max)}$

    d.   D+, D–, $V_{OH(min)}$ and $V_{OL(max)}$

15. USB networks have:

    a.   a host controller with a root hub and transceiver.

    b.   hubs or functions connected to the root hub.

    c.   a 90Ω transmission cable.

    d.   an upstream and downstream transmission direction.

    e.   all of the above.

    f.   only a and b above.

16. A USB transceiver has:

    a.   three op amps and a power amplifier.

    b.   two power amplifiers and three tuned-stage amplifiers.

    c.   a class C, a class A, a Class AB and a class B amplifier.

    d.   differential receivers, single-ended receivers, and output buffers.

17. A USB network:

    a.   identifies a 1 or 0 of data using single-ended signals.

    b.   identifies a 1 or 0 of data using differential signals.

    c.   identifies a 1 or 0 of data using class AB amplifiers.

    d.   none of the above.

18. In the USB network:

    a.   the transmitter sends a sync signal to synchronize the receiver.

    b.   the receiver runs freely at its own clock rate.

    c.   the receiver has a physical wire connection with the transmitter.

    d.   the USB device that is addressed selects itself.

    e.   a and b above.

    f.   a and d above.

19. At the end of a USB packet transmission:

    a.   the transceivers buffers are still actively connected to the bus.

    b.   the D+ and D– lines are both at a 1 level.

    c.   the output buffers in the transceivers are put into their high-impedance state.

    d.   the sync signal is inserted.

20. In a USB packet transmission:

    a.   there is no error-correction information.

    b.   there only is data in the packets.

    c.   the single-ended transmission is susceptible to noise.

    d.   seven 0s and a 1 are transmitted to develop sync pulses.

Answers: 1.b, 2.f, 3.f, 4.c, 5.a, 6.c, 7.a, 8.f, 9.d, 10.b, 11.c, 12.g, 13.c, 14.a, 15.e, 16.d, 17.b, 18.f, 19.c, 20.d.

System Power and Control

Introduction

All electronic systems require a source of power. In almost all cases the voltage and current values are specified. The current value is in amperes as a load on the supply, and the voltage value is to be held within a specified tolerance (usually a percentage of the nominal value) as the current value varies within specified limits as the load changes. The nominal value of voltage times the nominal value of current determines the watts of power required of the supply. In this chapter, not only will the source of the voltages and their regulation be discussed, but the way the supply voltages are distributed throughout a system. In addition, the sophisticated circuits that are now available to monitor, detect and protect systems from damage, errors and failure will be discussed.

AC to DC Power Supplies

Figure 9-1a shows a general AC to DC power supply. Its source is the alternating current voltage of 120VAC or 250VAC, 60 Hz that is distributed commercially by the local power company. The alternating voltage varying plus and minus around zero is rectified into voltages that vary only above zero. The

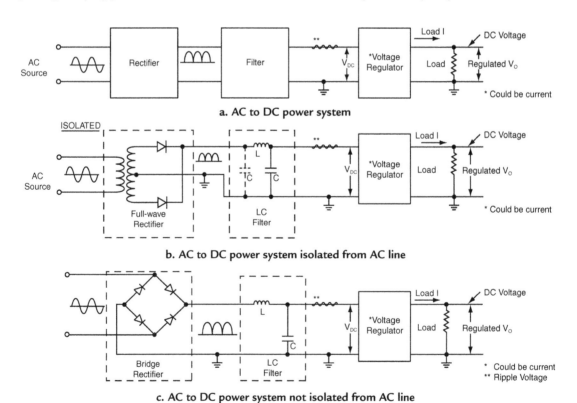

a. AC to DC power system

b. AC to DC power system isolated from AC line

c. AC to DC power system not isolated from AC line

Figure 9-1: Examples of AC to DC power supplies

half-alterations are passed through a filter that produces a DC voltage of designed amplitude. A small ripple voltage results from the amount of filtering compared to the input voltage variations. The ripple is superimposed on the DC voltage and represents a so-called noise. Because the output voltage must be controlled accurately within tight tolerances, a voltage regulator (or it could be a current regulator) is required. The voltage is held to within 1% to 10% of V_{OUT} over the specified load current and its changes depending on the application and type of regulator.

Many AC to DC power supplies must be isolated from the incoming AC line. *Figure 9-1b* shows such a design using a full-wave rectifier and transformer isolation. If the power supply need not be isolated, *Figure 9-1c* shows a design using a bridge rectifier supplied directly from the AC line.

Voltage Regulators

Zener Regulator

Figure 9-2 shows different versions of linear voltage regulators. The simplest of these is shown in *Figure 9-2a*. It consists of a zener diode and a resistor connected to a voltage V_{IN}. A zener diode is a semiconductor diode designed to operate in the reverse-biased avalanche region (similar to breakdown) that has the characteristic of maintaining a constant voltage across it as the current through it varies. It must have a minimum current, $I_{Z(min)}$, through it to operate properly, and because of a power dissipation temperature limit, it can

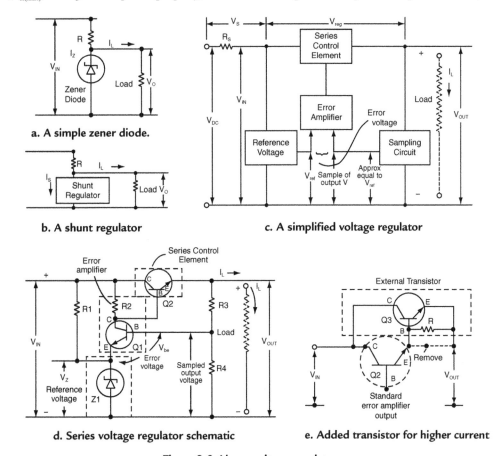

a. A simple zener diode.

b. A shunt regulator

c. A simplified voltage regulator

d. Series voltage regulator schematic

e. Added transistor for higher current

Figure 9-2: Linear voltage regulators

handle only up to a maximum current, $I_{Z(max)}$. Here is how it works as a regulator. I_Z can be any value from $I_{Z(min)}$ to $I_{Z(max)}$, and V_O will remain within a specified percentage of V_O. Initially with no-load, $I_L = 0$, the series resistor is set so that the current is $I_{Z(max)}$. When I_L is increased, I_Z decreases, but V_O will remain within specified limits until $I_Z = I_{Z(min)}$. Thus, I_L varies over its range but V_O remains within specified limits. The zener diode is not a regulator for wide variations in current; it is more a regulator for a constant load with little variations. Currents that it can handle are usually less than 100 milliamperes (100 mA).

Example 1. Zener Voltage Regulator

A Zener diode has the characteristics shown for points 1 and 2. What is the percent load regulation when the load changes between point 1 and point 2?

	V_Z	I_Z	
Point 1	6.0V	1 mA	(maximum current drawn by load)
Point 2	6.42V	100 mA	(minimum current drawn by load)

Solution:

% Load Regulation = $V_{NO\ LOAD} - V_{LOAD}/V_{LOAD} \times 100$

% Load Regulation = $6.42 - 6.0/6.0 \times 100 = 0.42/6 \times 100 = 0.07 \times 100$

% Load Regulation = 7%

Shunt Regulator

The shunt regulator shown in *Figure 9-2b* also shunts current from the load but is designed to handle much larger currents. It duplicates the zener diode regulator. Initially, I_S is a maximum through the shunt. As I_L increases to a maximum, I_S will decrease to a minimum. It is packaged to handle much greater power dissipation, since the device power dissipation is V_O times I_S.

Linear Series Voltage Regulators

A true feedback-type linear voltage regulator is shown in *Figure 9-2c*. All components operate in their linear mode. It is a simplified block diagram that does not have all the bells and whistles that are designed into IC regulators today, but the modern IC regulators are based on the same principles. The input voltage, a DC voltage, is separated from the load by a control element in series between V_{IN} and V_{OUT}. There is a voltage drop across the control element of V_{REG}. The series control element is controlled by an error amplifier. The error amplifier amplifies a voltage difference, called the error voltage, between a reference voltage and a sampled portion of the output voltage approximately equal to the reference voltage. Changes in the load current cause V_{OUT} to vary and the error voltage to change such that the series drop across the control element compensates for the change in V_{OUT}.

Load Variations

The regulation works as follows: If I_L increases it will tend to reduce V_{OUT}. The reduction in V_{OUT} is fed through the sampling circuit to an input of the error amplifier. The reference voltage is on the other input. Since the reference voltage is constant, the error voltage decreases and causes the voltage across the control element to decrease. As a result, V_{OUT} increases to compensate for the initial decrease.

Likewise, if I_L decreases, it tends to increase V_{OUT}. Increasing V_{OUT} increases the error voltage, which increases the control element voltage, V_{REG} and reduces V_{OUT} to compensate for the initial increase. The stable operating point of the system is such that with V_{IN} a particular value and V_{OUT} a specified value, the error voltage is tending toward zero.

Actual Linear Voltage Regulator Circuit

Figure 9-2d is a schematic of the interconnection of components for a linear series voltage regulator. The active devices shown are bipolar transistors, but MOS devices can (and are) used for the same design. NPN transistors and a positive output voltage are used in the design because the circuit is a bit easier to understand.

Note first that the series control element is just a NPN transistor. Note also that the reference voltage is really the zener voltage regulator that was discussed in *Figure 9-2a*. The only variations in the zener diode current will be those caused by variations in the input voltage, V_{IN}. Q_1 and R_2 form an inverting amplifier whose output drives the base of Q_2, the series control element. The input to the amplifier is the error voltage, V_{be} of Q_1. The sampled output voltage under the quiescent state is equal to a V_{be} voltage above V_Z, the reference voltage.

Load Variations

The regulation proceeds as follows: When I_L increases and V_{OUT} tends to reduce due to the increased drop across Q_2, the V_{be} error voltage on Q_1 reduces, reducing current through Q_1 and R_2. The rise in the collector voltage of Q_1 and base voltage of Q_2 raises the emitter voltage of Q_2, increasing V_{OUT} to compensate for the initial reduction.

Likewise, for a decrease in I_L, V_{OUT} tends to increase because of the decreased drop across Q_2. The V_{be} error voltage increases, which reduces the Q_1 collector voltage and base voltage of Q_2, which reduces the emitter voltage to compensate for the initial rise in V_{OUT}.

Line Variations

Similar regulation occurs for V_{IN} variations. If V_{IN} increases, V_{OUT} tends to increase, but the reference voltage, V_Z, changes very little. The error voltage increases because of an increase in V_{OUT}, which reduces the Q_1 collector voltage and base voltage of Q_2, which reduces the emitter voltage, V_{OUT}, to compensate for the increase. Similar to load variations, a decrease in V_{IN} will be met with a compensating increase in V_{OUT} to complete the regulation.

Higher-Current Regulators

In order to handle larger currents and, thus, more power dissipation for the regulator, external devices can be connected as shown in *Figure 9-2e*. Many IC regulators have the connections for the external devices provided in the design. Again, bipolar devices are used in the example, but MOS devices can be used just as well. Power devices with external heat sinks are usually required in order to satisfy the power dissipation requirements and keep the operating temperatures of the devices within specifications.

Voltage Regulation

Return now to *Figure 9-2c*. The input voltage V_{DC} is the voltage out of the rectifier and filter of *Figure 9-1*. There is an impedance associated with the rectifier and filter. It is R_S shown in *Figure 9-2c*. As a result, the output voltage, V_{OUT}, is equal to:

$$V_{OUT} = V_{DC} - I_L R_S - V_{REG}$$
$$= V_{DC} - V_S - V_{REG}$$
$$= V_{DC} - (V_S + V_{REG})$$

Using this information, the regulation can be explained as follows: V_{DC} is always considered constant. The regulator varies V_{REG} to keep V_{OUT} constant as I_L and V_S change. With an increase in I_L and thus V_S, V_{OUT} would tend to decrease; however, V_{REG} is reduced to compensate and V_{OUT} remains constant.

A decrease in I_L causes a decrease in V_S, but regulation compensates by increasing V_{REG} so that V_{OUT} remains constant. V_{DC} was considered constant but if V_{DC} changes, either up or down, regulation follows to compensate by increasing or decreasing V_{REG} to keep V_{OUT} constant.

Percent Regulation

The load regulation of a voltage regulator can be expressed as follows (where the load is some specified current value):

$$\% \text{ Load Regulation} = \frac{V_{NO\,LOAD} - V_{LOAD}}{V_{LOAD}} \times 100$$

If $V_{NO\,LOAD} = 11V$ and, at a specified current, $V_{LOAD} = 10V$ then

$$\% \text{ Load Regulation} = \frac{11 - 10}{10} \times 100$$

$$\% \text{ Load Regulation} = 10\%$$

Common percent load regulation for IC voltage regulators is from 1% to 5%.

Example 2. Voltage Regulation

To have 1% load regulation for the above supply where $V_{NO\,LOAD} = 11V$, at the specified load, what does the V_{LOAD} have to be?

$$1\% = \frac{11 - V_{LOAD}}{V_{LOAD}} \times 100$$

$$0.01 = \frac{11 - V_{LOAD}}{V_{LOAD}}$$

$$V_{LOAD}\,(1 + 0.01) = 11$$

$$V_{LOAD} = \frac{11}{1.01} = 10.89V$$

Power Dissipation

The power dissipation within an IC is a very important parameter because excessive temperature rise within a semiconductor junction can ruin the device. In linear series voltage regulators, the device that handles the most current is the series control element. The power dissipated in the control element is the product of the voltage across the unit times the current through the unit. The load is the current, I_{LOAD}, through the unit as shown in *Figure 9-2c and d*. The voltage across the series element is V_{REG}; therefore, the power dissipation in the series element is:

$$P_D = V_{REG} \times I_{LOAD}$$

V_{REG} is:

$$V_{REG} = V_{DC} - V_S - V_{OUT}$$

V_S is usually very small compared to V_{OUT}, therefore, the series element power dissipation can be expressed as:

$$P_D = (V_{DC} - V_{OUT})\,I_{LOAD}$$

Example 3. Power Dissipation

A voltage regulator has an input voltage of +12V and is regulating a +5V supply line. The load current is 100 mA. What is the power dissipation in the control element?

Solution:

$$P_D = (V_{DC} - V_{OUT})I_{LOAD}$$
$$P_D = (+12 - +5)V \times 0.1A$$
$$P_D = 7 \times 0.1 = 0.7 \text{ watts}$$

In many IC voltage regulators, especially the low-drop-out regulators, the V_{DC} is restricted to specified values so that the V_{REG} is not too great across the series element at the rated load current. This prevents exceeding the rated power dissipation of the device.

Switching Voltage Regulators

A regulator that has gained prominence as the requirements for load current increased is the switching voltage regulator. Standard linear regulators only have conversion efficiencies of less than 50%. Switching regulators can have efficiencies of up to 85%. This results in lower power dissipation, much smaller size components for a given power output, and operation over a wide range of voltage and current.

Figure 9-3 details a switching voltage regulator. One notes that there are similarities to a linear voltage regulator. There is the sampling of the output, the error amplifier, and the error voltage resulting from a comparison of a sample of the output voltage and the reference voltage.

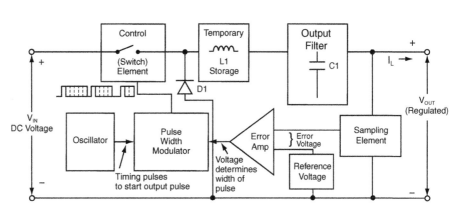

Figure 9-3: Switching voltage regulator (step-down)

Here are the differences between the two regulators:

1. The error amplifier output controls a switch whose ratio of open to closed is varied.

2. Since the control element is a switch rather than a linear element, there is considerable difference in the regulator action.

The Control Element

Instead of a series element operating in the linear mode, the control element is a switch that is in series with a temporary energy storage element, an inductor. The switch is opened and closed at a very rapid rate, and the ratio of the time it is closed to the time it is opened is varied to accomplish the regulation. There is no linear control element operation; it is all digital, either open or closed. When the switch is closed, it charges the inductor with energy by creating a field of magnetic flux around the inductor. When the switch is opened, the magnetic flux collapses across the inductor and returns the energy to the circuit. As the energy is returned, the inductor uses D1, shown in *Figure 9-3*, to complete the circuit and keep current, I_L, through the load.

Actual Regulation

Producing more or less voltage across the load is based upon modulating the time that the control element is closed. This is accomplished by the pulse-width modulator (PWM) driven by the error amplifier. An oscillator produces the start of pulses at a constant rate, but the end of the pulse is determined by the voltage supplied by the error amplifier. The relationship of the control voltage from the error amplifier to the pulse width that turns on the switch is shown in *Figure 9-4*. Note the center of the figure has a line that represents a constant level of the control voltage B that is the nominal voltage level at the rated current output. The pulse width for this control voltage is shown as width C.

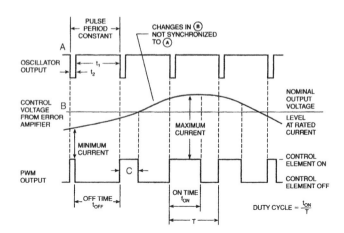

Figure 9-4: Switching regulator waveforms

Courtesy of Master Publishing, Inc.

When the demand for current increases, the pulse width increases because the ON time of the pulse is increased. More energy is stored in the inductor so that the increased current can be supplied and the voltage maintained. The integration of the current pulses by the output filter establishes the output voltage level. More ON time in the pulses produces a higher voltage, less ON time in the pulses produces a lower voltage. As shown in *Figure 9-4*, when minimum current is required the pulse width is narrow with a short ON time. Likewise, when maximum current is required the pulse width is wide with a long ON time.

Here is a description of the regulation in simple terms. When the load demands more current the output voltage tends to decrease. This voltage decrease is sampled and converted to an error voltage that increases the control voltage B and increases the ON time of the pulses. The increase in ON time supplies the increased current and raises the output voltage to its required value.

A load that demands less current would tend to increase the output voltage. The voltage increase is sampled and converted to an error voltage that decreases the control voltage B and decreases the ON time of the pulse. The decrease in ON time of the pulses lowers the voltage and satisfies the demand for less current.

Switching regulators operate at frequencies from 100 kHz to several million cycles/sec. Because of the range of frequencies and the switching action, there is some concern about RFI energy; and attention must be paid to the shielding of sensitive circuits.

Step-Up and Inverting Switching Regulators

The switching regulator shown in *Figure 9-3* is a step-down regulator—V_O is smaller in value than V_{IN}. *Figure 9-5* shows two other types of regulators, a step-up and an inverting regulator. The step-up regulator produces a regulated voltage V_O that is greater in value than V_{IN}, while the inverting regulator produces a V_O that is inverted in polarity from V_{IN}. A positive V_{IN} produces a negative V_O.

Switching Regulator Design

The design of switching regulators can be accomplished in a number of ways, but they all include the inductor as the temporary energy storage element and large storage capacitors. The inductor and capacitor(s) cannot be integrated into ICs; therefore they are external to any ICs used. Any of the other components,

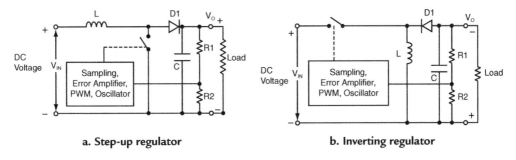

a. Step-up regulator   b. Inverting regulator

Figure 9-5: Different kinds of switching regulators

depending on the current and voltage requirements, can at least be partially integrated circuits. For example, if the current handling is within the range of 1–2 amperes, all of the error amplifier, PWM circuit, oscillator, and the control element can be one IC. With higher current requirements, external heat-sinked driver packages can be used for the control element. Resistor dividers are always used to sample the output voltage to feed back to the error amplifier.

Transformed PWM Regulators

In a different design than that shown in *Figure 9-3,* the PWM circuit, which contains the error amplifier, oscillator, voltage reference and some protection circuits, is used as an AC source. This AC source is transformed to the desired voltage, filtered, and fed back to the error amplifier to close the regulation loop. Such a regulator is similar to the ones described because it uses PWM pulses for regulation control, but it does not utilize the inductor as a temporary storage element. An increased pulse width (larger ON time) will increase the voltage out from the transformed source; while a decreased pulse width decreases the voltage output.

Summary of Regulators

Nothing has been said in the discussion on regulators about all of the protection techniques that can be used in the regulator circuit. For example, protection for maximum current, for short-circuits, for exceeding temperature limits, for over voltage, for under voltage, controlling the power-up or power-down sequence are all protection features that regulators may contain. Some of the features may be built into the IC regulator itself, while others may be separate ICs designed specifically to provide the protection function.

Many IC voltage regulators that handle low power requirements may have two separate individual regulators in a package, or the regulator may be one that has been designed to regulate two different voltages at the same time. As mentioned previously, many regulators have external connections provided so that higher current control elements can be driven as was shown in *Figure 9-2e.*

Switching regulators normally handle larger currents and voltage than fully-integrated regulators. Great care must be taken to keep regulators within temperature limits by the use of heat sinks and proper ventilation. The switching elements of switching regulators can be subjected to rather extreme current spikes and/or voltage spikes because of the nature of the operation; therefore, careful design is required to manage these concerns.

As mentioned previously, because switching is occurring at relatively high frequencies, and because the magnitude of currents switched are high, there is significant RF energy generated. Thus, a major concern is the circuit layouts and shielding of sensitive circuits due to the RF energy present.

Power Supply Distribution

Figure 9-6a shows an electronic system that needs regulated voltages of +5V, +3.3V, +3V, and +1.8V. The source for the system is a filtered DC of V_{DC} obtained from one of the rectifier and filter systems of *Figure 9-1*. One of the simplest ways to provide the system voltages is shown in *Figure 9-6b*. A regulated +5V line is used as the source, and resistor dividers with bypass capacitors are used to provide the required voltage supply lines. However, such a system is not satisfactory for many systems because the load current changes cause voltage variations that are not acceptable for proper operation of the system. The supply lines must be regulated voltages held to tight tolerances, much tighter tolerances than the resistor divider and capacitor by-pass can provide. In addition, the +5V regulator must be able to handle the total current required of the system.

A very acceptable system is shown in *Figure 9-6c*. Here individual linear regulators designed to operate with low V_{REG} voltages for low power dissipation (called LDO regulators) are used to provide the regulated voltages in steps. Regulator A provides a +5V rail from which regulator B and C provide +3.3V and +3.0V rails, respectively. Instead of regulator D deriving its source voltage from the +5V rail, it uses the +3.0 rail to provide the +1.8V rail. This keeps the power dissipation in each regulator at a low level, and also doubles the regulation for the stepped down rails. Regulator A must be able to supply the current used by the +3.3V, the +3.0V

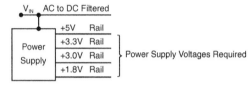

a. Overall system requirements

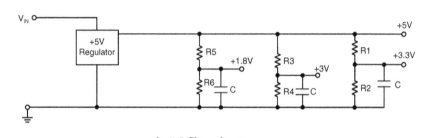

b. RC filtered system

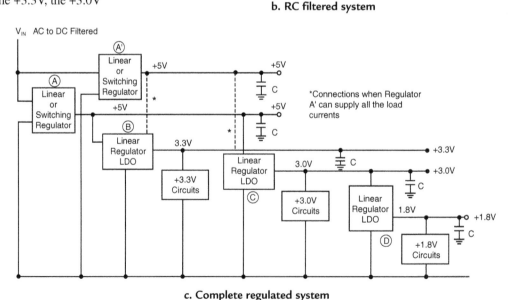

c. Complete regulated system

Figure 9-6: Power distribution system

and the +1.8V rails. This is the reason regulator A'
is included. It supplies all the remaining current
for the main +5V rail. If it is determined that the
current supplied by regulator A is small enough that
regulator A' would be able to handle the load and
still hold the regulation percentage, then regulator A
can be eliminated. Of course, on all rails, capacitor
bypass should be provided where the circuits tap off
for power.

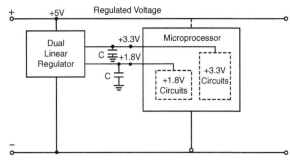

Dual-Output Regulators

Figure 9-7: Dual regulator for microprocessors

Many times circuits inside microprocessors operate
at different voltage levels to save power dissipation. *Figure 9-7* shows the use of a regulator that can supply two different voltages from the same input voltage. Obviously, two separate regulator packages could
be used, but with the one-chip package, variations with temperature will track better and "loss-of-power"
protection for circuits will be better.

DC/DC Converters

Figure 9-8 shows a
power distribution
system that requires,
beside the +5V for
logic circuits and other
computing circuits,
+12V, +24V and +3V.
DC/DC converters
are used for the +12V,
+24V and +3V rails.
The A and B convert-
ers are stepping up
the voltage from +5V,
while the C converter
steps down the volt-
age. The schematic of

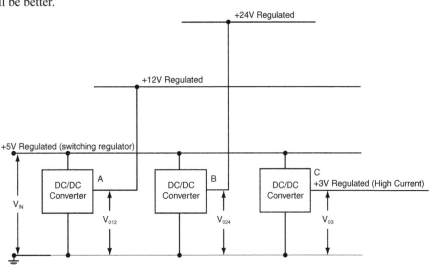

Figure 9-8: DC/DC converters for higher or lower voltages

the DC/DC converter is shown in *Figure 9-9*. The basic circuit of oscillator, pulse-width modulator, error
amplifier and reference voltage are used to provide a varying ON-time pulse to the control element, in this
case a power MOS. The varying pulse width pulses, transformed, rectified and filtered, provide the output
voltage V_O. The ratio of the secondary turns to the primary turns on the output transformer determine if the

voltage steps up or steps
down. The sensing of V_O
in this case is out at the
load, and current sensing
circuits provide protec-
tion if the current output
exceeds a set limit.

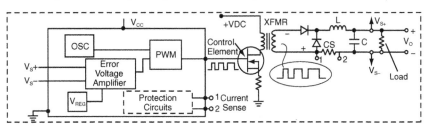

Figure 9-9: DC/DC converter schematic

Example 4. Power Distribution

Given the system specifications for the power supply rails required, design the power distribution for the system.

Voltage	Current	W	Comment
+12V	7 A	84	2% regulation, small current variations
+5V	2 A	10	1% regulation, moderate current changes
+3.3V	0.5 A	1.65	1% regulation, large % current changes
+3V	10 mA	0.03	2% regulation, no current variations

Solution:

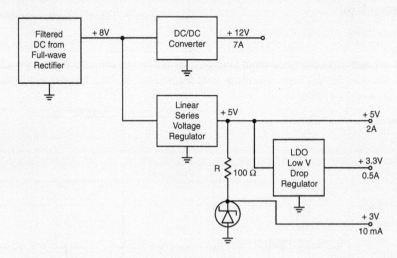

A V_{DC} of +8V from a filtered DC output of a full-wave rectifier is chosen as the main source. The +5V and +3.3V rails, because of the 1% regulation needed with moderate current changes, linear voltage regulators are selected. The +3.3V rail, since it has current changes that are a large % of the total current of 0.5A nominal value, uses the regulated +5V as its supply rail. The +5V LDO regulator must be able to supply the 2.5A requirements of the +5V and +3.3V rails. The high-current +12V rail needs only 2% regulation with small current variations; therefore, a DC/DC converter is used to supply the +12V. As a result, another source besides the +8V is not required. The +3V rail with 2% regulation and no current variations from the low current of 10 mA is regulated with a zener diode.

Power System Supervisors

Power system supervisors are circuits that watch over the power distribution system to detect variations in system power that may cause failure, faulty operation, or damaged circuitry. *Figure 9-10* shows a supervisor that is watching over three power rails. A is a main system processor. It is the heart of a system. If it were to lose its operating point, data and memory, it would be a catastrophe. The supervisor watches the supply voltage and anticipates power failure. By doing so, it allows the processor to terminate main operations, save data and enter a proper shut-down procedure.

At B are circuits that need time to stabilize—like a clock generator in a processor. The supervisor provides a time delay before the circuits operate which allows the power to be at it stabilized value and the circuits up and running properly.

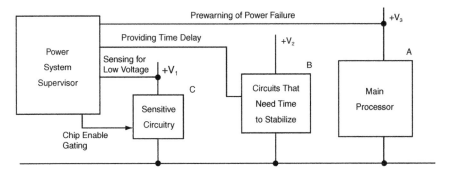

Figure 9-10: Power system supervisor

The supervisor for the circuits of C provides a gated chip enable. If the power supply voltage were to go below a critical value, sensitive circuits of C would not operate properly. The supervisor senses the power supply voltage, and if it goes below a critical set threshold, the supervisor disables the circuit with the chip enable line.

Summary

The basics of power supply regulation and some techniques used for distribution have been covered. A subject that is somewhat beyond the scope of this book is called "power management." It covers many exciting innovations such as, loadsharing, swapping boards in computers while power is ON, and many other protection techniques—for example, current limiting power switches, and watchdog timers. To equalize current in parallel power supplies, to have a technician replacing boards while the system runs, to turn ON or OFF a power supply at a current limit, and to reset a system if it doesn't respond after a given time should be subjects that wet one's interest to investigate further power management techniques.

Chapter 9 Quiz

1. An AC to DC regulated power supply for electronic systems consists of:
   a. a transformer, a rectifier and a filter.
   b. a rectifier, a filter and a voltage or current regulator.
   c. a transformer, a full-wave rectifier and a load.
   d. a bridge rectifier and a voltage or current regulator.

2. Common solid-state rectifiers for AC to DC power supplies are:
   a. single vacuum tube rectifiers.
   b. bridge rectifiers.
   c. full-wave rectifiers.
   d. all of above.
   e. b and c above.
   f. a above.

3. A zener diode is a semiconductor diode that:
   a. is forward biased in normal operation.
   b. can handle tens of amperes of current.
   c. doesn't have to be concerned about power dissipation.
   d. is designed to maintain a constant voltage across it as the current through it varies.

4. In linear series voltage regulators:
    a. all components operate within their linear range.
    b. the series control element operates in a switching mode.
    c. the error voltage is quite large at the stable operating point.
    d. the reference voltage source handles large currents.
5. The efficiency of linear series voltage regulators is:
    a. less than 50%.
    b. less than 75%.
    c. less than 35%
    d. less than 10%.
6. In a quiescent linear series voltage regulator using a zener diode reference voltage:
    a. the sampled output voltage is equal to the zener reference voltage.
    b. with V_{IN} constant, there will be large changes in the zener reference voltage.
    c. the sampled output voltage is a V_{be} voltage above the zener reference voltage.
    d. there is no limit on the current through the series control element.
7. A linear series voltage regulator can regulate output voltage:
    a. when load current changes occur.
    b. when input line variations occur.
    c. when the input voltage source is removed.
    d. when there is a short-circuit load.
    e. a and b only.
    f. none of the above.
8. To obtain higher current regulation from a linear series voltage regulator:
    a. more zener diodes are added.
    b. another higher-current transistor is added in parallel with the series control element.
    c. more heat sinks are added to the series control element.
    d. larger resistors are added to the sampled voltage resistor chain.
9. The component(s) in a linear series voltage regulator that must be protected from excessive temperature rise is(are):
    a. the zener diode.
    b. the error amplifier transistor(s).
    c. the series control element transistor(s).
    d. the sampled output voltage resistor string.
10. A switching voltage regulator is used instead of a linear series voltage regulator because:
    a. it has efficiencies up to 85%.
    b. it has lower power dissipation.
    c. it has smaller sized components for the output power required.
    d. all of the above.
    e. a only above.
    f. b and c only above.
11. What components, along with a switch, provide the basic circuit for a switching voltage regulator?
    a. an inductor, diode and capacitor.
    b. two inductors.
    c. an inductor and a capacitor.
    d. two capacitors.

12. The control element in a switching voltage regulator is:
    a. an inductor.
    b. a capacitor.
    c. a switch.
    d. a resistor.
13. An inverting switching voltage regulator:
    a. uses inverting logic circuits for its control.
    b. produces an output voltage that is inverted in polarity to that of the input voltage.
    c. uses a bridge rectifier in its control circuit.
    d. operates at 10 Hz.
14. Power distribution systems can:
    a. use linear series voltage regulators operating from a regulated rail to provide the desired output voltages.
    b. use switching voltage regulators operating from a regulated rail to provide the desired output voltages.
    c. use a combination of linear series and switching regulators to provide the desired output voltages.
    d. use DC to DC converters operating from a regulated rail to provide step-up and/or step-down regulated output voltages.
    e. all of the above.
    f. a only.
    g. a and c only.
15. Power system supervisors:
    a. contain circuits that watch over the power distribution system so variations do not occur that cause system failures.
    b. contain circuits that can detect low voltages, provide time delays, and provide signals that enable or disable system circuits.
    c. shuts down everything in the system without notice.
    d. all of the above.
    e. a and b only.
    f. a only.

Answers: 1.b, 2.e, 3.d, 4.a, 5.a, 6.c, 7.e, 8.b, 9.c, 10.d, 11.a, 12.c, 13.b, 14.e, 15.e.

A Microcontroller Application

Introduction

The thrust of this chapter is to provide the reader with the opportunity to actually build and implement an application using a microcontroller. The microcontroller that will be used is the same one that was chosen to describe assembly-language programming in *Chapter 7*. This should provide continuity from what was learned in *Chapter 7* to applying it to a working application. The microcontroller used is the MSP430F1232. The application is explained, and then the reader is given the opportunity to gather the parts, interconnect them, and with the help of a development kit for the microcontroller used, to program the microcontroller with the program provided to complete the application.

Application Block Diagram

The block diagram of the application is shown in *Figure 10-1*. It consists of a PT-100 resistive sensor whose resistance change with temperature is quite linear. The signal from the sensor is conditioned (amplified) by the TLV2451 operational amplifier. The output of the op amp is fed to the TLV1549, a 10-bit ADC, to convert the analog voltage into a digital code. The digital code representing the temperature sensed by the PT-100 is inputted to the MSP430F1232, the microcontroller. The microcontroller decodes the temperature code, and converts it to signals that drive the display to indicate the temperature in either °F and °C. In addition, the microcontroller uses its clock and counters to produce the timing pulses that produce the time, date, and year outputs that are then decoded for display. In sequence, the time, the date, the year, and the

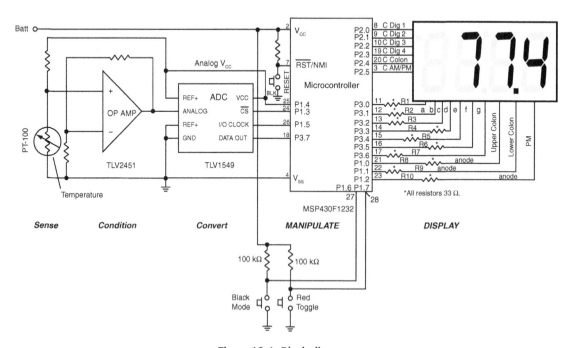

Figure 10-1: Block diagram

temperature in °F and then °C is displayed. The display is four digits with each digit having seven segments. Each segment is an LED diode that has an anode and a cathode and must be connected between V_{CC} and ground to pass current through the forward-biased diode. The microcontroller connects the cathode of the diodes to ground and the grounding is timed by the microcontroller to energize the correct digit as the selected anodes of the segments, a through g, are timed and energized by the microcontroller to produce the correct number or letter. Also included in the display are small LEDs that represent colons, decimal points, and AM and PM indicators. Their cathodes must be connected to ground and their anodes energized to produce their outputs at the correct time.

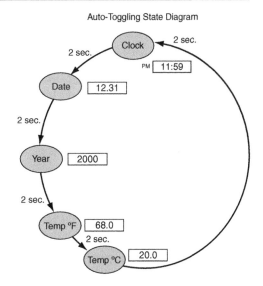

Figure 10-2: Auto toggling through states

The Auto-Toggling State Diagram

A diagram that shows the sequence of what is to be displayed is shown in *Figure 10-2*. It is called a state diagram. The sequence begins at **Clock** where the digits for the hour(s) and minute(s) are displayed with a colon between them. If the time is after 12:00 noon, the PM LED is energized. After two seconds, the **Date** is displayed. The month digits and the day digits are energized with a period between them. Two seconds pass and the **Year** digits are energized. Two seconds pass and the temperature in °F, **Temp °F**, is displayed, and two seconds later the temperature in °C, **Temp °C**, is displayed. If the temperature is negative, the g-segment LED is energized to produce a minus sign to indicate the negative value. Two seconds later the display is back at the start, **Clock**, displaying the time.

The Manual-Toggling State Diagram

In the block diagram of *Figure 10-1* there are three push-button switches. One is a black reset button; the other two—one red and the other black—are used in *Figure 10-3* while manually toggling through the same states as in *Figure 10-2*. The diagram shown in *Figure 10-3* is called the manual-toggling state diagram. The red button is used to move manually through the states shown in *Figure 10-2*. At each state, the black button is used to manually move to a quantity displayed in different modes in the particular state. Pushing the red button then increments the quantity. As a result, using these two buttons, the correct time, date, and year are set and the temperature checked for calibration at know points—freezing water and boiling water. They are also used to switch between the manual toggling of the states and the sequencing or auto-toggling of the states.

At startup, the system will be in the **Clock** state with the time 12:00 midnight. This is the start of the

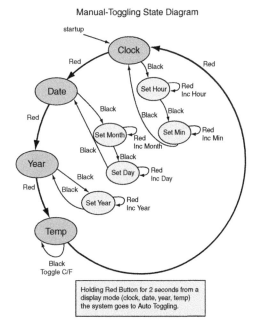

Figure 10-3: Manual toggling through states

manual-toggling. Pressing the black button will put the system in the "set hour" mode. Pressing the red button will increase the hour display from midnight to one o'clock. Further pressing advances the hour digits from one o'clock to 12 o'clock, and then the numbers start over for the PM hours. After the hour digits are set, pressing the black button puts the system in the "set minute" mode, and pressing the red button advances the minute digits from 1 to 59 and back to 1. Setting the minutes digits completes manually setting the modes in the clock state. Pressing the black button returns the system to the **Clock** state.

Pressing the red button will move the system to the **Date** state. Pressing the black button puts the system in the "set month" mode. Pressing the red button increases the month digits from 1 to 12 and back to 1. When the month digits are correct, pressing the black button puts the system in the "set day" mode. Here pressing the red button increases the day digits from 1 to 31 and back to 1 again until the day digits are set correctly. Now pressing the black button returns the system to the **Date** state.

Pressing the red button will move the system to the **Year** state. Pressing the black button puts the system in the "set year" mode, and pressing the red button advances the year. It starts at the initial year of 2000. After setting the year, pressing the black button returns the system to the **Year** state.

Pressing the red button moves the system to the **Temp** state, either **Temp °F** or **Temp °C,** and pressing the black button toggles the display from °F to °C. Temperature can be checked at the points of boiling water (212°F) and at freezing water (32°F). Pressing the red button returns the system to the **Clock** state. Thus, all the modes in each state that are to be manually set have been stepped through.

Switching Modes

Holding the red button for two seconds will put the system into auto-toggling. Manual toggling can be entered from any of the states that the system is in. If in the auto toggling sequence, which holds each state for two seconds before advancing to the next state, and in the **Date** state, pressing the red button will put the system in the **Year** state and manual setting mode so that the year can be set. If in the **Temp** state, pressing the red button will put the system in the **Clock** state and manual setting mode for the clock. Pressing the red button and holding the button for two seconds again puts the system into auto-toggling.

The Sleep Mode

There is a sleep mode used to conserve power and extend the life of the batteries. This is one of the significant advantages of using the MSP430 microcontroller. It operates at very low power and can be put into a sleep mode to significantly reduce the average power consumed. If the system is left on continuously, the battery drain is 25–30 mA, and the batteries would last only 2–3 days. Being able to put the system into a sleep mode and waking it to update the one second count, or waking it to sequence through time, date, year and temperatures reduces the current drain. This procedure can extend the battery life to around a year. *The sleep mode is entered if none of the buttons have been pressed for 15 seconds.*

The system is awakened by pressing the red or black button, or by a timer signal that is initiated every 20 minutes, on the 20-minute, 40-minute and 60-minute mark of the hour. The system goes through one cycle of the auto-toggle sequence and then goes back to sleep. When the system awakens, it returns to the state that it was in before it went to sleep. If it was sequencing, then it returns to sequencing; if it was in manual, then it returns to manual. When in the sleep mode, the 32,786 Hz crystal oscillator and the Timer_A used as a counter are the only circuits operating in the microcontroller.

In summary, starting with the system in the sleep mode, assuming it was in the auto-toggling mode when it went to sleep, pressing the red or black button awakens the system in the auto-toggling mode. Pressing the red button will put the system into manual toggling. Pressing the black button will put the system in a manual set mode; if in the **Clock** state, then the black button will move the system to be able to set the hour and minutes with the red button. Correspondingly for **Date** and **Year** states, setting month and day, and the year, respectively. Pressing the red button in any of these modes and holding it for two seconds puts the system in the auto-toggling mode. Pressing no buttons for 15 seconds puts the system in the sleep mode.

In the sleep mode, an internal timer is keeping track of time from the top of the hour. At the 20-minute after the hour point, or at the 40-minute point, or at the 60-minute point, a signal awakens the system, it cycles through one cycle of the auto-toggling mode, and returns to the sleep mode. The system will ignore the 20, 40, or 60-minute mark signal if the system is already awake.

System Schematic

A schematic is a diagram that shows all the package pins of the devices that are used in the system and all the electrical connections between the devices, resistors, capacitors, transistors, diodes and display elements that make up the system. It is shown in *Figure 10-4.* It represents the components with the respective accepted symbols, and all of the respective package connections are included, even the open or unconnected pins. It uses the block diagram of *Figure 10-1* as a base, but includes all the detail electrical connections including the components and connections needed to supply power to the system.

The Display

The Seven Segments

The system requires that four digits be displayed. The display shown in *Figure 10-1, Figure 10-4 and Figure 10-5* is made up of digits that have seven segments and some additional small LEDs in between.

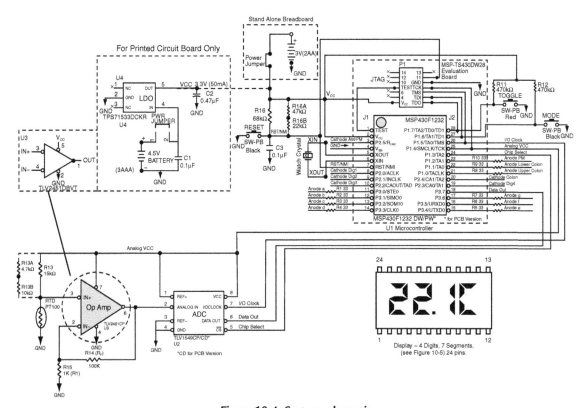

Figure 10-4: System schematic

The unit, which has four digits packaged together, has 24 pins, and *Figure 10-5* shows the 4-digit display connections. As shown in *Figure 10-6a*, each digit is made up of seven LED segments that each have an anode and a cathode. The segments are identified with the letter a through g, and will be energized when a positive voltage is applied to the anode and the cathode is grounded. The anodes of the segments are connected in parallel across the digits (a to a, b to b, and so on), while the cathodes for each digit are all tied together and identified with CDig1, CDig2, CDig3 and CDig4 for the four digit connections. This allows only one digit to be excited at a time.

As shown in *Figure 10-6b*, codes determine which segments are energized with a positive voltage. For example, the code for digit 1 segments a through g is 1101101 which means that segments a, b, d, e and g are energized when CDig1 is grounded. This displays 2 as digit 1. Since all the segments of all the digits are energized in parallel, the only digit that will be displayed is the one that has its cathode line grounded. The microcontroller controls the time that the cathodes of the desired digit to be displayed are grounded. Just to give a small example of the circuit connections to the display, *Figure 10-7* details a portion of the connections from the microcontroller to CDig1 and CDig2 to ground the cathodes for digit 1 and digit 2 and excite the anodes of segments a, b, c, d. Similar connections are made to CDig3, CDig4 and the segments e, f and g as shown in *Figure 10-4*.

Referring back to *Figure 10-6b*, the code for digit 2 changes to 1011011, which means that segments a, c, d, f and g are energized and digit 2 displays a 5 when its cathodes are grounded. Correspondingly,

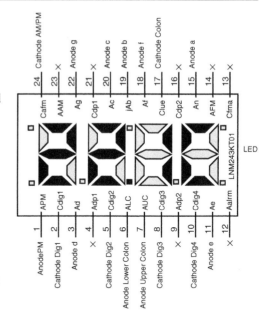

Figure 10-5: Connections to 7-segment display

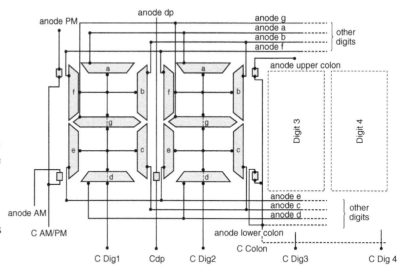

a. Diode interconnections

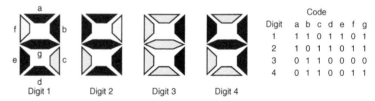

b. Segment codes

Figure 10-6: 7-segment LED detail

Digit	a	b	c	d	e	f	g
1	1	1	0	1	1	0	1
2	1	0	1	1	0	1	1
3	0	1	1	0	0	0	0
4	0	1	1	0	0	1	1

Code

digit 3 displays a 1 when the code 0110000 energizes the segments; and digit 4 displays a 4 when the code 0110011 energizes the segments. The microcontroller grounds the cathode connection and excites the anodes at the proper time to display the respective digit.

The Additional Small LEDs

There are several other small LEDs included in the display. When the system is in the **Clock** state, the hour digits are separated from the minutes digits by a colon (*Figure 10-2*). The colon is displayed by energizing the anodes of the LEDs identi-

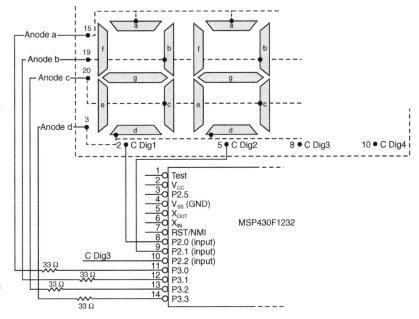

Figure 10-7: LED segment driver

fied as the upper colon and lower colon and grounding their cathodes. The cathodes are tied together at the connection "cathode colon." In similar fashion, when the system is in the **Date** state or the **Temp** state, a decimal point separates the month and day digits, or identifies tenths of a degree for temperature. The decimal point is displayed by energizing the anode of the lower colon, depending on the position required, at the same time grounding the cathodes through cathode colon. One other small LED that is used is the PM LED. The PM LED is energized when the hours exceed 12 noon, to indicate the hours displayed are in the PM. A positive voltage is placed on the PM anode while the cathodes of both the AM and PM LEDs, which are tied together, are grounded by the microcontroller. As noted previously, the segment g LED is energized when the temperature values are negative. The cathode, CDig1, is grounded by the microcontroller. Pins 4(Adp1), 9(Adp2), 12(Aalrm), 13(Cfma), 14(AFM), 16(Cdp2), 21(Ddp1) and 23(AAM) are not used.

The Microcontroller

Return to the system schematic in *Figure 10-4*. The microcontroller used is the MSP430F1232 out of the MSP430 family of microcontrollers, the same family used for *Chapter 7* to explain assembly-language programming. The MSP430F1232 has three I/Os, 8 kB + 256B of Flash memory, 256 bytes of SRAM, a watchdog timer, a Timer_A, and the brownout feature. It is packaged in a 28-pin package. It uses the same assembly-language instruction set used in *Chapter 7*, which is included in the Appendix.

The I/O Ports

The port 3 I/O pins P3.0, P3.1, P3.2, P3.3, P3.4, P3.5, P3.6, are configured as outputs and drive the anodes of display segments a through g through 33 Ω resistors between the output pin and the pin to the anode. The P3.0, P3.1, P3.2 and P3.3 connections were shown in *Figure 10-7*. Port 1 I/O pins P1.0, P1.1 and P1.2 are also configured as outputs to drive the anodes, through 33 Ω resistors, of display segments upper colon, lower colon and PM, respectively. The port 1 I/O pins P1.6 and P1.7 are configured as inputs and receive the signals from the red and black push buttons, respectively. A 470 kΩ resistor from V_{CC} connected to P1.6

or P1.7 is pulled to ground when the black or red button, respectively, is pushed. The ground signal on P1.6 or P 1.7 initiates microcontroller action as previously described in the discussion on the state diagrams. In addition, a black push button grounds pin 7, RST/NMI, through a 68 kΩ resistor to V_{CC} to reset the system. A detail of the push button connections is shown in *Figure 10-8*.

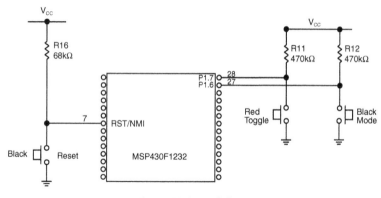

Figure 10-8: Push buttons

The port 2 I/O pins P2.0, P2.1, P2.2, P2.3, P2.4, P2.5 are configured as inputs and ground the cathodes CDig1, CDig2, CDig3, CDig4, Ccolon, and CAM/PM, respectively. The remaining port 1 and port 3 I/O pins are configured as follows:

Pin	Configured as:	Connected to:
P1.3	output	CS of TLV1549
P1.4	output	Analog V_{CC}
P1.5	output	I/O CLOCK of TLV1549
P3.7	input	DATA OUT of TLV1549

P1.3 outputs the low-level chip select to the TLV1549 to initiate the analog-to-digital conversion. P1.5 outputs a clock to the TLV1549. P3.7 inputs to the microcontroller the digital code data output from the TLV1549. P1.4 is configured as an output that provides the power to the analog portion of the system—the sensor, op amp and the ADC. When the system is awake, V_{CC} for these circuits is provided on P1.4 by the microcontroller; when the system is asleep, there is no power provided on P1.4. Thus, with this power control, as well as disconnecting power from circuits inside the microcontroller, the system consumes very little power in the sleep mode.

Here are the remaining connections to the microcontroller. Pin 1 is a TEST pin, pin 2 connects to the main V_{CC}, pin 4 is V_{SS} or main system GND, and pin 5 and pin 6 are the connections to the watch crystal (32,768 kHz) which provides the precision clocks generated in the microcontroller for the system.

The Analog Circuitry
The Sensor and Op Amp

The analog portion of the system schematic, as mentioned previously, consists of the PT-100 sensor, the TLV2451 op amp, the TLV1549 ADC. There is another separate analog circuit, U4, a low-voltage drop voltage regulator used for power control that will be discussed later. Pin 2 (inverting input) of the op amp has a 1 kΩ resistor to ground and a 100 kΩ feedback resistor from the output pin 6 to set the op amp gain. The noninverting input, pin 3, is connected to the intersection of the PT-100 sensor and a 15 kΩ resistor that connects to the power line Analog V_{CC}. The other end of the PT-100 sensor is connected to ground. The op amp power is supplied through pin 7, which is connected to Analog V_{CC}, and completed through pin 4 that is connected to ground.

ADC

The output of the op amp (pin 6) is connected to pin 2 (the analog input) of the TLV1549 ADC. The output digital data from the ADC is coupled to the microcontroller from output pin 6. The data is shifted out on the data line using the clock on pin 7, I/O Clock from Port 1 pin P1.5 of the microcontroller. The conversion process begins when the chip select input (pin 5) coupled from the microcontroller Port 1 pin P1.3 is brought low. Pin 1, REF+, and pin 8, V_{CC}, are connected to Analog V_{CC}. Pin 3, REF-, and pin 4, GND, are connected to GND.

Power and Power Control

The schematic of *Figure 10-4* shows several variations for supplying the main system V_{CC}. The first of these is when the complete system is assembled on a printed-circuit board (PCB). The V_{CC} of +3.3V is supplied from pin 5 of a TPS71533 low-voltage drop (LDO) voltage regulator, the U4 analog circuit mentioned previously. The input to the LDO on pin 4 is +4.5V supplied by three AAA batteries in series. The LDO keeps the +3.3V output constant as the batteries lose voltage, and thus, extends the battery life of the system. Pin 2 of the LDO is connected to GND, and pin 4 and pin 5 are bypassed by capacitors C1 (0.1 μF) and C2 (0.47 μF). There is a series jumper to disconnect the +4.5V battery source.

In the second variation, the system is in breadboard form and has the main system V_{CC} of +3V supplied by two AA batteries in series. There is no LDO. There is a series jumper to disconnect the battery source. The system stands alone operating from +3V battery power.

In the third variation, the main system V_{CC} is supplied from a PC (personal computer) through the JTAG connector on the evaluation board. There are no batteries and no voltage regulator. The system operates from the PC's V_{CC}. This is the usual connection used when programming the system.

Brown Out

The MSP430F1232 microcontroller has a system power protection circuit built in. This feature means that if the V_{CC} supply goes below a voltage level between 1.1V and 1.7V the system is reset. The actual trigger voltage varies with the device. When the V_{CC} voltage is restored above the trigger level, it will start at ground zero (time of 12:00) and run properly. Without brown out, if V_{CC} falls below the trigger level, the system would not operate properly and would be unpredictable when V_{CC} is restored. This feature requires the MSP430F1232, which has no comparator. If a comparator is required, the MSP430F123 is used without the brownout feature.

JTAG

There is a connector on the schematic that is on the evaluation board that sockets the microcontroller. It is a 14-pin connector called out as JTAG. It is used to connect the evaluation board to the PC development system, and supply V_{CC} to the breadboard. The development system is used to load the program software into the microcontroller and for debugging the system. Here are the connections from the JTAG connector to the microcontroller:

JTAG Pin	Microcontroller Pin
1 (TDO)	28 – TDO
2 (V_{CC})	2 – V_{CC}
3 (TDI)	27 – TDI
4	NC
5 (TMS)	26 – TMS
6	NC
7 (TCK)	25 – TCK
8 (TEST)	1 – TEST
9 (GND)	GND – V_{SS}
10	NC
11 (RST)	7 – RST/NMI
12 – 14	NC

Summary of Schematic

The sensor, PT-100, changes resistance linearly with temperature as discussed in *Chapter 3*, Sensors. At the junction of the sensor to ground and a resistor to V_{CC}, a voltage that varies with temperature is coupled to the noninverting input of the op amp. A positive input voltage produces a positive-going output voltage. The gain of the op amp, as discussed in *Chapter 4*, is set by the ratio of R_f/R_i (R14/R15), the ratio of the feedback resistor from the output to the inverting input, and the resistor from the inverting input to ground. The op amp output voltage is fed to the input of the ADC. The ADC, a switched capacitor, successive approximation ADC, produces a 10-bit digital code from its analog input. The 10-bit code is shifted out of the ADC into a data register in the microcontroller by a clock signal to the ADC from the microcontroller. The switched capacitor, successive approximation (SAR) ADC was discussed in *Chapter 5*. The shifting out of the data from the ADC to the microcontroller was discussed as one of the subprograms of *Chapter 7*.

The clocks are produced from a watch crystal oscillator with a frequency of 32,786 Hz. Counters in the microcontroller count the crystal oscillator clocks and produce a one-second pulse that then is counted to produce the time in minutes and hours, days, months and years. The counter initial conditions are set manually to adjust the system to present-day time. The program is installed in the microcontroller using a development system, and after all interconnections on the schematic are made, the system should operate properly.

System Development

There are three stages of system development—two breadboard stages and a final PCB stage.

1. The first stage is shown in *Figure 10-9*. The MSP430F1232 microcontroller is connected in a socket on the evaluation board (Texas Instruments MSP-TS430DW28) that is connected through the JTAG connector by cable to the Flash Emulator Tool. The Flash Emulator Tool is connected by cable to the parallel 25-pin input connector on a PC. This configuration, which uses the development system software in the PC to load the program software into the microcontroller and to debug the system, has the circuit interconnected in breadboard form. As shown in *Figure 10-9,* the breadboard consists of an analog board on one side of the evaluation board and a display board on the other side. Recall that the system is powered from the PC.

2. The second stage is shown in *Figure 10-10*. In this configuration, the breadboard stands alone powered by two AA batteries. The microcontroller is still plugged into the evaluation board, but the JTAG connector is disconnected. There is no longer any need for the PC development system; the system runs stand-alone.

3. The third stage is shown in *Figure 10-11*. This stage has the system assembled completely on a PCB. The parts used are now parts that are capable of being mounted on a PCB. There is an op amp in a smaller package, there is

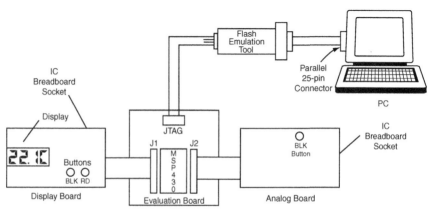

Figure 10-9: Breadboard powered by PC

the LDO voltage regulator, and there is a MSP430F1232 in a smaller package. The majority of the parts are assembled to the board using surface-mount technology. Power is supplied by three AAA batteries in a holder on the back of the PCB. It is a self-contained system. Such a system could be mounted in a separate case, or included as part of other equipment.

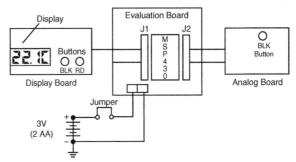

Figure 10-10: Stand-alone breadboard powered by +3V battery (two AA cells)

The Breadboard Circuit

The breadboard circuit shown in *Figure 10-9* and *Figure 10-10* is built using readily available parts. The main components of the breadboard are the MSP430 evaluation board, which contains a socket for the MSP430 microcontroller, and two IC breadboard sockets. One IC breadboard socket, called the display board, contains the 7-segment display, the red and black toggling and mode push button switches, and the associated resistors and connecting wires. The other IC breadboard, called the analog board, contains the analog devices and associated circuitry, the op amp, the ADC, the black RESET push button, and connecting wires. The evaluation board is in the development system from Texas Instruments (MSP-FET430P120), which contains the following:

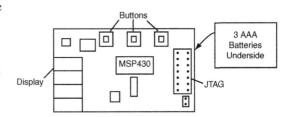

Figure 10-11: PCB powered by three AAA batteries (+4.5V)

1. The evaluation board (MSP-TS430DW28).

2. The microcontroller (MSP430F1232).

3. A MSP430 Flash Emulation Tool (MSP-FETP430IF).

4. Cable from JTAG to Flash Emulation Tool.

5. Cable from Flash Emulation Tool to parallel LPT (printer port) on PC.

6. Two 14-pin cable connectors for wiring to J1 and J2 on evaluation board.

7. PC software and instructions.

Here are the main sources for the parts:

a. Samples of the op amp and the ADC can be obtained online from Texas Instruments.

b. The 4-digit LED display and the 32-kHz crystal can be ordered online from Digi-Key.

c. The PT-100 sensor can be obtained online from Omega Engineering.

d. The IC breadboard sockets, the jumper wire kit, the push-button switches, and the resistors are all available from RadioShack or other electronic parts distributors. The resistors and capacitors can be ordered from Digi-Key Corporation at the same time as the display and the crystal if one wants to order online.

It is assumed that the reader will have available:

1. Long-nose pliers

2. Diagonal cutter

3. Wire stripper

4. Small soldering iron

5. Solder

If these are not available, they are easily purchased at RadioShack or any other electronic parts distributor.

Parts List for the Breadboard

Here are the parts required of the breadboard:

Part	Part No.	Quantity
MSP430 Development System	MSP-FET430P120	1
Sensor (PT100 – RD)	1PT100KN815	1
Op Amp ($V_{CC} = +3V$)	TLV2451CP	1
ADC (Analog-to-Digital Converter)	TLV1549CP	1
4-Digit 7-segment LED Display	LN543RKN8	1
32 kHz Watch Crystal	X802 – ND	1
IC Breadboard Socket	RS 276-175	2
Pair (Red and Black) SPST Momentary Push-button Sw.	RS 275-1556	2
Jumper Wire Kit	RS 276-173	1
Battery Clip for two AA Batteries	RS 270-0414	1
DIP Shunt Shorting Jumpers for AA battery connection	RS 276-1512	1
AA Battery (1.5V)	RS 23-872	2
R1 – R10 33 Ω 1/2W 5% resistor	RS 271-1104	10
R11, R12 470 kΩ 1/2W 5% resistor	RS 271-1133	2
R13A 4.7 kΩ 1/4W 5% resistor	RS 271-1330	1
R13B 10 kΩ 1/4W 5% resistor	RS 271-1335	1
R14 100 kΩ 1/4W 5% resistor	RS 271-1347	1
R15 1 kΩ 1/4W 5% resistor	RS 271-1321	1
R16A 47 kΩ 1/2W 5% resistor	RS 271-1130	1
R16B 22 kΩ 1/2W 5% resistor	RS 271-1128	1
C3 0.1 µF 50V capacitor	RS 272-135	1

*The 15 kΩ R13 is made up of a 4.7 kΩ and a 10 kΩ, standard values readily available.
**The 68 kΩ R16 is made up of a 47 kΩ and a 22 kΩ, standard values readily available.

Obtaining the Parts

The details that follow were current at the time of publication. Obviously, Web sites change, and so does the information presented and the procedures to be followed. However, as the reader becomes familiar with the Web sites and by checking Readme.txt files, they will be able to arrive at the desired outcome even though the Web site has changed.

MSP430 Development System—It may be purchased from Texas Instruments ($99.00) by calling the American Products Information Center at 972-644-5580. The board will be shipped directly to a home or business.

Op Amp and ADC—Samples of the Texas Instruments TLV2451CP and TLV1549CP can be obtained online via TI's Web site: http://www.ti.com. If this is the first time visiting the TI Web site, nonmembers of *my.ti* may have to register by using an e-mail address and choosing a password. On the Web site page there is a part number search window. Enter the part number—TLV2451CP—and click on "Go." A window displaying the results of the search shows a description of the TLV2451CP. Click on TLV2451 and a listing of TLV2451 products appears. On the left side of the product list there is a link that reads "samples." Click on it and a sample list of the TLV2451 products is displayed. Select the TLV2451CP and follow the prompts to register and fill out the shipping information. Repeat for the TLV1549CP. The parts should be received in a few days. When the TLV2451 or TLVR1549CP product list is displayed, if a complete data sheet is required, clicking on the "Data Sheet" box will print out the data sheet if an Acrobat Reader is available on your computer.

Watch Crystal and Display—The watch crystal and the 4-digit LED display can be purchased online from Digi-Key at the Web site http://www.digikey.com. Digi-Key Corporation is an online distributor of electronic parts. Enter the part number (LN543RKN8 for the display; X802-ND for the crystal) in the part number search window and click on "Go." Enter the quantity on the "Add to Order" line, then follow the prompts to supply shipping and payment information. The watch crystal is about $0.30 and the display is about $10.00, plus shipping.

Sensor—The sensor, PT100-RD, is purchased online from Omega Engineering at its Web site http://www.omega.com. Omega fulfills all their small purchases online. Enter the part number, 1PT100KN815, in the part number search window and click "Go." A part description window appears. Enter quantity and click "Add to Cart." Follow the prompts to supply shipping and payment information. The sensor should be under $20.00.

Remaining Parts—All the remaining parts are available from RadioShack or other electronic parts distributors. The part numbers given are RadioShack part numbers. Note that for R13 and R16 two resistors are placed in series to make up the total resistance because the values for the two-resistor combination are more readily available. The push-button switches come two in a pack—one black and one red. Two of the IC breadboard sockets are required for the system breadboard. As mentioned previously, resistors, capacitor(s) and push button switches may be obtained online from Digi-Key. RadioShack was chosen as the source because of the nationwide outlets, and immediate access to the parts.

Breadboard Construction—Powered by the PC

Cables for J1 and J2, the connections on the evaluation module, are constructed first. There are two connectors supplied in the development system for each of the J1 and J2 connections. One is a male, and the other a female. The female is attached to the PCB; the male is used as a jumper cable connector. Use the jumper wire kit to construct the cables. Use the solid-color red, green, yellow and orange wires and solder them to the male connectors provided as shown in *Figure 10-12*. The wire colors refer to the ones that are in the RadioShack kit. The reader may choose to use another wire supply with different colors. Use the text as a guide to assign the wire colors. If the RadioShack kit is used, the wires marked with an asterisk in *Figure 10-12* need to be extended by soldering a like color wire to them and insulating the solder joint with electrical tape. Solder the female connectors, also provided, to the evaluation module PCB in the J1 and J2 positions shown on the PCB. There are Rd/Blk and Or/Blk wires required. Take three red wires and two orange wires and mark across them with a marker to make the dual-colored wires. Make sure all connections are physically and electrically sound. When the cables are completed, plug the male connector into the female to connect the cables to the evaluation module.

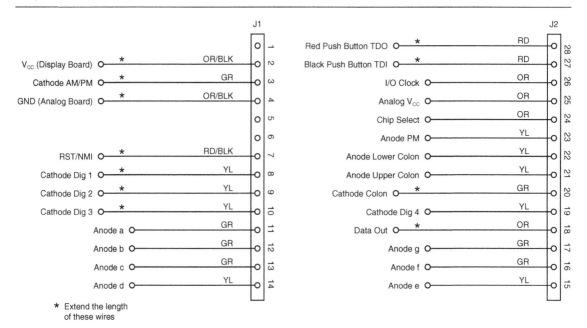

Figure 10-12: Cables for J1 and J2

Display Board

The display board is constructed using one of the IC breadboard sockets (RS 276-175). Plug the 4-digit LED display (LN543RKN8) into the breadboard socket in the position shown in *Figure 10-13*. The pins 13 through 24 plug into the bottom horizontal row of the upper portion of the IC breadboard, and the pins 1 through 12 plug into the second horizontal row of the lower portion of the breadboard. Pin 1 of the display is in vertical row 1 of the lower portion; pin 24 of the display is in vertical row 1 of the upper portion. The upper and lower portions of the breadboard are configured the same. There are 23 vertical rows with five connection points in each row. All five connection points, A, B, C, D, and E, in the upper portion vertical rows are connected together; all the connection points, F, G, H, I, and J, in the lower portion vertical rows are connected together. Each vertical row is electrically isolated from any of the other vertical rows. At the top of the upper portion is a horizontal row, X, of 20 connections all connected together. This row of connections will be used for V_{CC}. A similar horizontal row, Y, of 20 connections is at the bottom of the lower portion. This row of connections will be used for the ground (GND) of the circuit.

Note that in the block diagram of *Figure 10-1* and the schematic of *Figure 10-4* there are 10 33 Ω resistors that are connected from the microcontroller drive pin to the anode of the LED segments of the four digits of the display. Connect these resistors by pushing the wires of the resistors into the connection points as shown in *Figure 10-13*. For example, one end of R1 is connected to pin 15 of the display (vertical row 10 of upper portion), and the other end to vertical row 19. A green wire from pin 11 of the microcontroller connector J1, to be connected later, will connect to vertical row 19 to drive anode a. Plugging in R2 through R10 in a similar fashion will provide the drive for anodes b through g, PM, lower and upper colon.

To connect the push buttons, solder a short piece of wire to each push button terminal. Insert one terminal wire of the black push button into vertical row 20 of the lower portion of the socket and the other terminal wire into Y, the GND row of connections. Insert one end of R12 (470 kΩ) into the same vertical row 20 and the other end into the top row X of V_{CC} connections. Refer to *Figure 10-13* to check the connections. Later a red wire from pin 27 of J2, TD1, will also be connected to vertical row 20.

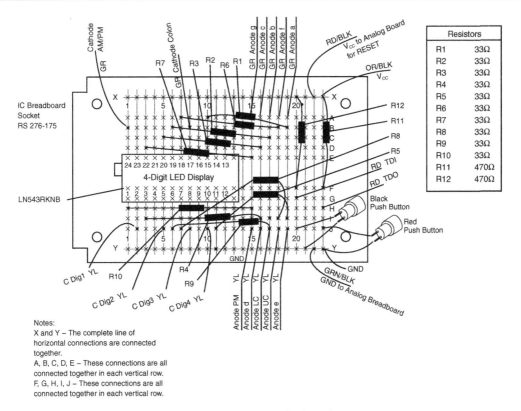

Figure 10-13: Display board

A similar connection pattern is followed for the red push button; one terminal wire is connected to vertical row 23 of the lower portion and the other terminal wire to GND. One end of R11 (470 kΩ) is inserted into vertical row 23 and the other end into the V_{CC} connections. Later a red wire from pin 28 of J2, TD0, will also be connected to vertical row 23. The grounding of TD0 and TD1 by the push buttons sends signals to the microcontroller to toggle, change modes, or to set a parameter.

The Analog Board

The analog board is constructed on the second IC breadboard socket as shown in *Figure 10-14*. Both the TLV2451CP op amp and the TLV1549CP ADC are packaged in 8-pin DIP packages. The are both plugged into horizontal connection rows, as shown in *Figure 10-14*, that span the upper and lower portions of the breadboard. For the TLV2451, vertical rows 1 through 4 of the lower portion are used for pin connections 1, 2, 3 and 4, respectively, and vertical rows 1 through 4 of the upper portion are used for pin connections 8, 7, 6, and 5, respectively. Pin 7 is connected to the top row, X, of connections, which are now identified as "Analog V_{CC}" because this V_{CC} is being switched on and off by the microcontroller.

The noninverting input, IN+, pin 3, (vertical row 3) has one end of the PT-100 sensor connected to it, as well as one end of R13B. The other end of R13B is connected in series with R13A to Analog V_{CC}. To make sure that the connections from the PT-100 sensor are good electrically, solder short pieces of wire to the very fine wire of the sensor before inserting the connections into the breadboard. It is even best to put a blob of silicone or plastic sealer (one that solidifies) around the wires to hold them in place, otherwise the

Resistors	
R13A	4.7 kΩ
R13B	10 kΩ
R14	100 kΩ
R15	1 kΩ
R16A	47 kΩ
R16B	22 kΩ

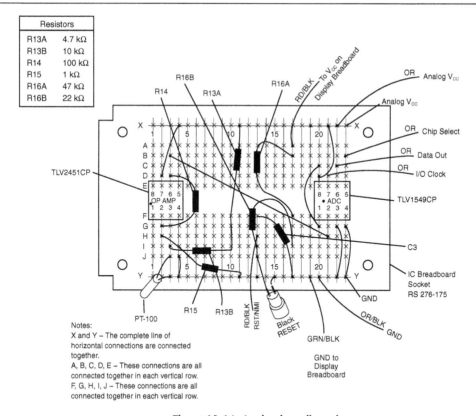

Figure 10-14: Analog breadboard

wires will break and ruin the PT100. One connection of the sensor is to vertical row 3 of the lower portion, and the other connection is to Y, the GND row of connections at the bottom of the breadboard.

Resistor R14 (100 kΩ) is connected from the output, pin 6, vertical row 3 of the upper portion, back to the inverting input, IN-, pin 2, vertical row 2 of the lower portion. Vertical row 2 also has one end of R15 (1 kΩ) connecting from it to the GND line of connections. The ratio of the 100 kΩ R14 to the 1 kΩ R15 sets the gain of the op amp at 100. The output of the op amp, pin 6, is coupled to the input of the ADC, pin 2.

The ADC's pin 1, 2, 3 and 4 are connected to vertical rows 20, 21, 22 and 23, respectively, of the lower portion of the breadboard. Vertical rows 20, 21, 22 and 23 of the upper portion are connected to pins 8, 7, 6 and 5, respectively. Pin 1 and pin 8 are connected to X the V_{CC} line "Analog V_{CC}." Pin 3 and pin 4 are connected to Y the GND line. I/O Clock, Data Out, and Chip Select will be connected from the microcontroller to complete the ADC connections.

The RESET circuit from the black RESET push-button switch is connected as follows: One end of resistor R16A (47 kΩ) is connected to vertical row 17 of the upper portion of the breadboard, and the other end is connected to vertical row 17 of the lower portion. One end of R16B (22 kΩ) is also connected to vertical row 17, lower portion. The other end of R16B is connected to vertical row 13, lower portion. Again, short wires are soldered to the terminals of the black RESET push button to make it easy to connect them to the breadboard. Insert one terminal wire into vertical row 13, lower portion and the other end into the bottom horizontal row GND line Y. A RST/NMI wire from the microcontroller will connect to vertical row 13, lower portion, to input the RESET signal to the microcontroller when the black button is pushed. To assure the RESET signal input is clear of noise, capacitor C3 is added from vertical row 13, lower portion to END at line Y.

Completing the System Connections

The system power connections come from the J1 connector on the evaluation board. V_{CC} is connected to the display board by inserting the OR/BLK wire from pin 2 of J1, shown in *Figure 10-12*, to the V_{CC} line X of the display board. GND is connected by inserting the OR/BLK wire from pin 4 of J1 into the Y GND line of the analog board. A GRN/BLK wire is connected from the Y GND line connections of the analog board to the Y GND line connections of the display board to complete the system GND.

Return to *Figure 10-12* and note that most of the remaining wires from J1 and J2 are either yellow or green wires. The green wires will be connected to the upper portion of the display board, and the yellow wires to the lower portion of the display board.

The Display Board

The green wires from J1 are connected first:

1. Pin 3 (Cathode AM/PM) is connected to pin 24 (vertical row 1) of the LED display.
2. Pin 11 (Anode a) is connected to vertical row 19, upper portion, one end of R1.
3. Pin 12 (Anode b) is connected to vertical row 17, upper portion, one end of R2.
4. Pin 13 (Anode c) is connected to vertical row 16, upper portion, one end of R3.

The green wires from J2 are connected next (also to upper portion):

1. Pin 16 (Anode f) is connected to vertical row 18, one end of R6.
2. Pin 17 (Anode g) is connected to vertical row 15, one end of R7.
3. Pin 20 (Cathode colon) is connected to vertical row 8, pin 17 of display.

The yellow wires from J1 are connected next (they all go to the lower portion of the breadboard):

1. Pin 8 (Cathode Dig1) is connected to vertical row 2, pin 2 of display.
2. Pin 9 (Cathode Dig2) is connected to vertical row 5, pin 5 of display.
3. Pin 10 (Cathode Dig3) is connected to vertical row 8, pin 8 of display.
4. Pin 14 (Anode d) is connected to vertical row 16, one end of R4.

The yellow wires from J2 are connected next (also to lower portion):

1. Pin 15 (Anode e) is connected to vertical row 19, one end of R5.
2. Pin 19 (Cathode Dig4) is connected to vertical row 10, pin 10 of display.
3. Pin 21 (Anode UC) is connected to vertical row 18, one end of R8.
4. Pin 22 (Anode LC) is connected to vertical row 17, one end of R9.
5. Pin 23 (Anode PM) is connected to vertical row 15, one end of R10.

The red-wire push-button inputs TD1 and TD0 from J2 are connected next:

1. Pin 27 (TD1) is connected to vertical row 20, one end of R12.
2. Pin 28 (TD0) is connected to vertical row 23, one end of R11.

All connections to the display board should be complete.

The Analog Board

The connections to the analog board start with connecting a RD/BLK wire from vertical row 17, upper portion, one end of R16A, to the V_{CC} line on the display board; and an orange wire (Analog V_{CC}) from pin 25 of J2 to the "Analog V_{CC}" line.

The remaining orange lines from J2 to the upper portion are connected next:

1. Pin 26 (I/O Clock) is connected to vertical row 21, pin 7 of TLV1549.

2. Pin 24 (Chip Select) is connected to vertical row 23, pin 5 of TLV1549.

3. Pin 18 (Data Out) is connected to vertical row 22, pin 6 of TLV1549.

4. One remaining wire—a RD/BLK wire from pin 7 of J1 (RST/NMI)—is connected to vertical row 13, lower portion to input the RESET push-button signal to the microcontroller. This completes the connections to the analog board.

Breadboard Construction Completion

The breadboard construction is essentially complete; however, several items need clarification. First is the connection of the watch crystal. The connections to the watch crystal, X_{OUT} and X_{IN}, are made on the evaluation board, as shown in *Figure 10-15*, rather than with wires from pin 5 and 6 of J1. The connections for the crystal on the evaluation board PCB are connected to pin 5 and pin 6 of J1. Carefully (the wires are very delicate) insert the wires into the holes in the evaluation board PCB, as shown in *Figure 10-15a*, and solder the wires to the PCB; then solder the case of the crystal to the PCB pad provided. Before soldering to the pad, tin the case of the crystal with a small dab of solder. Do this quickly so the crystal will not overheat.

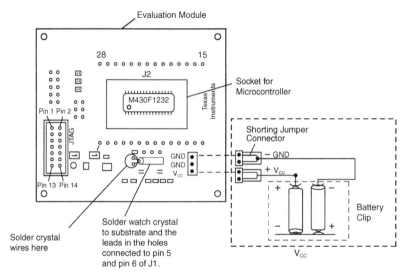

a. Watch crystal connections

b. Battery supply for stand-alone breadboard

Figure 10-15: Evaluation board connections for watch crystal and battery clip for stand-alone breadboard

Second, when the breadboard is to stand alone—disconnected from the PC—+3V power is provided by two AA batteries. The batteries are connected in series in a battery clip which is wired through shorting jumpers to V_{CC} and GND on the evaluation board as shown in *Figure 10-15b*. This power connection is normallyl used only when the JTAG connector from the development system is disconnected from the evaluation board. Shorting jumpers are used so that the batteries can be disconnected if the application is not being used. However, some computer power supplies cannot handle the load, especially when the red and black buttons are pushed. In these cases, connect the battery.

The Application Program

With the construction of the breadboard complete, the application program—a listing of the program in C language is included in the Appendix—must be loaded into the microcontroller. The most recent version is contained online; it may vary from what is printed in the Appendix. The development system is used for this purpose, and the connections for the system are shown in *Figure 10-9*. The first step is to load the software that comes with the development system, MSP-FET430P120.

Installing the Development System Software

Following are the steps to install the development system software:

1. Insert the CD-ROM that is contained in the MSP-430P120 development system in the CD-ROM drive of a computer. It should start automatically. If it does not, use a browser to open a file "index.htm" that is located in the root directory of the CD-ROM. The MSP430 start page will be displayed.
2. Click on "Tool Software."
3. Select the MSP430PI20 Flash Evaluation Tool.
4. Click "Save" to download the file "FET_RXXX.exe". The file gets updated as time progresses.
5. A "Save As" window appears. Select a directory path to store the download. For example: C:/My Computer/MSP430 Rel X.X (D:). Click "Save." Note: *The hard drive in this directory path is identified with the letter C. If another letter is used for the hard drive designation, substitute that letter in the directory path.*
6. Use Windows Explorer to follow the directory path to the window that contains and displays "FET_RXXX.exe". Click on it to execute the program.
7. The IAR Embedded Workbench window appears which has a welcome message to IAR Systems Product Setup. IAR Systems are the developers of the software.
8. Follow the prompts to install the FET software. The IAR Systems software licensing agreement must be accepted.
9. Turn off computer and reboot.
10. There should be an icon on your desktop that says "IAR Embedded Workbench."

Downloading the Application Software

The software program for the application can be downloaded from the Internet as follows. It is a file programmed in C:

1. Go to http://www.ti.com/MSP430university.
2. Click on "MSP430 Textbooks."
3. Click on *Analog and Digital Circuits for Control System Applications: Using the TI MSP430 Microcontroller* by Jerry Luecke. A web page for the book appears.
4. Select "TimeDateTemp Application" and a window opens that has "Save" or "Cancel" across the bottom. Click on "Save."
5. Choose a directory path to store the download. For example: C:/My Documents/MSP430Applications.
6. The file is a Zip file that contains the application program written in the C language. Name the file "TimeDateTemp.zip."
7. Click on "Save."

Unzipping the Application Software

1. Open Windows Explorer.
2. Navigate through the directory C:/MyDocuments/MSP430Applications to the MSP430Applications folder.
3. Select "Time,Date,Temp.zip."
4. Double clicking on "Time,Date,Temp.zip" will unzip the file if a WinZip program is installed on the computer used. If no WinZip program is installed, go to www.winzip.com on the Web and download the software to unzip the file.
5. In the WinZip window go to "Extract."
6. Select "all files" and extract to the "MSP430Applications" folder at C:/MyDocuments/ MSP430Applications.
7. In the "MSP430 Applications" folder there will be the "Time,Date,Temp.c" file and a "Readme.txt" file. The "Readme.txt" file will have the latest information on revisions or changes.

Loading Application Program in Microcontroller

Follow these steps to load the application program (TimeDateTemp.c) into the microcontroller: (Note: These instructions are based on the latest version of the IAR Workbench at the time of publication. Variations may occur in these instructions as the software is updated. See the "Readme.txt" for the latest information.)

Creating a Project in IAR Workbench©

1. On the PC desktop, click the IAR Embedded Workbench icon. A IAR Systems window appears.
2. Click File menu and then "New."
3. In the "New" window highlight "Workspace" and click "OK."
4. Choose directory C:/My Documents/430Applications and click "Open."
5. Enter TimeDateTemp.eww in the file window and click "Save."
6. A TimeDateTemp workspace window appears on the IAR workbench.
7. Click Project menu and then "Create New Project."
8. Choose directory C:/My Documents/430Applications and click "Open."
9. Enter TimeDateTemp.ewp in the file window and click "Create."
10. TimeDateTemp will appear in your workspace window.
11. Click on "TimeDateTemp _ Debug" in the workspace window then click the Project menu and then click "Add Files."
12. Choose directory C:/My Documents/430Applications and click "Open."
13. Select "TimeDateTemp.c" that you unzipped and click "Open."
14. "TimeDateTemp.c" will appear in the workspace window under the TimeDateTemp Project.
15. Click on "TimeDateTemp _ Debug" in the workspace window once again, then click the Project menu and now click "Options."
16. An "Options for node - TimeDateTemp _ Debug" window appears. Highlight the "General" category, then select msp430F1232 in the Target, Device box.
17. Click "OK" to set the options for the project.
18. Click on "TimeDateTemp _ Debug" in the workspace window once again, then click the Project menu and click "Options" again.

19. Highlight the "C-Spy" category, then select "Flash Emulation Tool" in the driver box and click on "OK" to set the options for the project.

Compiling the Program

20. Select Project menu and highlight "Build All." Click on it. The "Rebuild All" command executes compiling the C code file into assembly-language and links it with the specific MSP430 descriptor files which were defined by the project options.

21. A message window appears that gives the status of the programming. If it is successful, here is an example of a window that appears:

```
Messages

Rebuilding configuration: TimeDateTemp - Debug

0 file(s) deleted.

TimeDateTemp.c
icc430.exe -I C:\Program Files\IAR Systems\Embedded Workbench
3.2\430\INC\ -I C:\Program Files\IAR Systems\Embedded Workbench
3.2\430\INC\CLIB\ -o C:\
Documents and Settings\a0193378\My Documents\430Applications\Debug\Obj\
-z2 --no_cse --no_unroll --no_inline --no_code_motion --debug -e
--double=32 C:\
Documents and Settings\a0193378\My
Documents\430Applications\TimeDateTemp.c

    IAR MSP430 C Compiler V2.21A/W32 [Kickstart]
    Copyright 1996-2003 IAR Systems. All rights reserved.

3   258 bytes of CODE memory
    116 bytes of CONST memory
    168 bytes of DATA memory (+ 19 bytes shared)

Errors: none
Warnings: none


Linking
xlink.exe C:\Documents and Settings\a0193378\My
Documents\430Applications\Debug\Obj\TimeDateTemp.r43 -o C:\Documents and
Settings\a0193378\My
Documents\430Applications\Debug\Exe\TimeDateTemp.d43 -rt -IC:\Program
Files\IAR Systems\Embedded Workbench 3.2\430\LIB\ -f C:\Program
Files\IAR Systems\
Embedded Workbench 3.2\430\config\lnk430F123.xcl C:\Program Files\IAR
Systems\Embedded Workbench 3.2\430\lib\cl430f.r43
-e_small_write=_formatted_write
-e_medium_read=_formatted_read -f C:\Program Files\IAR Systems\Embedded
Workbench 3.2\430\config\compactmath.xcl

    IAR Universal Linker V4.56D/386
    Copyright 1987-2003 IAR Systems. All rights reserved.

3   412 bytes of CODE memory
```

```
      248 bytes of DATA memory (+ 19 absolute )
      144 bytes of CONST memory

  Errors: none
  Warnings: none

  Total number of errors: 0
  Total number of warnings: 0
```

Loading the Program

22. Select the Project menu, then click on "Debug." The program will be loaded through the JTAG connector onto the MSP430, and the Debugging interface will be shown.

23. Click "Debug," then "Go" and the program will begin to run on your breadboard.

The system comes up in the **Clock** state at 12:00. It can be set to the present hour and minutes by pushing the black button and adjusting the quantity with the red button. Pressing the black button after adjustments are finished returns to the **Clock** state. Pushing the red button toggles to the next state. Then the date and the year can be set in the same manner. Calibration can then occur for the temperature. Refer to the manual-toggling state diagram shown in *Figure 10-3* for guidance. Sometimes pushing the buttons causes current surges that take the system out of its sequence or shuts it down. If this persists, connect the battery shown in *Figure 10-15*.

Troubleshooting

If the breadboard does not function, look first at the construction. Most problems will occur because of a wrong connection or a connection that is not electrically sound—shorts, opens, broken wires, intermittent connections. One of the first considerations is that the MSP430 microcontroller is not securely in its socket. Check it carefully. It may require a magnifying glass to see it clearly. The second consideration is the interconnection of the wires. Be sure they are inserted properly and are making correct connection. It is less likely that the software is to blame, especially if the loading sequence has been followed and the windows and clicking followed religiously. The software has been and will be continually checked by the users to verify its authenticity. It may change but any changes should be in the "Readme.txt" file. However, if after following Steps 20 through 23, a "Failed to get target information" message is received, it indicates that the communication to the devise has failed. Do the following procedure after checking again the wiring connections and the socket connections.

1. Disconnect the J-TAG connector to the evaluator board.
2. Disconnect the battery if connected.
3. Short out the two power connections, "V_{CC}" and "GND." This discharges all capacitors and resets the microcontroller.
4. Reconnect the J-TAG connector and the battery, and redo Steps 1 through 23.

The Stand-Alone Breadboard

To make the breadboard stand alone, the separate battery supply shown in *Figure 10-15b* is added. Connect the shorting jumpers to the GND terminals and the V_{CC} terminal on the evaluation module. Disconnect the JTAG connector from the evaluation module *before* connecting the shorting jumpers. Disconnecting the JTAG connector removes the V_{CC} power that has been supplied by the PC. The program remains stored in the microcontroller and will not be destroyed while the stand-alone V_{CC} and GND are established.

To return to the JTAG connection and supplying power from the PC, remove the shorting jumpers from the AA batteries to the V_{CC} and GND terminals on the evaluation module. Reconnect the JTAG connector.

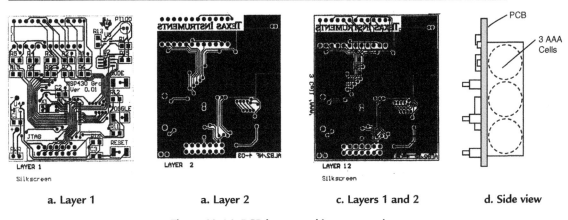

| a. Layer 1 | a. Layer 2 | c. Layers 1 and 2 | d. Side view |

Figure 10-16: PCB layout and interconnections

The PCB Circuit

For those readers that prefer to have a very compact, portable unit, the layers for a printed circuit layout are shown in *Figure 10-16*. There is a different parts list for the PCB because the circuit is laid out for surface-mount components. For example, the package notations change for the ICs to identify them for surface mounting. And recall that a different power supply for V_{CC} is added, so there are additional components. The same MSP430 Development System, MSP-FET430P120, is needed. Here is the parts list:

Part	Part No.	Quantity
Available from www.ti.com:		
Microcontroller	MSP430F1232PW	1
ADC Data converter (A – D)	TLV1549CD	1
Op Amp	TLV2451CDBVT	1
LDO (Voltage Regulator)	TPS71533DCKR	1
Available from www.omega.com:		
Sensor (PT-100 RTD)	1PT100KN815	1
Available from www.digikey.com:		
4-digit LED Display	LN543RKN8	1
32 kHz watch crystal	X802-ND	1
14-pin JTAG connector	H2902-ND	1
Push buttons	401-1096-2-ND	3
Jumper	99911-21011000000-ND	1
3-cell AAA battery holder	2479K-ND	1
C1, C3 0.1uF capacitor	C1206C104J5RACTU	2
C2 0.47uF capacitor	ECJ-3VB1C474K	1
R1 – R10 33-Ω resistor	9C12063A33R0FKHFT	10
R11, R12 470 KΩ resistor	9C12063A4703FKHFT	2
R13 15KΩ resistor	9C12063A1502FKHFT	1
R14 100KΩ resistor	9C12063A1003FKHF	1
R15 1KΩ resistor	9C12063A1001FKHFT	1
R16 68KΩ resistor	9C12063A6802FKHFT	1
Any Local Supply		
AAA battery		3

The parts may be obtained from the sources listed, or any other electronic parts distributor, and, especially, one that handles surface-mount components.

Refer back to the schematic, *Figure 10-4*, and note, again, that there are part changes and additions for the PCB circuit. The first is the op amp package. It is changed to a 5-pin surface-mount package TLV2451CDBVT. The pin connections are shown on *Figure 10-4*. The second is the addition of an LDO voltage regulator, TPS71533, so that the circuit has an extended battery life by operating from three AAA batteries. The regulator supplies +3.3V. In addition to the regulator, three capacitors—C_1(0.1 μF), C_2(0.47 μF) and C_3(0.1 μF) are added to the circuit.

One must have dexterity, patience, and skill to mount the components on the PCB. Search out a person who has experience with such surface-mount circuits to help in the construction. When completed, it results in a neat, self-contained unit that resides easily on the desk, in a briefcase, or on a shelf. *Figure 10-17a* is a photograph of a completed breadboard, and *Figure 10-17b* is the PCB circuit.

An additional word on programming the microcontroller when it is on the PCB. Make sure the power jumper shown on the *Figure 10-4* schematic that is in the LDO circuit block is disconnected while the microcontroller is being programmed. Let the PC power the circuit until the programming is complete, then reconnect the power jumper so the circuit can be powered from the AAA batteries through the LDO.

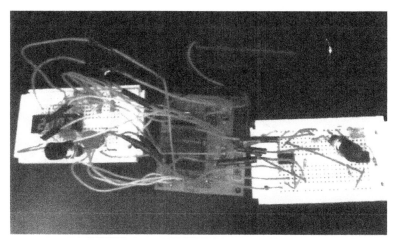

a. Breadboard circuit

b. PCB circuit

Figure 10-17: System in breadboard and PCB form

Summary

This completes the discussion of the functions and the electronic circuits that make up a system that inputs analog signals, converts them to digital, manipulates them digitally, converts them back from digital to analog and outputs them to productive tasks in the human world. Providing understanding and learning has been the major goal; it is hoped that this goal has been achieved.

Chapter 10 Quiz

1. The application that is implemented in this chapter:
   a. really doesn't do anything but provide exercises in programming.
   b. can't be built. It is just a paper exercise.
   c. requires no hardware, just software.
   d. displays the time, date, year and temperature in °F and °C in sequence.
2. For the application, a diagram that shows the sequence of what is displayed is called:
   a. a state diagram.
   b. a schematic.
   c. a logic diagram.
   d. a memory map.
3. In the application, the three modes of operation are:
   a. the running mode, the setting mode, and quiet mode.
   b. auto toggling, manual toggling and sleep modes.
   c. the ON, OFF and toggling modes.
   d. the logic, the arithmetic and the memory modes.
4. When the system is in the sleep mode:
   a. everything in the system is inactive.
   b. all circuits are active except the display.
   c. the crystal oscillator and Timer_A are the only circuits active.
   d. the microcontroller is fully active.
5. When in the sleep mode, a signal awakens the system:
   a. every 10-minute point in the hour.
   b. at the 20-minute, 40-minute and 60-minute point in the hour.
   c. every half hour.
   d. every hour.
6. What segments of the 7-segment LED display need to be energized to make the number 9?
   a. a, b, d, f and g.
   b. b, c, d, e and g.
   c. a, b, c, f and g.
   d. a, b, e, f and g.
7. Of the 24 pins on the 7-segment display, which ones are not used in the chapter application?
   a. 4, 9, 13, 14, 16, 21 and 23.
   b. 9, 12, 13, 14, 16, 21 and 23.
   c. 12, 13, 14, 16, 21, 23 and 24.
   d. 4, 9, 12, 13, 14, 16, 21 and 23.

8. On the schematic of *Figure 10-4*, what are the three push button switches used for?
   a. Red turns on power, black turns on power, and black resets the system.
   b. Red starts the microcontroller and black shuts it down.
   c. Red is for toggling, black is for changing modes, and black is for reset.
   d. Red selects the I/O, black selects the frequency of the clock, and black resets the system.
9. Which control register configures the Port 3 I/O pins as outputs?
   a. P3DIR.
   b. P3SEL.
   c. P3IN.
   d. P3OUT.
10. Which control register configures the Port 2 I/O pins as inputs?
    a. P2SEL.
    b. P2DIR.
    c. P2IN.
    d. P2OUT.
11. Which MSP430F1232 pin is configured as an output to provide the chip select –CS, to the TLV1549 ADC?
    a. P1.4.
    b. P2.6.
    c. P2.7.
    d. P1.3.
12. Which MSP430F1232 pin is configured as an input to receive the output data from the TLV1549 ADC?
    a. P2.6.
    b. P2.7.
    c. P3.7.
    d. P1.4.
13. MSP430F1232 pin 1.4 is configured as an output to supply V_{CC} to the analog circuits so that:
    a. the V_{CC} supplies are isolated.
    b. the system will consume very little power in the sleep mode.
    c. the V_{CC} to the analog circuits can be bypassed.
    d. noise is reduced on the analog V_{CC} line.
14. The breadboard system is made up of:
    a. the display board, the JTAG connector and a PC.
    b. the evaluation board and a PC.
    c. the evaluation board, the display board and the analog board.
    d. the analog board and the evaluation board.
15. In the breadboard system, the connector used to connect the system to a PC is:
    a. the MSP430F1232 socket.
    b. the LED display socket.
    c. the JTAG connector.
    d. the battery connector.
16. What are the three stages of developing the MSP430F1232 system?
    a. Breadboard powered by a PC, stand-alone breadboard powered by batteries, a PCB circuit powered by batteries.

b. Evaluation board powered by batteries, an analog V_{CC} breadboard, and a PCB circuit powered by batteries.

c. a stand-alone breadboard powered from a PC, a JTAG breadboard, and a PCB circuit powered by batteries.

d. a stand-alone breadboard powered by batteries, and two PCB circuits powered by batteries.

17. The analog board contains:

a. the display, the sensor, and the ADC.

b. the sensor, the signal-conditioning amplifier and the ADC.

c. the signal conditioning amplifier and the display.

d. Only the sensor and the display.

18. The display board contains:

a. the display and the resistors to energize the LED anodes.

b. the sensor, the signal-conditioning amplifier and the ADC.

c. the resistors to set the gain of the signal-conditioning amplifier.

d. none of the above.

19. The stand-alone breadboard:

a. operates by itself on battery power.

b. is made up of a display board, an evaluation board, and an analog board.

c. operates stand alone after being programmed from a PC.

d. all of the above.

e. only a above.

20. The PCB system:

a. is a complete system assembled on a printed-circuit board.

b. must be programmed by a PC after assembly on a PCB.

c. must operate from a PC at all times.

d. only c above.

e. only a and b above.

f. none of the above.

Answers: 1.d, 2.a, 3.b, 4.c, 5.b, 6.c, 7.d, 8.c, 9.a, 10.b, 11.d, 12.c, 13.b, 14.c, 15.c, 16.a, 17.b, 18.a, 19.d, 20.e.

The MSP430 Instruction Set

This list of the 51 instructions—27 core instructions and 24 emulated instructions is included to help in understanding and clarifying assembly-language programs for the MSP430 family of microcontrollers. They are printed here by permission and courtesy of Texas Instruments Incorporated and are part of a *MSP430x1xx Family User's Guide*.

3.4 Instruction Set

The complete MSP430 instruction set consists of 27 core instructions and 24 emulated instructions. The core instructions are instructions that have unique op-codes decoded by the CPU. The emulated instructions are instructions that make code easier to write and read, but do not have op-codes themselves, instead they are replaced automatically by the assembler with an equivalent core instruction. There is no code or performance penalty for using emulated instruction.

There are three core-instruction formats:

❑ Dual-operand

❑ Single-operand

❑ Jump

All single-operand and dual-operand instructions can be byte or word instructions by using .B or .W extensions. Byte instructions are used to access byte data or byte peripherals. Word instructions are used to access word data or word peripherals. If no extension is used, the instruction is a word instruction.

The source and destination of an instruction are defined by the following fields:

src	The source operand defined by As and S-reg
dst	The destination operand defined by Ad and D-reg
As	The addressing bits responsible for the addressing mode used for the source (src)
S-reg	The working register used for the source (src)
Ad	The addressing bits responsible for the addressing mode used for the destination (dst)
D-reg	The working register used for the destination (dst)
B/W	Byte or word operation: 0: word operation 1: byte operation

---

Note: Destination Address

Destination addresses are valid anywhere in the memory map. However, when using an instruction that modifies the contents of the destination, the user must ensure the destination address is writable. For example, a masked-ROM location would be a valid destination address, but the contents are not modifiable, so the results of the instruction would be lost.

3.4.1   Double-Operand (Format I) Instructions

Figure 3–9 illustrates the double-operand instruction format.

Figure 3–9. Double Operand Instruction Format

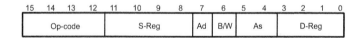

Table 3–11 lists and describes the double operand instructions.

Table 3–11. Double Operand Instructions

Mnemonic	S-Reg, D-Reg	Operation	Status Bits			
			V	N	Z	C
MOV(.B)	src,dst	src → dst	–	–	–	–
ADD(.B)	src,dst	src + dst → dst	*	*	*	*
ADDC(.B)	src,dst	src + dst + C → dst	*	*	*	*
SUB(.B)	src,dst	dst + .not.src + 1 → dst	*	*	*	*
SUBC(.B)	src,dst	dst + .not.src + C → dst	*	*	*	*
CMP(.B)	src,dst	dst – src	*	*	*	*
DADD(.B)	src,dst	src + dst + C → dst (decimally)	*	*	*	*
BIT(.B)	src,dst	src .and. dst	0	*	*	*
BIC(.B)	src,dst	.not.src .and. dst → dst	–	–	–	–
BIS(.B)	src,dst	src .or. dst → dst	–	–	–	–
XOR(.B)	src,dst	src .xor. dst → dst	*	*	*	*
AND(.B)	src,dst	src .and. dst → dst	0	*	*	*

*   The status bit is affected
–   The status bit is not affected
0   The status bit is cleared
1   The status bit is set

Note: Instructions CMP and SUB

The instructions CMP and SUB are identical except for the storage of the result. The same is true for the BIT and AND instructions.

3.4.2 Single-Operand (Format II) Instructions

Figure 3–10 illustrates the single-operand instruction format.

Figure 3–10. Single Operand Instruction Format

15 14 13 12 11 10 9 8 7	6	5 4	3 2 1 0
Op-code	B/W	Ad	D/S-Reg

Table 3–12 lists and describes the single operand instructions.

Table 3–12. Single Operand Instructions

Mnemonic	S-Reg, D-Reg	Operation	Status Bits V	N	Z	C
RRC(.B)	dst	C → MSB →.......LSB → C	*	*	*	*
RRA(.B)	dst	MSB → MSB →....LSB → C	0	*	*	*
PUSH(.B)	src	SP – 2 → SP, src → @SP	–	–	–	–
SWPB	dst	Swap bytes	–	–	–	–
CALL	dst	SP – 2 → SP, PC+2 → @SP	–	–	–	–
		dst → PC				
RETI		TOS → SR, SP + 2 → SP	*	*	*	*
		TOS → PC,SP + 2 → SP				
SXT	dst	Bit 7 → Bit 8........Bit 15	0	*	*	*

*   The status bit is affected
–   The status bit is not affected
0   The status bit is cleared
1   The status bit is set

All addressing modes are possible for the CALL instruction. If the symbolic mode (ADDRESS), the immediate mode (#N), the absolute mode (&EDE) or the indexed mode x(RN) is used, the word that follows contains the address information.

Instruction Set

3.4.3   Jumps

Figure 3–11 shows the conditional-jump instruction format.

Figure 3–11. Jump Instruction Format

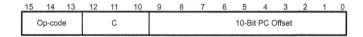

15	14	13	12	11	10	9	8	7	6	5	4	3	2	1	0
Op-code			C			10-Bit PC Offset									

Table 3–13 lists and describes the jump instructions.

Table 3–13. Jump Instructions

Mnemonic	S-Reg, D-Reg	Operation
JEQ/JZ	Label	Jump to label if zero bit is set
JNE/JNZ	Label	Jump to label if zero bit is reset
JC	Label	Jump to label if carry bit is set
JNC	Label	Jump to label if carry bit is reset
JN	Label	Jump to label if negative bit is set
JGE	Label	Jump to label if (N .XOR. V) = 0
JL	Label	Jump to label if (N .XOR. V) = 1
JMP	Label	Jump to label unconditionally

Conditional jumps support program branching relative to the PC and do not affect the status bits. The possible jump range is from −511 to +512 words relative to the PC value at the jump instruction. The 10-bit program-counter offset is treated as a signed 10-bit value that is doubled and added to the program counter:

$$PC_{new} = PC_{old} + 2 + PC_{offset} \times 2$$

ADC[.W]	Add carry to destination
ADC.B	Add carry to destination

Syntax   ADC   dst   or   ADC.W   dst
         ADC.B   dst

Operation   dst + C –> dst

Emulation   ADDC   #0,dst
         ADDC.B   #0,dst

Description   The carry bit (C) is added to the destination operand. The previous contents of the destination are lost.

Status Bits   N: Set if result is negative, reset if positive
         Z: Set if result is zero, reset otherwise
         C: Set if dst was incremented from 0FFFFh to 0000, reset otherwise
              Set if dst was incremented from 0FFh to 00, reset otherwise
         V: Set if an arithmetic overflow occurs, otherwise reset

Mode Bits   OSCOFF, CPUOFF, and GIE are not affected.

Example   The 16-bit counter pointed to by R13 is added to a 32-bit counter pointed to by R12.
         ADD   @R13,0(R12)   ; Add LSDs
         ADC   2(R12)   ; Add carry to MSD

Example   The 8-bit counter pointed to by R13 is added to a 16-bit counter pointed to by R12.
         ADD.B   @R13,0(R12)   ; Add LSDs
         ADC.B   1(R12)   ; Add carry to MSD

Instruction Set

ADD[.W]   Add source to destination
ADD.B   Add source to destination

Syntax   ADD   src,dst   or   ADD.W   src,dst
                    ADD.B   src,dst

Operation   src + dst –> dst

Description   The source operand is added to the destination operand. The source operand
                    is not affected. The previous contents of the destination are lost.

Status Bits   N: Set if result is negative, reset if positive
                    Z: Set if result is zero, reset otherwise
                    C: Set if there is a carry from the result, cleared if not
                    V: Set if an arithmetic overflow occurs, otherwise reset

Mode Bits   OSCOFF, CPUOFF, and GIE are not affected.

Example   R5 is increased by 10. The jump to TONI is performed on a carry.

                    ADD   #10,R5
                    JC   TONI   ; Carry occurred
                    ......   ; No carry

Example   R5 is increased by 10. The jump to TONI is performed on a carry.

                    ADD.B   #10,R5   ; Add 10 to Lowbyte of R5
                    JC   TONI   ; Carry occurred, if (R5) ≥ 246 [0Ah+0F6h]
                    ......   ; No carry

3-22   *RISC 16–Bit CPU*

ADDC[.W]	Add source and carry to destination
ADDC.B	Add source and carry to destination

Syntax   ADDC   src,dst   or   ADDC.W   src,dst
ADDC.B   src,dst

Operation   src + dst + C –> dst

Description   The source operand and the carry bit (C) are added to the destination operand. The source operand is not affected. The previous contents of the destination are lost.

Status Bits   N: Set if result is negative, reset if positive
Z: Set if result is zero, reset otherwise
C: Set if there is a carry from the MSB of the result, reset otherwise
V: Set if an arithmetic overflow occurs, otherwise reset

Mode Bits   OSCOFF, CPUOFF, and GIE are not affected.

Example   The 32-bit counter pointed to by R13 is added to a 32-bit counter, eleven words (20/2 + 2/2) above the pointer in R13.

```
ADD       @R13+,20(R13)   ; ADD LSDs with no carry in
ADDC      @R13+,20(R13)   ; ADD MSDs with carry
...                       ; resulting from the LSDs
```

Example   The 24-bit counter pointed to by R13 is added to a 24-bit counter, eleven words above the pointer in R13.

```
ADD.B     @R13+,10(R13)   ; ADD LSDs with no carry in
ADDC.B    @R13+,10(R13)   ; ADD medium Bits with carry
ADDC.B    @R13+,10(R13)   ; ADD MSDs with carry
...                       ; resulting from the LSDs
```

Instruction Set

AND[.W]　　　　　Source AND destination
AND.B　　　　　　Source AND destination

Syntax　　　　AND　　　　src,dst　　or　AND.W src,dst
　　　　　　　　　AND.B　　　src,dst

Operation　　　src .AND. dst –> dst

Description　　　The source operand and the destination operand are logically ANDed. The result is placed into the destination.

Status Bits　　　N:　Set if result MSB is set, reset if not set
　　　　　　　　　Z:　Set if result is zero, reset otherwise
　　　　　　　　　C:　Set if result is not zero, reset otherwise (= .NOT. Zero)
　　　　　　　　　V:　Reset

Mode Bits　　　OSCOFF, CPUOFF, and GIE are not affected.

Example　　　　The bits set in R5 are used as a mask (#0AA55h) for the word addressed by TOM. If the result is zero, a branch is taken to label TONI.

　　　　　　　　MOV　　　#0AA55h,R5　　　; Load mask into register R5
　　　　　　　　AND　　　R5,TOM　　　　　; mask word addressed by TOM with R5
　　　　　　　　JZ　　　　TONI　　　　　　;
　　　　　　　　......　　　　　　　　　　　; Result is not zero
　　　　　　　　;
　　　　　　　　;
　　　　　　　　;
　　　　　　　　;　　　　　　or
　　　　　　　　;
　　　　　　　　;
　　　　　　　　AND　　　#0AA55h,TOM
　　　　　　　　JZ　　　　TONI

Example　　　　The bits of mask #0A5h are logically ANDed with the low byte TOM. If the result is zero, a branch is taken to label TONI.

　　　　　　　　AND.B　　#0A5h,TOM　　　; mask Lowbyte TOM with 0A5h
　　　　　　　　JZ　　　　TONI　　　　　　;
　　　　　　　　......　　　　　　　　　　　; Result is not zero

BIC[.W]	Clear bits in destination
BIC.B	Clear bits in destination

Syntax   BIC   src,dst   or   BIC.W src,dst
                 BIC.B   src,dst

Operation   .NOT.src .AND. dst –> dst

Description The inverted source operand and the destination operand are logically ANDed. The result is placed into the destination. The source operand is not affected.

Status Bits Status bits are not affected.

Mode Bits   OSCOFF, CPUOFF, and GIE are not affected.

Example   The six MSBs of the RAM word LEO are cleared.

                 BIC   #0FC00h,LEO   ; Clear 6 MSBs in MEM(LEO)

Example   The five MSBs of the RAM byte LEO are cleared.

                 BIC.B   #0F8h,LEO   ; Clear 5 MSBs in Ram location LEO

BIS[.W]   Set bits in destination
BIS.B   Set bits in destination

Syntax   BIS   src,dst   or BIS.W   src,dst
                    BIS.B   src,dst

Operation   src .OR. dst –> dst

Description   The source operand and the destination operand are logically ORed. The result is placed into the destination. The source operand is not affected.

Status Bits   Status bits are not affected.

Mode Bits   OSCOFF, CPUOFF, and GIE are not affected.

Example   The six LSBs of the RAM word TOM are set.

                    BIS   #003Fh,TOM; set the six LSBs in RAM location TOM

Example   The three MSBs of RAM byte TOM are set.

                    BIS.B   #0E0h,TOM   ; set the 3 MSBs in RAM location TOM

BIT[.W]	Test bits in destination
BIT.B	Test bits in destination

Syntax   BIT   src,dst   or   BIT.W   src,dst

Operation   src .AND. dst

Description   The source and destination operands are logically ANDed. The result affects only the status bits. The source and destination operands are not affected.

Status Bits   N: Set if MSB of result is set, reset otherwise
Z: Set if result is zero, reset otherwise
C: Set if result is not zero, reset otherwise (.NOT. Zero)
V: Reset

Mode Bits   OSCOFF, CPUOFF, and GIE are not affected.

Example   If bit 9 of R8 is set, a branch is taken to label TOM.

```
BIT        #0200h,R8       ; bit 9 of R8 set?
JNZ        TOM             ; Yes, branch to TOM
...                        ; No, proceed
```

Example   If bit 3 of R8 is set, a branch is taken to label TOM.

```
BIT.B      #8,R8
JC         TOM
```

Example   A serial communication receive bit (RCV) is tested. Because the carry bit is equal to the state of the tested bit while using the BIT instruction to test a single bit, the carry bit is used by the subsequent instruction; the read information is shifted into register RECBUF.

```
;
; Serial communication with LSB is shifted first:
                           ; xxxx    xxxx    xxxx    xxxx
BIT.B      #RCV,RCCTL      ; Bit info into carry
RRC        RECBUF          ; Carry –> MSB of RECBUF
                           ; cxxx    xxxx
......                     ; repeat previous two instructions
......                     ; 8 times
                           ; cccc    cccc
                           ; ^       ^
                           ; MSB     LSB
; Serial communication with MSB shifted first:
BIT.B      #RCV,RCCTL      ; Bit info into carry
RLC.B      RECBUF          ; Carry –> LSB of RECBUF
                           ; xxxx    xxxc
......                     ; repeat previous two instructions
......                     ; 8 times
                           ; cccc    cccc
                           ; |       LSB
                           ; MSB
```

* BR, BRANCH	Branch to .... destination	
Syntax	BR	dst
Operation	dst –> PC	
Emulation	MOV	dst,PC

Description An unconditional branch is taken to an address anywhere in the 64K address space. All source addressing modes can be used. The branch instruction is a word instruction.

Status Bits Status bits are not affected.

Example Examples for all addressing modes are given.

BR	#EXEC	;Branch to label EXEC or direct branch (e.g. #0A4h) ; Core instruction MOV @PC+,PC
BR	EXEC	; Branch to the address contained in EXEC ; Core instruction MOV X(PC),PC ; Indirect address
BR	&EXEC	; Branch to the address contained in absolute ; address EXEC ; Core instruction MOV X(0),PC ; Indirect address
BR	R5	; Branch to the address contained in R5 ; Core instruction MOV R5,PC ; Indirect R5
BR	@R5	; Branch to the address contained in the word ; pointed to by R5. ; Core instruction MOV @R5,PC ; Indirect, indirect R5
BR	@R5+	; Branch to the address contained in the word pointed ; to by R5 and increment pointer in R5 afterwards. ; The next time—S/W flow uses R5 pointer—it can ; alter program execution due to access to ; next address in a table pointed to by R5 ; Core instruction MOV @R5,PC ; Indirect, indirect R5 with autoincrement
BR	X(R5)	; Branch to the address contained in the address ; pointed to by R5 + X (e.g. table with address ; starting at X). X can be an address or a label ; Core instruction MOV X(R5),PC ; Indirect, indirect R5 + X

CALL	Subroutine		
Syntax	CALL	dst	
Operation	dst	–> tmp	dst is evaluated and stored
	SP – 2	–> SP	
	PC	–> @SP	PC updated to TOS
	tmp	–> PC	dst saved to PC

Description A subroutine call is made to an address anywhere in the 64K address space. All addressing modes can be used. The return address (the address of the following instruction) is stored on the stack. The call instruction is a word instruction.

Status Bits Status bits are not affected.

Example Examples for all addressing modes are given.

    CALL   #EXEC   ; Call on label EXEC or immediate address (e.g. #0A4h)
                    ; SP–2 → SP, PC+2 → @SP, @PC+ → PC

    CALL   EXEC   ; Call on the address contained in EXEC
                    ; SP–2 → SP, PC+2 → @SP, X(PC) → PC
                    ; Indirect address

    CALL   &EXEC   ; Call on the address contained in absolute address
                    ; EXEC
                    ; SP–2 → SP, PC+2 → @SP, X(0) → PC
                    ; Indirect address

    CALL   R5   ; Call on the address contained in R5
                    ; SP–2 → SP, PC+2 → @SP, R5 → PC
                    ; Indirect R5

    CALL   @R5   ; Call on the address contained in the word
                    ; pointed to by R5
                    ; SP–2 → SP, PC+2 → @SP, @R5 → PC
                    ; Indirect, indirect R5

    CALL   @R5+   ; Call on the address contained in the word
                    ; pointed to by R5 and increment pointer in R5.
                    ; The next time—S/W flow uses R5 pointer—
                    ; it can alter the program execution due to
                    ; access to next address in a table pointed to by R5
                    ; SP–2 → SP, PC+2 → @SP, @R5 → PC
                    ; Indirect, indirect R5 with autoincrement

    CALL   X(R5)   ; Call on the address contained in the address pointed
                    ; to by R5 + X (e.g. table with address starting at X)
                    ; X can be an address or a label
                    ; SP–2 → SP, PC+2 → @SP, X(R5) → PC
                    ; Indirect, indirect R5 + X

Instruction Set

* CLR[.W]	Clear destination
* CLR.B	Clear destination

Syntax   CLR   dst   or   CLR.W dst
                 CLR.B   dst

Operation   0 –> dst

Emulation   MOV   #0,dst
                 MOV.B   #0,dst

Description The destination operand is cleared.

Status Bits Status bits are not affected.

Example   RAM word TONI is cleared.

                 CLR   TONI   ; 0 –> TONI

Example   Register R5 is cleared.

                 CLR   R5

Example   RAM byte TONI is cleared.

                 CLR.B   TONI   ; 0 –> TONI

* CLRC	Clear carry bit
Syntax	CLRC
Operation	0 –> C
Emulation	BIC #1,SR
Description	The carry bit (C) is cleared. The clear carry instruction is a word instruction.
Status Bits	N: Not affected Z: Not affected C: Cleared V: Not affected
Mode Bits	OSCOFF, CPUOFF, and GIE are not affected.
Example	The 16-bit decimal counter pointed to by R13 is added to a 32-bit counter pointed to by R12.

```
CLRC                        ; C=0: defines start
DADD    @R13,0(R12)  ; add 16-bit counter to low word of 32-bit counter
DADC    2(R12)          ; add carry to high word of 32-bit counter
```

Instruction Set

* CLRN	Clear negative bit
Syntax	CLRN
Operation	$0 \rightarrow N$ or (.NOT.src .AND. dst –> dst)
Emulation	BIC #4,SR
Description	The constant 04h is inverted (0FFFBh) and is logically ANDed with the destination operand. The result is placed into the destination. The clear negative bit instruction is a word instruction.
Status Bits	N: Reset to 0 Z: Not affected C: Not affected V: Not affected
Mode Bits	OSCOFF, CPUOFF, and GIE are not affected.
Example	The Negative bit in the status register is cleared. This avoids special treatment with negative numbers of the subroutine called.

```
        CLRN
        CALL      SUBR
        ......
        ......
SUBR    JN        SUBRET    ; If input is negative: do nothing and return
        ......
        ......
        ......
SUBRET  RET
```

* CLRZ	Clear zero bit
Syntax	CLRZ
Operation	$0 \rightarrow Z$ or (.NOT.src .AND. dst –> dst)
Emulation	BIC #2,SR
Description	The constant 02h is inverted (0FFFDh) and logically ANDed with the destination operand. The result is placed into the destination. The clear zero bit instruction is a word instruction.
Status Bits	N: Not affected Z: Reset to 0 C: Not affected V: Not affected
Mode Bits	OSCOFF, CPUOFF, and GIE are not affected.
Example	The zero bit in the status register is cleared. CLRZ

CMP[.W]   Compare source and destination
CMP.B   Compare source and destination

Syntax   CMP   src,dst   or   CMP.W   src,dst
                   CMP.B   src,dst

Operation   dst + .NOT.src + 1
                   or
                   (dst – src)

Description   The source operand is subtracted from the destination operand. This is
                   accomplished by adding the 1s complement of the source operand plus 1. The
                   two operands are not affected and the result is not stored; only the status bits
                   are affected.

Status Bits   N: Set if result is negative, reset if positive (src >= dst)
                   Z: Set if result is zero, reset otherwise (src = dst)
                   C: Set if there is a carry from the MSB of the result, reset otherwise
                   V: Set if an arithmetic overflow occurs, otherwise reset

Mode Bits   OSCOFF, CPUOFF, and GIE are not affected.

Example   R5 and R6 are compared. If they are equal, the program continues at the label
                   EQUAL.

                   CMP   R5,R6   ; R5 = R6?
                   JEQ   EQUAL   ; YES, JUMP

Example   Two RAM blocks are compared. If they are not equal, the program branches
                   to the label ERROR.

                          MOV   #NUM,R5   ; number of words to be compared
                          MOV   #BLOCK1,R6   ; BLOCK1 start address in R6
                          MOV   #BLOCK2,R7   ; BLOCK2 start address in R7
                   L$1   CMP   @R6+,0(R7)   ; Are Words equal? R6 increments
                          JNZ   ERROR   ; No, branch to ERROR
                          INCD R7   ; Increment R7 pointer
                          DEC   R5   ; Are all words compared?
                          JNZ   L$1   ; No, another compare

Example   The RAM bytes addressed by EDE and TONI are compared. If they are equal,
                   the program continues at the label EQUAL.

                          CMP.B EDE,TONI   ; MEM(EDE) = MEM(TONI)?
                          JEQ   EQUAL   ; YES, JUMP

* DADC[.W]	Add carry decimally to destination
* DADC.B	Add carry decimally to destination

Syntax
DADC   dst   or   DADC.W   src,dst
DADC.B   dst

Operation
dst + C –> dst (decimally)

Emulation
DADD   #0,dst
DADD.B   #0,dst

Description
The carry bit (C) is added decimally to the destination.

Status Bits
N: Set if MSB is 1
Z: Set if dst is 0, reset otherwise
C: Set if destination increments from 9999 to 0000, reset otherwise
     Set if destination increments from 99 to 00, reset otherwise
V: Undefined

Mode Bits
OSCOFF, CPUOFF, and GIE are not affected.

Example
The four-digit decimal number contained in R5 is added to an eight-digit deci-mal number pointed to by R8.

```
CLRC                    ; Reset carry
                        ; next instruction's start condition is defined
DADD      R5,0(R8)      ; Add LSDs + C
DADC      2(R8)         ; Add carry to MSD
```

Example
The two-digit decimal number contained in R5 is added to a four-digit decimal number pointed to by R8.

```
CLRC                    ; Reset carry
                        ; next instruction's start condition is defined
DADD.B    R5,0(R8)      ; Add LSDs + C
DADC      1(R8)         ; Add carry to MSDs
```

Instruction Set

DADD[.W]	Source and carry added decimally to destination
DADD.B	Source and carry added decimally to destination

Syntax DADD src,dst or DADD.W src,dst
DADD.B src,dst

Operation src + dst + C –> dst (decimally)

Description The source operand and the destination operand are treated as four binary coded decimals (BCD) with positive signs. The source operand and the carry bit (C) are added decimally to the destination operand. The source operand is not affected. The previous contents of the destination are lost. The result is not defined for non-BCD numbers.

Status Bits N: Set if the MSB is 1, reset otherwise
Z: Set if result is zero, reset otherwise
C: Set if the result is greater than 9999
Set if the result is greater than 99
V: Undefined

Mode Bits OSCOFF, CPUOFF, and GIE are not affected.

Example The eight-digit BCD number contained in R5 and R6 is added decimally to an eight-digit BCD number contained in R3 and R4 (R6 and R4 contain the MSDs).

```
CLRC                    ; clear carry
DADD        R5,R3       ; add LSDs
DADD        R6,R4       ; add MSDs with carry
JC          OVERFLOW    ; If carry occurs go to error handling routine
```

Example The two-digit decimal counter in the RAM byte CNT is incremented by one.

```
CLRC                    ; clear carry
DADD.B      #1,CNT      ; increment decimal counter
```

or

```
SETC
DADD.B      #0,CNT      ; ≡ DADC.B      CNT
```

3-36   *RISC 16–Bit CPU*

* DEC[.W]	Decrement destination
* DEC.B	Decrement destination

Syntax DEC   dst   or   DEC.W   dst
            DEC.B   dst

Operation dst − 1 –> dst

Emulation SUB   #1,dst
Emulation SUB.B #1,dst

Description The destination operand is decremented by one. The original contents are lost.

Status Bits N: Set if result is negative, reset if positive
               Z: Set if dst contained 1, reset otherwise
               C: Reset if dst contained 0, set otherwise
               V: Set if an arithmetic overflow occurs, otherwise reset.
                   Set if initial value of destination was 08000h, otherwise reset.
                   Set if initial value of destination was 080h, otherwise reset.

Mode Bits OSCOFF, CPUOFF, and GIE are not affected.

Example R10 is decremented by 1

            DEC   R10   ; Decrement R10

```
; Move a block of 255 bytes from memory location starting with EDE to memory location starting with
;TONI. Tables should not overlap: start of destination address TONI must not be within the range EDE
; to EDE+0FEh
;
                MOV     #EDE,R6
                MOV     #255,R10
L$1             MOV.B   @R6+,TONI−EDE−1(R6)
                DEC     R10
                JNZ     L$1
```

; Do not transfer tables using the routine above with the overlap shown in Figure 3–12.

Figure 3–12. Decrement Overlap

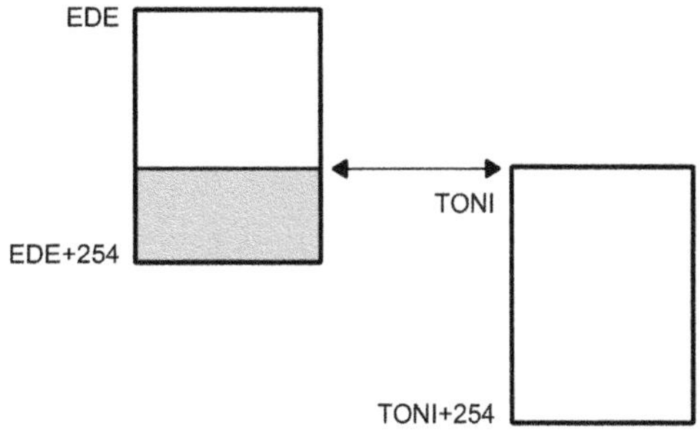

Instruction Set

* DECD[.W]	Double-decrement destination
* DECD.B	Double-decrement destination

Syntax	DECD dst or DECD.W dst
	DECD.B dst

Operation	dst – 2 –> dst

Emulation	SUB #2,dst
Emulation	SUB.B #2,dst

Description	The destination operand is decremented by two. The original contents are lost.

Status Bits	N: Set if result is negative, reset if positive
	Z: Set if dst contained 2, reset otherwise
	C: Reset if dst contained 0 or 1, set otherwise
	V: Set if an arithmetic overflow occurs, otherwise reset.
	Set if initial value of destination was 08001 or 08000h, otherwise reset.
	Set if initial value of destination was 081 or 080h, otherwise reset.

Mode Bits	OSCOFF, CPUOFF, and GIE are not affected.

Example	R10 is decremented by 2.

```
                DECD        R10         ; Decrement R10 by two
```

```
; Move a block of 255 words from memory location starting with EDE to memory location
; starting with TONI
; Tables should not overlap: start of destination address TONI must not be within the
; range EDE to EDE+0FEh
;
                MOV         #EDE,R6
                MOV         #510,R10
L$1             MOV         @R6+,TONI–EDE–2(R6)
                DECD        R10
                JNZ         L$1
```

Example	Memory at location LEO is decremented by two.

```
                DECD.B      LEO         ; Decrement MEM(LEO)
```

Decrement status byte STATUS by two.

```
                DECD.B      STATUS
```

* DINT	Disable (general) interrupts
Syntax	DINT
Operation	0 → GIE or (0FFF7h .AND. SR → SR / .NOT.src .AND. dst –> dst)
Emulation	BIC #8,SR
Description	All interrupts are disabled. The constant 08h is inverted and logically ANDed with the status register (SR). The result is placed into the SR.
Status Bits	Status bits are not affected.
Mode Bits	GIE is reset. OSCOFF and CPUOFF are not affected.
Example	The general interrupt enable (GIE) bit in the status register is cleared to allow a nondisrupted move of a 32-bit counter. This ensures that the counter is not modified during the move by any interrupt.

```
DINT                          ; All interrupt events using the GIE bit are disabled
NOP
MOV     COUNTHI,R5   ; Copy counter
MOV     COUNTLO,R6
EINT                          ; All interrupt events using the GIE bit are enabled
```

> **Note: Disable Interrupt**
>
> If any code sequence needs to be protected from interruption, the DINT should be executed at least one instruction before the beginning of the uninterruptible sequence, or should be followed by a NOP instruction.

Instruction Set

* EINT	Enable (general) interrupts
Syntax	EINT
Operation	1 → GIE or (0008h .OR. SR –> SR / .src .OR. dst –> dst)
Emulation	BIS #8,SR
Description	All interrupts are enabled. The constant #08h and the status register SR are logically ORed. The result is placed into the SR.
Status Bits	Status bits are not affected.
Mode Bits	GIE is set. OSCOFF and CPUOFF are not affected.
Example	The general interrupt enable (GIE) bit in the status register is set.

```
; Interrupt routine of ports P1.2 to P1.7
; P1IN is the address of the register where all port bits are read. P1IFG is the address of
; the register where all interrupt events are latched.
;
                PUSH.B   &P1IN
                BIC.B    @SP,&P1IFG    ; Reset only accepted flags
                EINT                   ; Preset port 0 interrupt flags stored on stack
                                       ; other interrupts are allowed
                BIT      #Mask,@SP
                JEQ      MaskOK        ; Flags are present identically to mask: jump
                ......
MaskOK          BIC      #Mask,@SP
                ......
                INCD     SP            ; Housekeeping: inverse to PUSH instruction
                                       ; at the start of interrupt subroutine. Corrects
                                       ; the stack pointer.
                RETI
```

> **Note: Enable Interrupt**
>
> The instruction following the enable interrupt instruction (EINT) is always executed, even if an interrupt service request is pending when the interrupts are enable.

* INC[.W]	Increment destination
* INC.B	Increment destination

Syntax   INC   dst   or   INC.W dst
             INC.B   dst

Operation   dst + 1 –> dst

Emulation   ADD   #1,dst

Description The destination operand is incremented by one. The original contents are lost.

Status Bits   N: Set if result is negative, reset if positive
             Z: Set if dst contained 0FFFFh, reset otherwise
                 Set if dst contained 0FFh, reset otherwise
             C: Set if dst contained 0FFFFh, reset otherwise
                 Set if dst contained 0FFh, reset otherwise
             V: Set if dst contained 07FFFh, reset otherwise
                 Set if dst contained 07Fh, reset otherwise

Mode Bits   OSCOFF, CPUOFF, and GIE are not affected.

Example   The status byte, STATUS, of a process is incremented. When it is equal to 11, a branch to OVFL is taken.

                    INC.B   STATUS
                    CMP.B   #11,STATUS
                    JEQ   OVFL

Instruction Set

* INCD[.W]	Double-increment destination
* INCD.B	Double-increment destination

Syntax INCD dst or INCD.W dst
INCD.B dst

Operation dst + 2 –> dst

Emulation ADD #2,dst
Emulation ADD.B #2,dst

Example The destination operand is incremented by two. The original contents are lost.

Status Bits N: Set if result is negative, reset if positive
Z: Set if dst contained 0FFFEh, reset otherwise
Set if dst contained 0FEh, reset otherwise
C: Set if dst contained 0FFFEh or 0FFFFh, reset otherwise
Set if dst contained 0FEh or 0FFh, reset otherwise
V: Set if dst contained 07FFEh or 07FFFh, reset otherwise
Set if dst contained 07Eh or 07Fh, reset otherwise

Mode Bits OSCOFF, CPUOFF, and GIE are not affected.

Example The item on the top of the stack (TOS) is removed without using a register.

```
.......
PUSH      R5          ; R5 is the result of a calculation, which is stored
                      ; in the system stack
INCD      SP          ; Remove TOS by double-increment from stack
                      ; Do not use INCD.B, SP is a word-aligned
                      ; register
RET
```

Example The byte on the top of the stack is incremented by two.

```
INCD.B    0(SP)       ; Byte on TOS is increment by two
```

3-42   *RISC 16–Bit CPU*

* INV[.W]	Invert destination	
* INV.B	Invert destination	

Syntax   INV   dst
                 INV.B   dst

Operation   .NOT.dst –> dst

Emulation   XOR   #0FFFFh,dst
Emulation   XOR.B   #0FFh,dst

Description The destination operand is inverted. The original contents are lost.

Status Bits N: Set if result is negative, reset if positive
                 Z: Set if dst contained 0FFFFh, reset otherwise
                     Set if dst contained 0FFh, reset otherwise
                 C: Set if result is not zero, reset otherwise (= .NOT. Zero)
                     Set if result is not zero, reset otherwise (= .NOT. Zero)
                 V: Set if initial destination operand was negative, otherwise reset

Mode Bits   OSCOFF, CPUOFF, and GIE are not affected.

Example   Content of R5 is negated (twos complement).
                 MOV   #00AEh,R5   ;   R5 = 000AEh
                 INV   R5   ; Invert R5,   R5 = 0FF51h
                 INC   R5   ; R5 is now negated,   R5 = 0FF52h

Example   Content of memory byte LEO is negated.

                 MOV.B   #0AEh,LEO   ;   MEM(LEO) = 0AEh
                 INV.B   LEO   ; Invert LEO,   MEM(LEO) = 051h
                 INC.B   LEO   ; MEM(LEO) is negated, MEM(LEO) = 052h

Instruction Set

JC   Jump if carry set
JHS   Jump if higher or same

Syntax   JC   label
                JHS   label

Operation   If C = 1: PC + 2 × offset –> PC
                If C = 0: execute following instruction

Description The status register carry bit (C) is tested. If it is set, the 10-bit signed offset contained in the instruction LSBs is added to the program counter. If C is reset, the next instruction following the jump is executed. JC (jump if carry/higher or same) is used for the comparison of unsigned numbers (0 to 65536).

Status Bits Status bits are not affected.

Example   The P1IN.1 signal is used to define or control the program flow.

                BIT   #01h,&P1IN   ; State of signal –> Carry
                JC   PROGA   ; If carry=1 then execute program routine A
                ......   ; Carry=0, execute program here

Example   R5 is compared to 15. If the content is higher or the same, branch to LABEL.

                CMP   #15,R5
                JHS   LABEL   ; Jump is taken if R5 ≥ 15
                ......   ; Continue here if R5 < 15

JEQ, JZ	Jump if equal, jump if zero
Syntax	JEQ label, JZ label
Operation	If Z = 1: PC + 2 × offset –> PC If Z = 0: execute following instruction
Description	The status register zero bit (Z) is tested. If it is set, the 10-bit signed offset contained in the instruction LSBs is added to the program counter. If Z is not set, the instruction following the jump is executed.
Status Bits	Status bits are not affected.
Example	Jump to address TONI if R7 contains zero.

```
TST     R7          ; Test R7
JZ      TONI        ; if zero: JUMP
```

Example   Jump to address LEO if R6 is equal to the table contents.

```
CMP     R6,Table(R5)    ; Compare content of R6 with content of
                        ; MEM (table address + content of R5)
JEQ     LEO             ; Jump if both data are equal
......                  ; No, data are not equal, continue here
```

Example   Branch to LABEL if R5 is 0.

```
TST     R5
JZ      LABEL
......
```

JGE   Jump if greater or equal

Syntax   JGE   label

Operation   If (N .XOR. V) = 0 then jump to label: PC + 2 × offset –> PC
                           If (N .XOR. V) = 1 then execute the following instruction

Description   The status register negative bit (N) and overflow bit (V) are tested. If both N
                           and V are set or reset, the 10-bit signed offset contained in the instruction LSBs
                           is added to the program counter. If only one is set, the instruction following the
                           jump is executed.

                           This allows comparison of signed integers.

Status Bits   Status bits are not affected.

Example   When the content of R6 is greater or equal to the memory pointed to by R7,
                           the program continues at label EDE.

                           CMP   @R7,R6   ; R6 ≥ (R7)?, compare on signed numbers
                           JGE   EDE   ; Yes, R6 ≥ (R7)
                           ......   ; No, proceed
                           ......
                           ......

JL	Jump if less
Syntax	JL label
Operation	If (N .XOR. V) = 1 then jump to label: PC + 2 × offset –> PC If (N .XOR. V) = 0 then execute following instruction
Description	The status register negative bit (N) and overflow bit (V) are tested. If only one is set, the 10-bit signed offset contained in the instruction LSBs is added to the program counter. If both N and V are set or reset, the instruction following the jump is executed. This allows comparison of signed integers.
Status Bits	Status bits are not affected.
Example	When the content of R6 is less than the memory pointed to by R7, the program continues at label EDE.

```
CMP     @R7,R6      ; R6 < (R7)?,  compare on signed numbers
JL      EDE         ; Yes, R6 < (R7)
......               ; No, proceed
......
......
```

Instruction Set

JMP	Jump unconditionally
Syntax	JMP label
Operation	PC + 2 × offset –> PC
Description	The 10-bit signed offset contained in the instruction LSBs is added to the program counter.
Status Bits	Status bits are not affected.
Hint:	This one-word instruction replaces the BRANCH instruction in the range of −511 to +512 words relative to the current program counter.

JN	Jump if negative
Syntax	JN label
Operation	if N = 1: PC + 2 × offset –> PC if N = 0: execute following instruction
Description	The negative bit (N) of the status register is tested. If it is set, the 10-bit signed offset contained in the instruction LSBs is added to the program counter. If N is reset, the next instruction following the jump is executed.
Status Bits	Status bits are not affected.
Example	The result of a computation in R5 is to be subtracted from COUNT. If the result is negative, COUNT is to be cleared and the program continues execution in another path.

```
              SUB       R5,COUNT    ; COUNT – R5 –> COUNT
              JN        L$1         ; If negative continue with COUNT=0 at PC=L$1
              ......                ; Continue with COUNT≥0
              ......
              ......
              ......
L$1           CLR       COUNT
              ......
              ......
              ......
```

JNC	Jump if carry not set
JLO	Jump if lower

Syntax	JNC	label
	JLO	label

Operation   if C = 0: PC + 2 × offset –> PC
if C = 1: execute following instruction

Description   The status register carry bit (C) is tested. If it is reset, the 10-bit signed offset contained in the instruction LSBs is added to the program counter. If C is set, the next instruction following the jump is executed. JNC (jump if no carry/lower) is used for the comparison of unsigned numbers (0 to 65536).

Status Bits   Status bits are not affected.

Example   The result in R6 is added in BUFFER. If an overflow occurs, an error handling routine at address ERROR is used.

```
        ADD     R6,BUFFER   ; BUFFER + R6 –> BUFFER
        JNC     CONT        ; No carry, jump to CONT
ERROR   ......              ; Error handler start
        ......
        ......
        ......
CONT    ......              ; Continue with normal program flow
        ......
        ......
```

Example   Branch to STL2 if byte STATUS contains 1 or 0.

```
        CMP.B   #2,STATUS
        JLO     STL2        ; STATUS < 2
        ......              ; STATUS ≥ 2, continue here
```

JNE   Jump if not equal
JNZ   Jump if not zero

Syntax   JNE   label
                 JNZ   label

Operation   If Z = 0: PC + 2 × offset –> PC
                 If Z = 1: execute following instruction

Description The status register zero bit (Z) is tested. If it is reset, the 10-bit signed offset contained in the instruction LSBs is added to the program counter. If Z is set, the next instruction following the jump is executed.

Status Bits Status bits are not affected.

Example   Jump to address TONI if R7 and R8 have different contents.

                 CMP   R7,R8   ; COMPARE R7 WITH R8
                 JNE   TONI   ; if different: jump
                 ......   ; if equal, continue

Instruction Set

MOV[.W]   Move source to destination
MOV.B   Move source to destination

Syntax   MOV   src,dst   or   MOV.W   src,dst
                     MOV.B   src,dst

Operation   src –> dst

Description   The source operand is moved to the destination.
                     The source operand is not affected. The previous contents of the destination
                     are lost.

Status Bits   Status bits are not affected.

Mode Bits   OSCOFF, CPUOFF, and GIE are not affected.

Example   The contents of table EDE (word data) are copied to table TOM. The length
                     of the tables must be 020h locations.

```
        MOV   #EDE,R10              ; Prepare pointer
        MOV   #020h,R9              ; Prepare counter
Loop    MOV   @R10+,TOM–EDE–2(R10)  ; Use pointer in R10 for both tables
        DEC   R9                    ; Decrement counter
        JNZ   Loop                  ; Counter ≠ 0, continue copying
        ......                      ; Copying completed
        ......
        ......
```

Example   The contents of table EDE (byte data) are copied to table TOM. The length of
                     the tables should be 020h locations

```
        MOV   #EDE,R10              ; Prepare pointer
        MOV   #020h,R9              ; Prepare counter
Loop    MOV.B @R10+,TOM–EDE–1(R10)  ; Use pointer in R10 for
                                    ; both tables
        DEC   R9                    ; Decrement counter
        JNZ   Loop                  ; Counter ≠ 0, continue
                                    ; copying
        ......                      ; Copying completed
        ......
        ......
```

* NOP	No operation
Syntax	NOP
Operation	None
Emulation	MOV #0, R3
Description	No operation is performed. The instruction may be used for the elimination of instructions during the software check or for defined waiting times.
Status Bits	Status bits are not affected.

The NOP instruction is mainly used for two purposes:

❏ To fill one, two, or three memory words
❏ To adjust software timing

Note: Emulating No-Operation Instruction

Other instructions can emulate the NOP function while providing different numbers of instruction cycles and code words. Some examples are:

Examples:

```
MOV    #0,R3          ; 1 cycle, 1 word
MOV    0(R4),0(R4)    ; 6 cycles, 3 words
MOV    @R4,0(R4)      ; 5 cycles, 2 words
BIC    #0,EDE(R4)     ; 4 cycles, 2 words
JMP    $+2            ; 2 cycles, 1 word
BIC    #0,R5          ; 1 cycle, 1 word
```

However, care should be taken when using these examples to prevent unintended results. For example, if MOV 0(R4), 0(R4) is used and the value in R4 is 120h, then a security violation will occur with the watchdog timer (address 120h) because the security key was not used.

Instruction Set

---

* POP[.W]	Pop word from stack to destination
* POP.B	Pop byte from stack to destination

Syntax　　　　POP　　　　dst
　　　　　　　　　POP.B　　　dst

Operation　　@SP　–> temp
　　　　　　　　SP + 2　–> SP
　　　　　　　　temp –> dst

Emulation　　MOV　　　　@SP+,dst　　or　　MOV.W　　@SP+,dst
Emulation　　MOV.B　　　@SP+,dst

Description　　The stack location pointed to by the stack pointer (TOS) is moved to the destination. The stack pointer is incremented by two afterwards.

Status Bits　　Status bits are not affected.

Example　　The contents of R7 and the status register are restored from the stack.

　　　　　　POP　　　　R7　　　　; Restore R7
　　　　　　POP　　　　SR　　　　; Restore status register

Example　　The contents of RAM byte LEO is restored from the stack.

　　　　　　POP.B　　　LEO　　　; The low byte of the stack is moved to LEO.

Example　　The contents of R7 is restored from the stack.

　　　　　　POP.B　　　R7　　　　; The low byte of the stack is moved to R7,
　　　　　　　　　　　　　　　　 ; the high byte of R7 is 00h

Example　　The contents of the memory pointed to by R7 and the status register are restored from the stack.

　　　　　　POP.B　　　0(R7)　　; The low byte of the stack is moved to the
　　　　　　　　　　　　　　　　 ; the byte which is pointed to by R7
　　　　　　　　　　　　　　　　 : Example:　　R7 = 203h
　　　　　　　　　　　　　　　　 ;　　　　　　　　Mem(R7) = low byte of system stack
　　　　　　　　　　　　　　　　 : Example:　　R7 = 20Ah
　　　　　　　　　　　　　　　　 ;　　　　　　　　Mem(R7) = low byte of system stack
　　　　　　POP　　　　SR　　　　; Last word on stack moved to the SR

Note:　The System Stack Pointer

The system stack pointer (SP) is always incremented by two, independent of the byte suffix.

PUSH[.W]	Push word onto stack
PUSH.B	Push byte onto stack

Syntax   PUSH   src   or   PUSH.W   src
                PUSH.B   src

Operation   SP – 2 → SP
                src → @SP

Description   The stack pointer is decremented by two, then the source operand is moved to the RAM word addressed by the stack pointer (TOS).

Status Bits   Status bits are not affected.

Mode Bits   OSCOFF, CPUOFF, and GIE are not affected.

Example   The contents of the status register and R8 are saved on the stack.

      PUSH   SR   ; save status register
      PUSH   R8   ; save R8

Example   The contents of the peripheral TCDAT is saved on the stack.

      PUSH.B   &TCDAT   ; save data from 8-bit peripheral module,
                             ; address TCDAT, onto stack

Note: The System Stack Pointer

The system stack pointer (SP) is always decremented by two, independent of the byte suffix.

Instruction Set

* RET	Return from subroutine
Syntax	RET
Operation	@SP→ PC SP + 2 → SP
Emulation	MOV @SP+,PC
Description	The return address pushed onto the stack by a CALL instruction is moved to the program counter. The program continues at the code address following the subroutine call.
Status Bits	Status bits are not affected.

RETI	Return from interrupt

Syntax   RETI

Operation
TOS   → SR
SP + 2   → SP
TOS   → PC
SP + 2   → SP

Description   The status register is restored to the value at the beginning of the interrupt service routine by replacing the present SR contents with the TOS contents. The stack pointer (SP) is incremented by two.

The program counter is restored to the value at the beginning of interrupt service. This is the consecutive step after the interrupted program flow. Restoration is performed by replacing the present PC contents with the TOS memory contents. The stack pointer (SP) is incremented.

Status Bits   N: restored from system stack
Z: restored from system stack
C: restored from system stack
V: restored from system stack

Mode Bits   OSCOFF, CPUOFF, and GIE are restored from system stack.

Example   Figure 3–13 illustrates the main program interrupt.

Figure 3–13. Main Program Interrupt

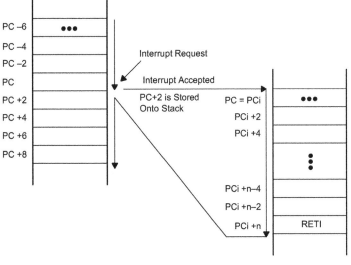

Instruction Set

* RLA[.W]	Rotate left arithmetically
* RLA.B	Rotate left arithmetically

Syntax RLA dst or RLA.W dst
RLA.B dst

Operation C <– MSB <– MSB–1 .... LSB+1 <– LSB <– 0

Emulation ADD dst,dst
ADD.B dst,dst

Description The destination operand is shifted left one position as shown in Figure 3–14. The MSB is shifted into the carry bit (C) and the LSB is filled with 0. The RLA instruction acts as a signed multiplication by 2.

An overflow occurs if dst ≥ 04000h and dst < 0C000h before operation is performed: the result has changed sign.

Figure 3–14. Destination Operand—Arithmetic Shift Left

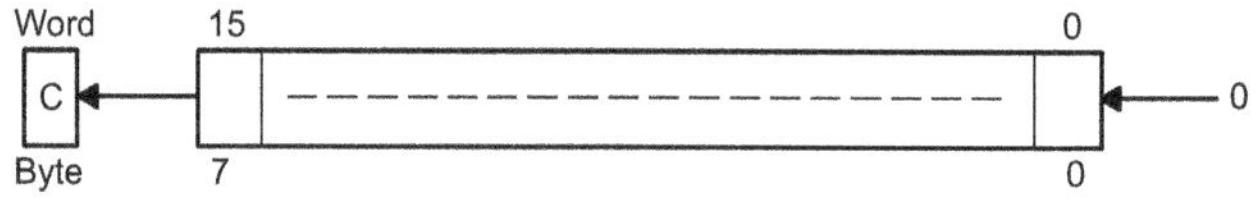

An overflow occurs if dst ≥ 040h and dst < 0C0h before the operation is performed: the result has changed sign.

Status Bits N: Set if result is negative, reset if positive
Z: Set if result is zero, reset otherwise
C: Loaded from the MSB
V: Set if an arithmetic overflow occurs:
the initial value is 04000h ≤ dst < 0C000h; reset otherwise
Set if an arithmetic overflow occurs:
the initial value is 040h ≤ dst < 0C0h; reset otherwise

Mode Bits OSCOFF, CPUOFF, and GIE are not affected.

Example R7 is multiplied by 2.

RLA R7 ; Shift left R7 (× 2)

Example The low byte of R7 is multiplied by 4.

RLA.B R7 ; Shift left low byte of R7 (× 2)
RLA.B R7 ; Shift left low byte of R7 (× 4)

Note: RLA Substitution

The assembler does not recognize the instruction:

RLA @R5+ nor RLA.B @R5+.

It must be substituted by:

ADD @R5+,–2(R5) or ADD.B @R5+,–1(R5).

---

* RLC[.W]	Rotate left through carry
* RLC.B	Rotate left through carry

Syntax   RLC   dst   or   RLC.W   dst
                 RLC.B   dst

Operation   C <– MSB <– MSB–1 .... LSB+1 <– LSB <– C

Emulation   ADDC   dst,dst

Description The destination operand is shifted left one position as shown in Figure 3–15. The carry bit (C) is shifted into the LSB and the MSB is shifted into the carry bit (C).

Figure 3–15. Destination Operand—Carry Left Shift

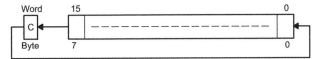

Status Bits N: Set if result is negative, reset if positive
                 Z: Set if result is zero, reset otherwise
                 C: Loaded from the MSB
                 V: Set if an arithmetic overflow occurs
                     the initial value is 04000h $\leq$ dst < 0C000h; reset otherwise
                     Set if an arithmetic overflow occurs:
                     the initial value is 040h $\leq$ dst < 0C0h; reset otherwise

Mode Bits   OSCOFF, CPUOFF, and GIE are not affected.

Example   R5 is shifted left one position.

                 RLC   R5   ; (R5 x 2) + C –> R5

Example   The input P1IN.1 information is shifted into the LSB of R5.

                 BIT.B   #2,&P1IN   ; Information –> Carry
                 RLC   R5   ; Carry=P0in.1 –> LSB of R5

Example   The MEM(LEO) content is shifted left one position.

                 RLC.B   LEO   ; Mem(LEO) x 2 + C –> Mem(LEO)

Note: RLC and RLC.B Substitution

The assembler does not recognize the instruction:
    RLC   @R5+.
It must be substituted by:
    ADDC   @R5+,–2(R5).

Instruction Set

RRA[.W]   Rotate right arithmetically
RRA.B   Rotate right arithmetically

Syntax   RRA   dst   or   RRA.W   dst
          RRA.B   dst

Operation   MSB –> MSB, MSB –> MSB–1, ... LSB+1 –> LSB,   LSB –> C

Description   The destination operand is shifted right one position as shown in Figure 3–16. The MSB is shifted into the MSB, the MSB is shifted into the MSB–1, and the LSB+1 is shifted into the LSB.

Figure 3–16. Destination Operand—Arithmetic Right Shift

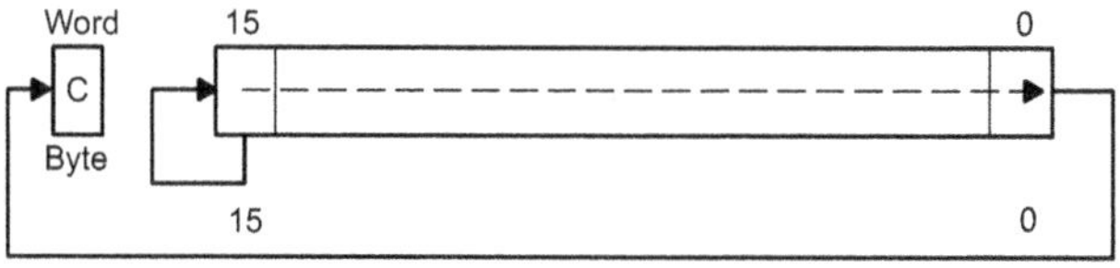

Status Bits   N: Set if result is negative, reset if positive
          Z: Set if result is zero, reset otherwise
          C: Loaded from the LSB
          V: Reset

Mode Bits   OSCOFF, CPUOFF, and GIE are not affected.

Example   R5 is shifted right one position. The MSB retains the old value. It operates equal to an arithmetic division by 2.

          RRA   R5   ; R5/2 –> R5

;   The value in R5 is multiplied by 0.75 (0.5 + 0.25).
;
;
          PUSH   R5   ; Hold R5 temporarily using stack
          RRA   R5   ; R5 × 0.5 –> R5
          ADD   @SP+,R5   ; R5 × 0.5 + R5 = 1.5 × R5 –> R5
          RRA   R5   ; (1.5 × R5) × 0.5 = 0.75 × R5 –> R5
          ......

Example   The low byte of R5 is shifted right one position. The MSB retains the old value. It operates equal to an arithmetic division by 2.

          RRA.B   R5   ; R5/2 –> R5: operation is on low byte only
                    ; High byte of R5 is reset
          PUSH.B   R5   ; R5 × 0.5 –> TOS
          RRA.B   @SP   ; TOS × 0.5 = 0.5 × R5 × 0.5 = 0.25 × R5 –> TOS
          ADD.B   @SP+,R5   ; R5 × 0.5 + R5 × 0.25 = 0.75 × R5 –> R5
          ......

| **RRC[.W]** | Rotate right through carry |
| **RRC.B** | Rotate right through carry |

Syntax   RRC   dst   or   RRC.W   dst
                RRC   dst

Operation   C –> MSB –> MSB–1 .... LSB+1 –> LSB –> C

Description   The destination operand is shifted right one position as shown in Figure 3–17.
The carry bit (C) is shifted into the MSB, the LSB is shifted into the carry bit (C).

Figure 3–17. Destination Operand—Carry Right Shift

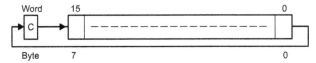

Status Bits   N: Set if result is negative, reset if positive
Z: Set if result is zero, reset otherwise
C: Loaded from the LSB
V: Set if initial destination is positive and initial carry is set, otherwise reset

Mode Bits   OSCOFF, CPUOFF, and GIE are not affected.

Example   R5 is shifted right one position. The MSB is loaded with 1.

      SETC   ; Prepare carry for MSB
      RRC   R5   ; R5/2 + 8000h –> R5

Example   R5 is shifted right one position. The MSB is loaded with 1.

      SETC   ; Prepare carry for MSB
      RRC.B   R5   ; R5/2 + 80h –> R5; low byte of R5 is used

Instruction Set

* SBC[.W]	Subtract source and borrow/.NOT. carry from destination
* SBC.B	Subtract source and borrow/.NOT. carry from destination

Syntax SBC dst or SBC.W dst
SBC.B dst

Operation dst + 0FFFFh + C –> dst
dst + 0FFh + C –> dst

Emulation SUBC #0,dst
SUBC.B #0,dst

Description The carry bit (C) is added to the destination operand minus one. The previous contents of the destination are lost.

Status Bits N: Set if result is negative, reset if positive
Z: Set if result is zero, reset otherwise
C: Set if there is a carry from the MSB of the result, reset otherwise.
  Set to 1 if no borrow, reset if borrow.
V: Set if an arithmetic overflow occurs, reset otherwise.

Mode Bits OSCOFF, CPUOFF, and GIE are not affected.

Example The 16-bit counter pointed to by R13 is subtracted from a 32-bit counter pointed to by R12.

SUB @R13,0(R12) ; Subtract LSDs
SBC 2(R12) ; Subtract carry from MSD

Example The 8-bit counter pointed to by R13 is subtracted from a 16-bit counter pointed to by R12.

SUB.B @R13,0(R12) ; Subtract LSDs
SBC.B 1(R12) ; Subtract carry from MSD

Note: Borrow Implementation.

The borrow is treated as a .NOT. carry :

	Borrow	Carry bit
	Yes	0
	No	1

RISC 16–Bit CPU

* SETC	Set carry bit
Syntax	SETC
Operation	1 –> C
Emulation	BIS #1,SR
Description	The carry bit (C) is set.
Status Bits	N: Not affected
	Z: Not affected
	C: Set
	V: Not affected
Mode Bits	OSCOFF, CPUOFF, and GIE are not affected.
Example	Emulation of the decimal subtraction:
	Subtract R5 from R6 decimally
	Assume that R5 = 03987h and R6 = 04137h

DSUB	ADD	#06666h,R5	; Move content R5 from 0–9 to 6–0Fh
			; R5 = 03987h + 06666h = 09FEDh
	INV	R5	; Invert this (result back to 0–9)
			; R5 = .NOT. R5 = 06012h
	SETC		; Prepare carry = 1
	DADD	R5,R6	; Emulate subtraction by addition of:
			; (010000h – R5 – 1)
			; R6 = R6 + R5 + 1
			; R6 = 0150h

* SETN	Set negative bit
Syntax	SETN
Operation	1 –> N
Emulation	BIS #4,SR
Description	The negative bit (N) is set.
Status Bits	N: Set
	Z: Not affected
	C: Not affected
	V: Not affected
Mode Bits	OSCOFF, CPUOFF, and GIE are not affected.

* SETZ	Set zero bit
Syntax	SETZ
Operation	1 –> Z
Emulation	BIS #2,SR
Description	The zero bit (Z) is set.
Status Bits	N: Not affected Z: Set C: Not affected V: Not affected
Mode Bits	OSCOFF, CPUOFF, and GIE are not affected.

Instruction Set

SUB[.W]   Subtract source from destination
SUB.B   Subtract source from destination

Syntax   SUB   src,dst   or   SUB.W   src,dst
                  SUB.B   src,dst

Operation   dst + .NOT.src + 1 –> dst
               or
               [(dst – src –> dst)]

Description   The source operand is subtracted from the destination operand by adding the source operand's 1s complement and the constant 1. The source operand is not affected. The previous contents of the destination are lost.

Status Bits   N: Set if result is negative, reset if positive
               Z: Set if result is zero, reset otherwise
               C: Set if there is a carry from the MSB of the result, reset otherwise.
                   Set to 1 if no borrow, reset if borrow.
               V: Set if an arithmetic overflow occurs, otherwise reset

Mode Bits   OSCOFF, CPUOFF, and GIE are not affected.

Example   See example at the SBC instruction.

Example   See example at the SBC.B instruction.

Note: Borrow Is Treated as a .NOT.

The borrow is treated as a .NOT. carry :

	Borrow	Carry bit
	Yes	0
	No	1

SUBC[.W]SBB[.W]	Subtract source and borrow/.NOT. carry from destination
SUBC.B,SBB.B	Subtract source and borrow/.NOT. carry from destination

Syntax
    SUBC   src,dst   or   SUBC.W   src,dst   or
    SBB   src,dst   or   SBB.W   src,dst
    SUBC.B src,dst   or   SBB.B   src,dst

Operation
dst + .NOT.src + C –> dst
or
(dst – src – 1 + C –> dst)

Description
The source operand is subtracted from the destination operand by adding the source operand's 1s complement and the carry bit (C). The source operand is not affected. The previous contents of the destination are lost.

Status Bits
N: Set if result is negative, reset if positive.
Z: Set if result is zero, reset otherwise.
C: Set if there is a carry from the MSB of the result, reset otherwise.
    Set to 1 if no borrow, reset if borrow.
V: Set if an arithmetic overflow occurs, reset otherwise.

Mode Bits
OSCOFF, CPUOFF, and GIE are not affected.

Example
Two floating point mantissas (24 bits) are subtracted.
LSBs are in R13 and R10, MSBs are in R12 and R9.

    SUB.W   R13,R10 ; 16-bit part, LSBs
    SUBC.B R12,R9   ; 8-bit part, MSBs

Example
The 16-bit counter pointed to by R13 is subtracted from a 16-bit counter in R10 and R11(MSD).

    SUB.B   @R13+,R10   ; Subtract LSDs without carry
    SUBC.B @R13,R11   ; Subtract MSDs with carry
    ...   ; resulting from the LSDs

Note: **Borrow Implementation**

The borrow is treated as a .NOT. carry :	Borrow	Carry bit
	Yes	0
	No	1

Instruction Set

SWPB	Swap bytes
Syntax	SWPB dst
Operation	Bits 15 to 8 <–> bits 7 to 0
Description	The destination operand high and low bytes are exchanged as shown in Figure 3–18.
Status Bits	Status bits are not affected.
Mode Bits	OSCOFF, CPUOFF, and GIE are not affected.

Figure 3–18. Destination Operand Byte Swap

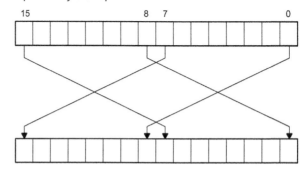

Example

```
MOV     #040BFh,R7      ; 0100000010111111 -> R7
SWPB    R7             ; 1011111101000000 in R7
```

Example   The value in R5 is multiplied by 256. The result is stored in R5,R4.

```
SWPB    R5             ;
MOV     R5,R4          ;Copy the swapped value to R4
BIC     #0FF00h,R5     ;Correct the result
BIC     #00FFh,R4      ;Correct the result
```

SXT   Extend Sign

Syntax   SXT   dst

Operation   Bit 7 –> Bit 8 .... Bit 15

Description   The sign of the low byte is extended into the high byte as shown in Figure 3–19.

Status Bits   N: Set if result is negative, reset if positive
                      Z: Set if result is zero, reset otherwise
                      C: Set if result is not zero, reset otherwise (.NOT. Zero)
                      V: Reset

Mode Bits   OSCOFF, CPUOFF, and GIE are not affected.

Figure 3–19. Destination Operand Sign Extension

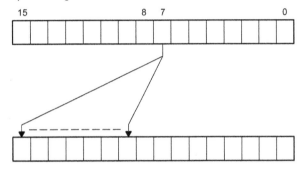

Example   R7 is loaded with the P1IN value. The operation of the sign-extend instruction
                      expands bit 8 to bit 15 with the value of bit 7.
                      R7 is then added to R6.

```
MOV.B    &P1IN,R7      ; P1IN = 080h:     . . . . . . . . 1000 0000
SXT      R7            ; R7 = 0FF80h:     1111 1111 1000 0000
```

| *** TST[.W]** | Test destination |
| *** TST.B** | Test destination |

Syntax   TST   dst   or   TST.W dst
               TST.B   dst

Operation   dst + 0FFFFh + 1
               dst + 0FFh + 1

Emulation   CMP   #0,dst
               CMP.B   #0,dst

Description The destination operand is compared with zero. The status bits are set according to the result. The destination is not affected.

Status Bits N: Set if destination is negative, reset if positive
               Z: Set if destination contains zero, reset otherwise
               C: Set
               V: Reset

Mode Bits   OSCOFF, CPUOFF, and GIE are not affected.

Example   R7 is tested. If it is negative, continue at R7NEG; if it is positive but not zero, continue at R7POS.

```
                      TST      R7          ; Test R7
                      JN       R7NEG       ; R7 is negative
                      JZ       R7ZERO      ; R7 is zero
            R7POS     ......               ; R7 is positive but not zero
            R7NEG     ......               ; R7 is negative
            R7ZERO    ......               ; R7 is zero
```

Example   The low byte of R7 is tested. If it is negative, continue at R7NEG; if it is positive but not zero, continue at R7POS.

```
                      TST.B    R7          ; Test low byte of R7
                      JN       R7NEG       ; Low byte of R7 is negative
                      JZ       R7ZERO      ; Low byte of R7 is zero
            R7POS     ......               ; Low byte of R7 is positive but not zero
            R7NEG     .....                ; Low byte of R7 is negative
            R7ZERO    ......               ; Low byte of R7 is zero
```

XOR[.W]	Exclusive OR of source with destination
XOR.B	Exclusive OR of source with destination

Syntax   XOR   src,dst   or   XOR.W   src,dst
               XOR.B   src,dst

Operation   src .XOR. dst –> dst

Description   The source and destination operands are exclusive ORed. The result is placed into the destination. The source operand is not affected.

Status Bits   N: Set if result MSB is set, reset if not set
                    Z: Set if result is zero, reset otherwise
                    C: Set if result is not zero, reset otherwise (= .NOT. Zero)
                    V: Set if both operands are negative

Mode Bits   OSCOFF, CPUOFF, and GIE are not affected.

Example   The bits set in R6 toggle the bits in the RAM word TONI.

          XOR   R6,TONI   ; Toggle bits of word TONI on the bits set in R6

Example   The bits set in R6 toggle the bits in the RAM byte TONI.

          XOR.B   R6,TONI   ; Toggle bits of byte TONI on the bits set in
                                          ; low byte of R6

Example   Reset to 0 those bits in low byte of R7 that are different from bits in RAM byte EDE.

          XOR.B   EDE,R7   ; Set different bit to "1s"
          INV.B   R7   ; Invert Lowbyte, Highbyte is 0h

3.4.4 Instruction Cycles and Lengths

The number of CPU clock cycles required for an instruction depends on the instruction format and the addressing modes used - not the instruction itself. The number of clock cycles refers to the MCLK.

Interrupt and Reset Cycles

Table 3–14 lists the CPU cycles for interrupt overhead and reset.

Table 3–14. Interrupt and Reset Cycles

Action	No. of Cycles	Length of Instruction
Return from interrupt (RETI)	5	1
Interrupt accepted	6	–
WDT reset	4	–
Reset (RST/NMI)	4	–

Format-II (Single Operand) Instruction Cycles and Lengths

Table 3–15 lists the length and CPU cycles for all addressing modes of format-II instructions.

Table 3–15. Format-II Instruction Cycles and Lengths

Addressing Mode	No. of Cycles RRA, RRC SWPB, SXT	PUSH	CALL	Length of Instruction	Example
Rn	1	3	4	1	SWPB R5
@Rn	3	4	4	1	RRC @R9
@Rn+	3	4	5	1	SWPB @R10+
#N	(See note)	4	5	2	CALL #81H
X(Rn)	4	5	5	2	CALL 2(R7)
EDE	4	5	5	2	PUSH EDE
&EDE	4	5	5	2	SXT &EDE

Note: Instruction Format II Immediate Mode

Do not use instructions RRA, RRC, SWPB, and SXT with the immediate mode in the destination field. Use of these in the immediate mode results in an unpredictable program operation.

Format-III (Jump) Instruction Cycles and Lengths

All jump instructions require one code word, and take two CPU cycles to execute, regardless of whether the jump is taken or not.

Format-I (Double Operand) Instruction Cycles and Lengths

Table 3–16 lists the length and CPU cycles for all addressing modes of format-I instructions.

Table 3–16. Format 1 Instruction Cycles and Lengths

Addressing Mode		No. of Cycles	Length of Instruction	Example	
Src	Dst				
Rn	Rm	1	1	MOV	R5,R8
	PC	2	1	BR	R9
	x(Rm)	4	2	ADD	R5,3(R6)
	EDE	4	2	XOR	R8,EDE
	&EDE	4	2	MOV	R5,&EDE
@Rn	Rm	2	1	AND	@R4,R5
	PC	3	1	BR	@R8
	x(Rm)	5	2	XOR	@R5,8(R6)
	EDE	5	2	MOV	@R5,EDE
	&EDE	5	2	XOR	@R5,&EDE
@Rn+	Rm	2	1	ADD	@R5+,R6
	PC	3	1	BR	@R9+
	x(Rm)	5	2	XOR	@R5,8(R6)
	EDE	5	2	MOV	@R9+,EDE
	&EDE	5	2	MOV	@R9+,&EDE
#N	Rm	2	2	MOV	#20,R9
	PC	3	2	BR	#2AEh
	x(Rm)	5	3	MOV	#0300h,0(SP)
	EDE	5	3	ADD	#33,EDE
	&EDE	5	3	ADD	#33,&EDE
x(Rn)	Rm	3	2	MOV	2(R5),R7
	PC	3	2	BR	2(R6)
	TONI	6	3	MOV	4(R7),TONI
	x(Rm)	6	3	ADD	3(R4),6(R9)
	&TONI	6	3	MOV	3(R4),&TONI
EDE	Rm	3	2	AND	EDE,R6
	PC	3	2	BR	EDE
	TONI	6	3	CMP	EDE,TONI
	x(Rm)	6	3	MOV	EDE,0(SP)
	&TONI	6	3	MOV	EDE,&TONI
&EDE	Rm	3	2	MOV	&EDE,R8
	PC	3	2	BRA	&EDE
	TONI	6	3	MOV	&EDE,TONI
	x(Rm)	6	3	MOV	&EDE,0(SP)
	&TONI	6	3	MOV	&EDE,&TONI

Instruction Set

3.4.5 Instruction Set Description

The instruction map is shown in Figure 3–20 and the complete instruction set is summarized in Table 3–17.

Figure 3–20. Core Instruction Map

	000	040	080	0C0	100	140	180	1C0	200	240	280	2C0	300	340	380	3C0
0xxx																
4xxx																
8xxx																
Cxxx																
1xxx	RRC	RRC.B	SWPB		RRA	RRA.B	SXT		PUSH	PUSH.B	CALL		RETI			
14xx																
18xx																
1Cxx																
20xx	JNE/JNZ															
24xx	JEQ/JZ															
28xx	JNC															
2Cxx	JC															
30xx	JN															
34xx	JGE															
38xx	JL															
3Cxx	JMP															
4xxx	MOV, MOV.B															
5xxx	ADD, ADD.B															
6xxx	ADDC, ADDC.B															
7xxx	SUBC, SUBC.B															
8xxx	SUB, SUB.B															
9xxx	CMP, CMP.B															
Axxx	DADD, DADD.B															
Bxxx	BIT, BIT.B															
Cxxx	BIC, BIC.B															
Dxxx	BIS, BIS.B															
Exxx	XOR, XOR.B															
Fxxx	AND, AND.B															

Table 3–17. MSP430 Instruction Set

Mnemonic		Description		V	N	Z	C
ADC(.B)†	dst	Add C to destination	dst + C → dst	*	*	*	*
ADD(.B)	src,dst	Add source to destination	src + dst → dst	*	*	*	*
ADDC(.B)	src,dst	Add source and C to destination	src + dst + C → dst	*	*	*	*
AND(.B)	src,dst	AND source and destination	src .and. dst → dst	0	*	*	*
BIC(.B)	src,dst	Clear bits in destination	.not.src .and. dst → dst	–	–	–	–
BIS(.B)	src,dst	Set bits in destination	src .or. dst → dst	–	–	–	–
BIT(.B)	src,dst	Test bits in destination	src .and. dst	0	*	*	*
BR†	dst	Branch to destination	dst → PC	–	–	–	–
CALL	dst	Call destination	PC+2 → stack, dst → PC	–	–	–	–
CLR(.B)†	dst	Clear destination	0 → dst	–	–	–	–
CLRC†		Clear C	0 → C	–	–	–	0
CLRN†		Clear N	0 → N	–	0	–	–
CLRZ†		Clear Z	0 → Z	–	–	0	–
CMP(.B)†	src,dst	Compare source and destination	dst – src	*	*	*	*
DADC(.B)†	dst	Add C decimally to destination	dst + C → dst (decimally)	*	*	*	*
DADD(.B)	src,dst	Add source and C decimally to dst.	src + dst + C → dst (decimally)	*	*	*	*
DEC(.B)†	dst	Decrement destination	dst – 1 → dst	*	*	*	*
DECD(.B)†	dst	Double-decrement destination	dst – 2 → dst	*	*	*	*
DINT†		Disable interrupts	0 → GIE	–	–	–	–
EINT†		Enable interrupts	1 → GIE	–	–	–	–
INC(.B)†	dst	Increment destination	dst +1 → dst	*	*	*	*
INCD(.B)†	dst	Double-increment destination	dst+2 → dst	*	*	*	*
INV(.B)†	dst	Invert destination	.not.dst → dst	*	*	*	*
JC/JHS	label	Jump if C set/Jump if higher or same		–	–	–	–
JEQ/JZ	label	Jump if equal/Jump if Z set		–	–	–	–
JGE	label	Jump if greater or equal		–	–	–	–
JL	label	Jump if less		–	–	–	–
JMP	label	Jump	PC + 2 x offset → PC	–	–	–	–
JN	label	Jump if N set		–	–	–	–
JNC/JLO	label	Jump if C not set/Jump if lower		–	–	–	–
JNE/JNZ	label	Jump if not equal/Jump if Z not set		–	–	–	–
MOV(.B)	src,dst	Move source to destination	src → dst	–	–	–	–
NOP†		No operation		–	–	–	–
POP(.B)†	dst	Pop item from stack to destination	@SP → dst, SP+2 → SP	–	–	–	–
PUSH(.B)	src	Push source onto stack	SP – 2 → SP, src → @SP	–	–	–	–
RET†		Return from subroutine	@SP → PC, SP + 2 → SP	–	–	–	–
RETI		Return from interrupt		*	*	*	*
RLA(.B)†	dst	Rotate left arithmetically		*	*	*	*
RLC(.B)†	dst	Rotate left through C		*	*	*	*
RRA(.B)	dst	Rotate right arithmetically		0	*	*	*
RRC(.B)	dst	Rotate right through C		*	*	*	*
SBC(.B)†	dst	Subtract not(C) from destination	dst + 0FFFFh + C → dst	*	*	*	*
SETC†		Set C	1 → C	–	–	–	1
SET†		Set N	1 → N	–	1	–	–
SETZ†		Set Z	1 → C	–	–	1	–
SUB(.B)	src,dst	Subtract source from destination	dst + .not.src + 1 → dst	*	*	*	*
SUBC(.B)	src,dst	Subtract source and not(C) from dst.	dst + .not.src + C → dst	*	*	*	*
SWPB	dst	Swap bytes		–	–	–	–
SXT	dst	Extend sign		0	*	*	*
TST(.B)†	dst	Test destination	dst + 0FFFFh + 1	0	*	*	1
XOR(.B)	src,dst	Exclusive OR source and destination	src .xor. dst → dst	*	*	*	*

† Emulated Instruction

Standard Register and Bit Definitions for the MSP430 Microcontrollers

This reference list of standard register and bit definitions for the MSP430 microcontrollers is the basis of much of the syntaxic substitution by the assembler when assembling and compiling an assembly-language program. It is also very useful when developing programs for the MSP430 family of microcontrollers using the C language.

```
/********************************************************************
*
* Standard register and bit definitions for the Texas Instruments
* MSP430 microcontroller.
*
* This file supports assembler and C development for
* MSP430x12x devices.
*
* Texas Instruments, Version 2.1
*
* Rev. 1.1, Corrected LPMx_EXIT to reference new intrinsic _BIC_SR_IRQ
*           Changed TAIV to be read-only
*
* Rev. 1.2, Enclose all #define statements with parentheses
*
* Rev. 1.3, Defined vectors for USART (in addition to UART)
*
* Rev. 1.4, Added USART special function labels (UxME, UxIE, UxIFG)
*
* Rev. 2.1, Alignment of defintions in Users Guide and of version numbers
*
********************************************************************/

#ifndef __msp430x12x
#define __msp430x12x

#if (((__TID__ >> 8) & 0x7F) != 0x2b)      /* 0x2b = 43 dec */
#error MSP430X44X.H file for use with ICC430/A430 only
#endif


#ifdef __IAR_SYSTEMS_ICC__
#include <in430.h>
#pragma language=extended

#define DEFC(name, address) __no_init volatile unsigned char name @ address;
#define DEFW(name, address) __no_init volatile unsigned short name @ address;
```

```
#endif  /* __IAR_SYSTEMS_ICC__  */


#ifdef __IAR_SYSTEMS_ASM__
#define DEFC(name, address) sfrb name = address;
#define DEFW(name, address) sfrw name = address;

#endif /* __IAR_SYSTEMS_ASM__*/

#ifdef __cplusplus
#define READ_ONLY
#else
#define READ_ONLY const
#endif

/***********************************************************
* STANDARD BITS
***********************************************************/

#define BIT0                (0x0001)
#define BIT1                (0x0002)
#define BIT2                (0x0004)
#define BIT3                (0x0008)
#define BIT4                (0x0010)
#define BIT5                (0x0020)
#define BIT6                (0x0040)
#define BIT7                (0x0080)
#define BIT8                (0x0100)
#define BIT9                (0x0200)
#define BITA                (0x0400)
#define BITB                (0x0800)
#define BITC                (0x1000)
#define BITD                (0x2000)
#define BITE                (0x4000)
#define BITF                (0x8000)

/***********************************************************
* STATUS REGISTER BITS
***********************************************************/

#define C                   (0x0001)
#define Z                   (0x0002)
#define N                   (0x0004)
#define V                   (0x0100)
#define GIE                 (0x0008)
#define CPUOFF              (0x0010)
#define OSCOFF              (0x0020)
#define SCG0                (0x0040)
#define SCG1                (0x0080)

/* Low Power Modes coded with Bits 4-7 in SR */
```

```
#ifndef __IAR_SYSTEMS_ICC /* Begin #defines for assembler */
#define LPM0                (CPUOFF)
#define LPM1                (SCG0+CPUOFF)
#define LPM2                (SCG1+CPUOFF)
#define LPM3                (SCG1+SCG0+CPUOFF)
#define LPM4                (SCG1+SCG0+OSCOFF+CPUOFF)
/* End #defines for assembler */

#else /* Begin #defines for C */
#define LPM0_bits           (CPUOFF)
#define LPM1_bits           (SCG0+CPUOFF)
#define LPM2_bits           (SCG1+CPUOFF)
#define LPM3_bits           (SCG1+SCG0+CPUOFF)
#define LPM4_bits           (SCG1+SCG0+OSCOFF+CPUOFF)

#include <In430.h>

#define LPM0      _BIS_SR(LPM0_bits)      /* Enter Low Power Mode 0 */
#define LPM0_EXIT _BIC_SR_IRQ(LPM0_bits) /* Exit Low Power Mode 0 */
#define LPM1      _BIS_SR(LPM1_bits)      /* Enter Low Power Mode 1 */
#define LPM1_EXIT _BIC_SR_IRQ(LPM1_bits) /* Exit Low Power Mode 1 */
#define LPM2      _BIS_SR(LPM2_bits)      /* Enter Low Power Mode 2 */
#define LPM2_EXIT _BIC_SR_IRQ(LPM2_bits) /* Exit Low Power Mode 2 */
#define LPM3      _BIS_SR(LPM3_bits)      /* Enter Low Power Mode 3 */
#define LPM3_EXIT _BIC_SR_IRQ(LPM3_bits) /* Exit Low Power Mode 3 */
#define LPM4      _BIS_SR(LPM4_bits)      /* Enter Low Power Mode 4 */
#define LPM4_EXIT _BIC_SR_IRQ(LPM4_bits) /* Exit Low Power Mode 4 */
#endif /* End #defines for C */

/*************************************************************
* PERIPHERAL FILE MAP
*************************************************************/

/*************************************************************
* SPECIAL FUNCTION REGISTER ADDRESSES + CONTROL BITS
*************************************************************/

#define IE1_                (0x0000)  /* Interrupt Enable 1 */
DEFC( IE1                , IE1_)
#define WDTIE               (0x01)
#define OFIE                (0x02)
#define NMIIE               (0x10)
#define ACCVIE              (0x20)

#define IFG1_               (0x0002)  /* Interrupt Flag 1 */
DEFC( IFG1               , IFG1_)
#define WDTIFG              (0x01)
#define OFIFG               (0x02)
#define NMIIFG              (0x10)

#define IE2_                (0x0001)  /* Interrupt Enable 2 */
DEFC( IE2                , IE2_)
```

```
#define U0IE                  IE2        /* UART0 Interrupt Enable Register */
#define URXIE0                (0x01)
#define UTXIE0                (0x02)


#define IFG2_                 (0x0003)   /* Interrupt Flag 2 */
DEFC( IFG2             , IFG2_)
#define U0IFG                 IFG2       /* UART0 Interrupt Flag Register */
#define URXIFG0               (0x01)
#define UTXIFG0               (0x02)


#define ME2_                  (0x0005)   /* Module Enable 2 */
DEFC( ME2              , ME2_)
#define U0ME                  ME2        /* UART0 Module Enable Register */
#define URXE0                 (0x01)
#define USPIE0                (0x01)
#define UTXE0                 (0x02)


/**************************************************************
* WATCHDOG TIMER
**************************************************************/

#define WDTCTL_               (0x0120)   /* Watchdog Timer Control */
DEFW( WDTCTL           , WDTCTL_)
/* The bit names have been prefixed with "WDT" */
#define WDTIS0                (0x0001)
#define WDTIS1                (0x0002)
#define WDTSSEL               (0x0004)
#define WDTCNTCL              (0x0008)
#define WDTTMSEL              (0x0010)
#define WDTNMI                (0x0020)
#define WDTNMIES              (0x0040)
#define WDTHOLD               (0x0080)


#define WDTPW                 (0x5A00)


/* WDT-interval times [1ms] coded with Bits 0-2 */
/* WDT is clocked by fMCLK (assumed 1MHz) */
#define WDT_MDLY_32           (WDTPW+WDTTMSEL+WDTCNTCL)                    /*
32ms interval (default) */
#define WDT_MDLY_8            (WDTPW+WDTTMSEL+WDTCNTCL+WDTIS0)             /*
8ms      " */
#define WDT_MDLY_0_5          (WDTPW+WDTTMSEL+WDTCNTCL+WDTIS1)            /*
0.5ms   " */
#define WDT_MDLY_0_064        (WDTPW+WDTTMSEL+WDTCNTCL+WDTIS1+WDTIS0)     /*
0.064ms   " */
/* WDT is clocked by fACLK (assumed 32KHz) */
#define WDT_ADLY_1000         (WDTPW+WDTTMSEL+WDTCNTCL+WDTSSEL)           /*
1000ms " */
#define WDT_ADLY_250          (WDTPW+WDTTMSEL+WDTCNTCL+WDTSSEL+WDTIS0)    /*
250ms   " */
#define WDT_ADLY_16           (WDTPW+WDTTMSEL+WDTCNTCL+WDTSSEL+WDTIS1)    /*
16ms    " */
```

```
#define WDT_ADLY_1_9          (WDTPW+WDTTMSEL+WDTCNTCL+WDTSSEL+WDTIS1+WDTIS0)    /*
1.9ms  " */
/* Watchdog mode -> reset after expired time */
/* WDT is clocked by fMCLK (assumed 1MHz) */
#define WDT_MRST_32           (WDTPW+WDTCNTCL)                                   /*
32ms interval (default) */
#define WDT_MRST_8            (WDTPW+WDTCNTCL+WDTIS0)                            /*
8ms    " */
#define WDT_MRST_0_5          (WDTPW+WDTCNTCL+WDTIS1)                            /*
0.5ms  " */
#define WDT_MRST_0_064        (WDTPW+WDTCNTCL+WDTIS1+WDTIS0)                     /*
0.064ms   " */
/* WDT is clocked by fACLK (assumed 32KHz) */
#define WDT_ARST_1000         (WDTPW+WDTCNTCL+WDTSSEL)                           /*
1000ms    " */
#define WDT_ARST_250          (WDTPW+WDTCNTCL+WDTSSEL+WDTIS0)                    /*
250ms     " */
#define WDT_ARST_16           (WDTPW+WDTCNTCL+WDTSSEL+WDTIS1)                    /*
16ms      " */
#define WDT_ARST_1_9          (WDTPW+WDTCNTCL+WDTSSEL+WDTIS1+WDTIS0)             /*
1.9ms     " */


/* INTERRUPT CONTROL */
/* These two bits are defined in the Special Function Registers */
/* #define WDTIE              0x01 */
/* #define WDTIFG             0x01 */


/*************************************************************
* DIGITAL I/O Port1/2
*************************************************************/

#define P1IN_                 (0x0020)  /* Port 1 Input */
READ_ONLY DEFC( P1IN          , P1IN_)
#define P1OUT_                (0x0021)  /* Port 1 Output */
DEFC( P1OUT                   , P1OUT_)
#define P1DIR_                (0x0022)  /* Port 1 Direction */
DEFC( P1DIR                   , P1DIR_)
#define P1IFG_                (0x0023)  /* Port 1 Interrupt Flag */
DEFC( P1IFG                   , P1IFG_)
#define P1IES_                (0x0024)  /* Port 1 Interrupt Edge Select */
DEFC( P1IES                   , P1IES_)
#define P1IE_                 (0x0025)  /* Port 1 Interrupt Enable */
DEFC( P1IE                    , P1IE_)
#define P1SEL_                (0x0026)  /* Port 1 Selection */
DEFC( P1SEL                   , P1SEL_)

#define P2IN_                 (0x0028)  /* Port 2 Input */
READ_ONLY DEFC( P2IN          , P2IN_)
#define P2OUT_                (0x0029)  /* Port 2 Output */
DEFC( P2OUT                   , P2OUT_)
#define P2DIR_                (0x002A)  /* Port 2 Direction */
DEFC( P2DIR                   , P2DIR_)
```

```
#define P2IFG_              (0x002B)  /* Port 2 Interrupt Flag */
DEFC( P2IFG             , P2IFG_)
#define P2IES_              (0x002C)  /* Port 2 Interrupt Edge Select */
DEFC( P2IES             , P2IES_)
#define P2IE_               (0x002D)  /* Port 2 Interrupt Enable */
DEFC( P2IE              , P2IE_)
#define P2SEL_              (0x002E)  /* Port 2 Selection */
DEFC( P2SEL             , P2SEL_)


/*************************************************************
* DIGITAL I/O Port3
*************************************************************/

#define P3IN_               (0x0018)  /* Port 3 Input */
READ_ONLY DEFC( P3IN        , P3IN_)
#define P3OUT_              (0x0019)  /* Port 3 Output */
DEFC( P3OUT             , P3OUT_)
#define P3DIR_              (0x001A)  /* Port 3 Direction */
DEFC( P3DIR             , P3DIR_)
#define P3SEL_              (0x001B)  /* Port 3 Selection */
DEFC( P3SEL             , P3SEL_)


/*************************************************************
* USART
*************************************************************/

/* UxCTL */
#define PENA                (0x80)        /* Parity enable */
#define PEV                 (0x40)        /* Parity 0:odd / 1:even */
#define SPB                 (0x20)        /* Stop Bits 0:one / 1: two */
#define CHAR                (0x10)        /* Data 0:7-bits / 1:8-bits */
#define LISTEN              (0x08)        /* Listen mode */
#define SYNC                (0x04)        /* UART / SPI mode */
#define MM                  (0x02)        /* Master Mode off/on */
#define SWRST               (0x01)        /* USART Software Reset */

/* UxTCTL */
#define CKPH                (0x80)        /* SPI: Clock Phase */
#define CKPL                (0x40)        /* Clock Polarity */
#define SSEL1               (0x20)        /* Clock Source Select 1 */
#define SSEL0               (0x10)        /* Clock Source Select 0 */
#define URXSE               (0x08)        /* Receive Start edge select */
#define TXWAKE              (0x04)        /* TX Wake up mode */
#define STC                 (0x02)        /* SPI: STC enable 0:on / 1:off */
#define TXEPT               (0x01)        /* TX Buffer empty */

/* UxRCTL */
#define FE                  (0x80)        /* Frame Error */
#define PE                  (0x40)        /* Parity Error */
#define OE                  (0x20)        /* Overrun Error */
#define BRK                 (0x10)        /* Break detected */
#define URXEIE              (0x08)        /* RX Error interrupt enable */
```

```
#define URXWIE                (0x04)        /* RX Wake up interrupt enable */
#define RXWAKE                (0x02)        /* RX Wake up detect */
#define RXERR                 (0x01)        /* RX Error Error */


/***************************************************************
* USART 0
***************************************************************/

#define U0CTL_                (0x0070)  /* USART 0 Control */
DEFC( U0CTL            , U0CTL_)
#define U0TCTL_               (0x0071)  /* USART 0 Transmit Control */
DEFC( U0TCTL           , U0TCTL_)
#define U0RCTL_               (0x0072)  /* USART 0 Receive Control */
DEFC( U0RCTL           , U0RCTL_)
#define U0MCTL_               (0x0073)  /* USART 0 Modulation Control */
DEFC( U0MCTL           , U0MCTL_)
#define U0BR0_                (0x0074)  /* USART 0 Baud Rate 0 */
DEFC( U0BR0            , U0BR0_)
#define U0BR1_                (0x0075)  /* USART 0 Baud Rate 1 */
DEFC( U0BR1            , U0BR1_)
#define U0RXBUF_              (0x0076)  /* USART 0 Receive Buffer */
READ_ONLY DEFC( U0RXBUF       ,    U0RXBUF_)
#define U0TXBUF_              (0x0077)  /* USART 0 Transmit Buffer */
DEFC( U0TXBUF          , U0TXBUF_)


/* Alternate register names */

#define UCTL0                 U0CTL      /* USART 0 Control */
#define UTCTL0                U0TCTL     /* USART 0 Transmit Control */
#define URCTL0                U0RCTL     /* USART 0 Receive Control */
#define UMCTL0                U0MCTL     /* USART 0 Modulation Control */
#define UBR00                 U0BR0      /* USART 0 Baud Rate 0 */
#define UBR10                 U0BR1      /* USART 0 Baud Rate 1 */
#define RXBUF0                U0RXBUF    /* USART 0 Receive Buffer */
#define TXBUF0                U0TXBUF    /* USART 0 Transmit Buffer */
#define UCTL0_                U0CTL_     /* USART 0 Control */
#define UTCTL0_               U0TCTL_    /* USART 0 Transmit Control */
#define URCTL0_               U0RCTL_    /* USART 0 Receive Control */
#define UMCTL0_               U0MCTL_    /* USART 0 Modulation Control */
#define UBR00_                U0BR0_     /* USART 0 Baud Rate 0 */
#define UBR10_                U0BR1_     /* USART 0 Baud Rate 1 */
#define RXBUF0_               U0RXBUF_   /* USART 0 Receive Buffer */
#define TXBUF0_               U0TXBUF_   /* USART 0 Transmit Buffer */
#define UCTL_0                U0CTL      /* USART 0 Control */
#define UTCTL_0               U0TCTL     /* USART 0 Transmit Control */
#define URCTL_0               U0RCTL     /* USART 0 Receive Control */
#define UMCTL_0               U0MCTL     /* USART 0 Modulation Control */
#define UBR0_0                U0BR0      /* USART 0 Baud Rate 0 */
#define UBR1_0                U0BR1      /* USART 0 Baud Rate 1 */
#define RXBUF_0               U0RXBUF    /* USART 0 Receive Buffer */
#define TXBUF_0               U0TXBUF    /* USART 0 Transmit Buffer */
#define UCTL_0_               U0CTL_     /* USART 0 Control */
```

```
#define UTCTL_0_              U0TCTL_   /* USART 0 Transmit Control */
#define URCTL_0_              U0RCTL_   /* USART 0 Receive Control */
#define UMCTL_0_              U0MCTL_   /* USART 0 Modulation Control */
#define UBR0_0_               U0BR0_    /* USART 0 Baud Rate 0 */
#define UBR1_0_               U0BR1_    /* USART 0 Baud Rate 1 */
#define RXBUF_0_              U0RXBUF_  /* USART 0 Receive Buffer */
#define TXBUF_0_              U0TXBUF_  /* USART 0 Transmit Buffer */
/********************************************************
* Timer A3
********************************************************/

#define TAIV_                 (0x012E)  /* Timer A Interrupt Vector Word */
READ_ONLY DEFW( TAIV          , TAIV_)
#define TACTL_                (0x0160)  /* Timer A Control */
DEFW( TACTL       , TACTL_)
#define TACCTL0_              (0x0162)  /* Timer A Capture/Compare Control 0 */
DEFW( TACCTL0     , TACCTL0_)
#define TACCTL1_              (0x0164)  /* Timer A Capture/Compare Control 1 */
DEFW( TACCTL1     , TACCTL1_)
#define TACCTL2_              (0x0166)  /* Timer A Capture/Compare Control 2 */
DEFW( TACCTL2     , TACCTL2_)
#define TAR_                  (0x0170)  /* Timer A */
DEFW( TAR         , TAR_)
#define TACCR0_               (0x0172)  /* Timer A Capture/Compare 0 */
DEFW( TACCR0      , TACCR0_)
#define TACCR1_               (0x0174)  /* Timer A Capture/Compare 1 */
DEFW( TACCR1      , TACCR1_)
#define TACCR2_               (0x0176)  /* Timer A Capture/Compare 2 */
DEFW( TACCR2      , TACCR2_)

/* Alternate register names */
#define CCTL0                 TACCTL0   /* Timer A Capture/Compare Control 0 */
#define CCTL1                 TACCTL1   /* Timer A Capture/Compare Control 1 */
#define CCTL2                 TACCTL2   /* Timer A Capture/Compare Control 2 */
#define CCR0                  TACCR0    /* Timer A Capture/Compare 0 */
#define CCR1                  TACCR1    /* Timer A Capture/Compare 1 */
#define CCR2                  TACCR2    /* Timer A Capture/Compare 2 */
#define CCTL0_                TACCTL0_  /* Timer A Capture/Compare Control 0 */
#define CCTL1_                TACCTL1_  /* Timer A Capture/Compare Control 1 */
#define CCTL2_                TACCTL2_  /* Timer A Capture/Compare Control 2 */
#define CCR0_                 TACCR0_   /* Timer A Capture/Compare 0 */
#define CCR1_                 TACCR1_   /* Timer A Capture/Compare 1 */
#define CCR2_                 TACCR2_   /* Timer A Capture/Compare 2 */

#define TASSEL2               (0x0400)  /* unused */        /* to distinguish from
USART SSELx */
#define TASSEL1               (0x0200)  /* Timer A clock source select 0 */
#define TASSEL0               (0x0100)  /* Timer A clock source select 1 */
#define ID1                   (0x0080)  /* Timer A clock input devider 1 */
#define ID0                   (0x0040)  /* Timer A clock input devider 0 */
#define MC1                   (0x0020)  /* Timer A mode control 1 */
#define MC0                   (0x0010)  /* Timer A mode control 0 */
```

```
#define TACLR              (0x0004)    /* Timer A counter clear */
#define TAIE               (0x0002)    /* Timer A counter interrupt enable */
#define TAIFG              (0x0001)    /* Timer A counter interrupt flag */


#define MC_0               (0*0x10u)   /* Timer A mode control: 0 - Stop */
#define MC_1               (1*0x10u)   /* Timer A mode control: 1 - Up to CCR0 */
#define MC_2               (2*0x10u)   /* Timer A mode control: 2 - Continous up */
#define MC_3               (3*0x10u)   /* Timer A mode control: 3 - Up/Down */
#define ID_0               (0*0x40u)   /* Timer A input divider: 0 - /1 */
#define ID_1               (1*0x40u)   /* Timer A input divider: 1 - /2 */
#define ID_2               (2*0x40u)   /* Timer A input divider: 2 - /4 */
#define ID_3               (3*0x40u)   /* Timer A input divider: 3 - /8 */
#define TASSEL_0           (0*0x100u)  /* Timer A clock source select: 0 - TACLK */
#define TASSEL_1           (1*0x100u)  /* Timer A clock source select: 1 - ACLK  */
#define TASSEL_2           (2*0x100u)  /* Timer A clock source select: 2 - SMCLK */
#define TASSEL_3           (3*0x100u)  /* Timer A clock source select: 3 - INCLK */


#define CM1                (0x8000)    /* Capture mode 1 */
#define CM0                (0x4000)    /* Capture mode 0 */
#define CCIS1              (0x2000)    /* Capture input select 1 */
#define CCIS0              (0x1000)    /* Capture input select 0 */
#define SCS                (0x0800)    /* Capture sychronize */
#define SCCI               (0x0400)    /* Latched capture signal (read) */
#define CAP                (0x0100)    /* Capture mode: 1 /Compare mode : 0 */
#define OUTMOD2            (0x0080)    /* Output mode 2 */
#define OUTMOD1            (0x0040)    /* Output mode 1 */
#define OUTMOD0            (0x0020)    /* Output mode 0 */
#define CCIE               (0x0010)    /* Capture/compare interrupt enable */
#define CCI                (0x0008)    /* Capture input signal (read) */
#define OUT                (0x0004)    /* PWM Output signal if output mode 0 */
#define COV                (0x0002)    /* Capture/compare overflow flag */
#define CCIFG              (0x0001)    /* Capture/compare interrupt flag */


#define OUTMOD_0           (0*0x20u)   /* PWM output mode: 0 - output only */
#define OUTMOD_1           (1*0x20u)   /* PWM output mode: 1 - set */
#define OUTMOD_2           (2*0x20u)   /* PWM output mode: 2 - PWM toggle/reset */
#define OUTMOD_3           (3*0x20u    /* PWM output mode: 3 - PWM set/reset */
#define OUTMOD_4           (4*0x20u)   /* PWM output mode: 4 - toggle */
#define OUTMOD_5           (5*0x20u)   /* PWM output mode: 5 - Reset */
#define OUTMOD_6           (6*0x20u)   /* PWM output mode: 6 - PWM toggle/set */
#define OUTMOD_7           (7*0x20u)   /* PWM output mode: 7 - PWM reset/set */
#define CCIS_0             (0*0x1000u) /* Capture input select: 0 - CCIxA */
#define CCIS_1             (1*0x1000u) /* Capture input select: 1 - CCIxB */
#define CCIS_2             (2*0x1000u) /* Capture input select: 2 - GND */
#define CCIS_3             (3*0x1000u) /* Capture input select: 3 - Vcc */
#define CM_0               (0*0x4000u) /* Capture mode: 0 - disabled */
#define CM_1               (1*0x4000u) /* Capture mode: 1 - pos. edge */
#define CM_2               (2*0x4000u) /* Capture mode: 1 - neg. edge */
#define CM_3               (3*0x4000u) /* Capture mode: 1 - both edges */
```

```
/**********************************************************
* Basic Clock Module
**********************************************************/

#define DCOCTL_              (0x0056)   /* DCO Clock Frequency Control */
DEFC( DCOCTL            , DCOCTL_)
#define BCSCTL1_             (0x0057)   /* Basic Clock System Control 1 */
DEFC( BCSCTL1           , BCSCTL1_)
#define BCSCTL2_             (0x0058)   /* Basic Clock System Control 2 */
DEFC( BCSCTL2           , BCSCTL2_)

#define MOD0                 (0x01)     /* Modulation Bit 0 */
#define MOD1                 (0x02)     /* Modulation Bit 1 */
#define MOD2                 (0x04)     /* Modulation Bit 2 */
#define MOD3                 (0x08)     /* Modulation Bit 3 */
#define MOD4                 (0x10)     /* Modulation Bit 4 */
#define DCO0                 (0x20)     /* DCO Select Bit 0 */
#define DCO1                 (0x40)     /* DCO Select Bit 1 */
#define DCO2                 (0x80)     /* DCO Select Bit 2 */

#define RSEL0                (0x01)     /* Resistor Select Bit 0 */
#define RSEL1                (0x02)     /* Resistor Select Bit 1 */
#define RSEL2                (0x04)     /* Resistor Select Bit 2 */
#define XT5V                 (0x08)     /* XT5V should always be reset */
#define DIVA0                (0x10)     /* ACLK Divider 0 */
#define DIVA1                (0x20)     /* ACLK Divider 1 */
#define XTS                  (0x40)     /* LFXTCLK 0:Low Freq. / 1: High Freq. */
#define XT2OFF               (0x80)     /* Enable XT2CLK */

#define DIVA_0               (0x00)     /* ACLK Divider 0: /1 */
#define DIVA_1               (0x10)     /* ACLK Divider 1: /2 */
#define DIVA_2               (0x20)     /* ACLK Divider 2: /4 */
#define DIVA_3               (0x30)     /* ACLK Divider 3: /8 */

#define DCOR                 (0x01)     /* Enable External Resistor : 1 */
#define DIVS0                (0x02)     /* SMCLK Divider 0 */
#define DIVS1                (0x04)     /* SMCLK Divider 1 */
#define SELS                 (0x08)     /* SMCLK Source Select 0:DCOCLK / 1:XT2CLK/
LFXTCLK */
#define DIVM0                (0x10)     /* MCLK Divider 0 */
#define DIVM1                (0x20)     /* MCLK Divider 1 */
#define SELM0                (0x40)     /* MCLK Source Select 0 */
#define SELM1                (0x80)     /* MCLK Source Select 1 */

#define DIVS_0               (0x00)     /* SMCLK Divider 0: /1 */
#define DIVS_1               (0x02)     /* SMCLK Divider 1: /2 */
#define DIVS_2               (0x04)     /* SMCLK Divider 2: /4 */
#define DIVS_3               (0x06)     /* SMCLK Divider 3: /8 */

#define DIVM_0               (0x00)     /* MCLK Divider 0: /1 */
#define DIVM_1               (0x10)     /* MCLK Divider 1: /2 */
#define DIVM_2               (0x20)     /* MCLK Divider 2: /4 */
```

```
#define DIVM_3                (0x30)     /* MCLK Divider 3: /8 */

#define SELM_0                (0x00)     /* MCLK Source Select 0: DCOCLK */
#define SELM_1                (0x40)     /* MCLK Source Select 1: DCOCLK */
#define SELM_2                (0x80)     /* MCLK Source Select 2: XT2CLK/LFXTCLK */
#define SELM_3                (0xC0)     /* MCLK Source Select 3: LFXTCLK */

/*************************************************************
* Flash Memory
*************************************************************/

#define  FCTL1_               (0x0128)  /* FLASH Control 1 */
DEFW(    FCTL1           , FCTL1_)
#define  FCTL2_               (0x012A)  /* FLASH Control 2 */
DEFW(    FCTL2           , FCTL2_)
#define  FCTL3_               (0x012C)  /* FLASH Control 3 */
DEFW(    FCTL3           , FCTL3_)

#define  FRKEY               (0x9600)  /* Flash key returned by read */
#define  FWKEY               (0xA500)  /* Flash key for write */
#define  FXKEY               (0x3300)  /* for use with XOR instruction */

#define  ERASE               (0x0002)  /* Enable bit for Flash segment erase */
#define  MERAS               (0x0004)  /* Enable bit for Flash mass erase */
#define  WRT                 (0x0040)  /* Enable bit for Flash write */
#define  BLKWRT              (0x0080)  /* Enable bit for Flash segment write */
#define  SEGWRT              (0x0080)  /* old definition */ /* Enable bit for Flash
segment write */

#define  FN0                 (0x0001)  /* Devide Flash clock by 1 to 64 using FN0
to FN5 according to: */
#define  FN1                 (0x0002   /*  32*FN5 + 16*FN4 + 8*FN3 + 4*FN2 + 2*FN1
+ FN0 + 1 */
#ifndef FN2
#define  FN2                 (0x0004)
#endif
#ifndef FN3
#define  FN3                 (0x0008)
#endif
#ifndef FN4
#define  FN4                 (0x0010)
#endif
#define  FN5                 (0x0020)
#define  FSSEL0              (0x0040)  /* Flash clock select 0 */        /* to
distinguish from USART SSELx */
#define  FSSEL1              (0x0080)  /* Flash clock select 1 */

#define  FSSEL_0             (0x0000)  /* Flash clock select: 0 - ACLK */
#define  FSSEL_1             (0x0040)  /* Flash clock select: 1 - MCLK */
#define  FSSEL_2             (0x0080)  /* Flash clock select: 2 - SMCLK */
#define  FSSEL_3             (0x00C0)  /* Flash clock select: 3 - SMCLK */
```

```
#define   BUSY                (0x0001)   /* Flash busy: 1 */
#define   KEYV                (0x0002)   /* Flash Key violation flag */
#define   ACCVIFG             (0x0004)   /* Flash Access violation flag */
#define   WAIT                (0x0008)   /* Wait flag for segment write */
#define   LOCK                (0x0010)   /* Lock bit: 1 - Flash is locked (read
only) */
#define   EMEX                (0x0020)   /* Flash Emergency Exit */


/*************************************************************
* Comparator A
*************************************************************/

#define CACTL1_             (0x0059)   /* Comparator A Control 1 */
DEFC(   CACTL1            , CACTL1_)
#define CACTL2_             (0x005A)   /* Comparator A Control 2 */
DEFC(   CACTL2            , CACTL2_)
#define CAPD_               (0x005B)   /* Comparator A Port Disable */
DEFC(   CAPD              , CAPD_)

#define CAIFG              (0x01)     /* Comp. A Interrupt Flag */
#define CAIE               (0x02)     /* Comp. A Interrupt Enable */
#define CAIES              (0x04)     /* Comp. A Int. Edge Select: 0:rising / 1:
falling */
#define CAON               (0x08)     /* Comp. A enable */
#define CAREF0             (0x10)     /* Comp. A Internal Reference Select 0 */
#define CAREF1             (0x20)     /* Comp. A Internal Reference Select 1 */
#define CARSEL             (0x40)     /* Comp. A Internal Reference Enable */
#define CAEX               (0x80)     /* Comp. A Exchange Inputs */

#define CAREF_0            (0x00)     /* Comp. A Int. Ref. Select 0 : Off */
#define CAREF_1            (0x10)     /* Comp. A Int. Ref. Select 1 : 0.25*Vcc */
#define CAREF_2            (0x20)     /* Comp. A Int. Ref. Select 2 : 0.5*Vcc */
#define CAREF_3            (0x30)     /* Comp. A Int. Ref. Select 3 : Vt*/

#define CAOUT              (0x01)     /* Comp. A Output */
#define CAF                (0x02)     /* Comp. A Enable Output Filter */
#define P2CA0              (0x04)     /* Comp. A Connect External Signal to CA0 :
1 */
#define P2CA1              (0x08)     /* Comp. A Connect External Signal to CA1 :
1 */
#define CACTL24            (0x10)
#define CACTL25            (0x20)
#define CACTL26            (0x40)
#define CACTL27            (0x80)

#define CAPD0              (0x01)     /* Comp. A Disable Input Buffer of Port
Register .0 */
#define CAPD1              (0x02)     /* Comp. A Disable Input Buffer of Port
Register .1 */
#define CAPD2              (0x04)     /* Comp. A Disable Input Buffer of Port
Register .2 */
#define CAPD3              (0x08)     /* Comp. A Disable Input Buffer of Port
```

271

```
Register .3 */
#define CAPD4                 (0x10)    /* Comp. A Disable Input Buffer of Port
Register .4 */
#define CAPD5                 (0x20)    /* Comp. A Disable Input Buffer of Port
Register .5 */
#define CAPD6                 (0x40)    /* Comp. A Disable Input Buffer of Port
Register .6 */
#define CAPD7                 (0x80)    /* Comp. A Disable Input Buffer of Port
Register .7 */


/**************************************************************
* Interrupt Vectors (offset from 0xFFE0)
**************************************************************/

#define PORT1_VECTOR          (2 * 2u)  /* 0xFFE4 Port 1 */
#define PORT2_VECTOR          (3 * 2u)  /* 0xFFE6 Port 2 */
#define USART0TX_VECTOR       (6 * 2u)  /* 0xFFEC USART 0 Transmit */
#define USART0RX_VECTOR       (7 * 2u)  /* 0xFFEE USART 0 Receive */
#define TIMERA1_VECTOR        (8 * 2u)  /* 0xFFF0 Timer A CC1-2, TA */
#define TIMERA0_VECTOR        (9 * 2u)  /* 0xFFF2 Timer A CC0 */
#define WDT_VECTOR            (10 * 2u) /* 0xFFF4 Watchdog Timer */
#define COMPARATORA_VECTOR    (11 * 2u) /* 0xFFF6 Comparator A */
#define NMI_VECTOR            (14 * 2u) /* 0xFFFC Non-maskable */
#define RESET_VECTOR          (15 * 2u) /* 0xFFFE Reset [Highest Priority] */

#define UART0TX_VECTOR        USART0TX_VECTOR
#define UART0RX_VECTOR        USART0RX_VECTOR


/**************************************************************
* End of Modules
**************************************************************/
#pragma language=default

#endif /* #ifndef __msp430x12x */
```

Application Program for Use in Chapter 10

The following is a copy of the application program written in the C language for use in the project of *Chapter 10*. The latest version of this program should be downloaded from Texas Instruments Incorporated Web site at http://www.ti.com. The instructions for doing this are contained in *Chapter 10*.

The program was written by Neal Frager of Texas Instruments Incorporated, and is used by permission and courtesy of Texas Instruments Incorporated.

```
                              TimeDateTemp.c
//****************************************************************************
**
//  MSP-FET430P120 - Temp Sensor + Clock
//
//  N. Frager
//  Texas Instruments, Inc
//  February 2003
//  Built with IAR Embedded Workbench Version: 1.26A
//  Version for MSP-FET430P120
//****************************************************************************
#include   <msp430x12x.h>

// Define segments on LED display
#define a 0x01
#define b 0x02
#define c 0x04
#define d 0x08
#define e 0x10
#define f 0x20
#define g 0x40


// Define numbers on LED display
#define zero        a+b+c+d+e+f
#define one         b+c
#define two         a+b+d+e+g
#define three       a+b+c+d+g
#define four        b+c+f+g
#define five        a+c+d+f+g
#define six         a+c+d+e+f+g
#define seven       a+b+c
#define eight       a+b+c+d+e+f+g
```

```
#define nine           a+b+c+f+g
#define celcius        d+e+g
#define fahrenheit     a+e+f+g
#define blank          0x00

// Define button press numbers
#define NOPUSH 0
#define MODE   1
#define TOGGLE 2

// Define state values
#define SHOW_TIME     0
#define SHOW_DATE     1
#define SHOW_YEAR     2
#define SHOW_TEMP     3
#define SET_HOUR      4
#define SET_MIN       5
#define SET_MONTH     6
#define SET_DAY       7
#define SET_YEAR      8
#define AUTO_TOGGLE   9

// Define Extra Auto Toggle States
#define SHOW_TEMP_F 3
#define SHOW_TEMP_C 4

// Define Clock and Date values
#define JAN 1
#define FEB 2
#define MAR 3
#define APR 4
#define MAY 5
#define JUN 6
#define JUL 7
#define AUG 8
#define SEP 9
#define OCT 10
#define NOV 11
#define DEC 12

#define CLK_PER_TEMP 512
#define CLK_PER_SEC 512
#define SEC_PER_MIN 60
#define MIN_PER_HR  60

#define MIN_YEAR 2000
#define MAX_YEAR 2100
```

```
#define TRUE 1
#define FALSE 0

// System Routines
void initialize(void);    // initialize ports and variables
void clocktick(void);     // run the system clock
void display(void);       // display driver
void displaydigit(int);   // digit display routine
void fillbuffer(void);    // fill 4-digit buffer

// Global variables
unsigned int digcount, dig1, dig2, dig3, dig4;  // digit buffer
unsigned int clock_count, sec, min, hour, ampm;  // clock values
unsigned int month, day, year;  // date values
unsigned int mainstate;  // main state machine
unsigned int buttonpress, buttonpush;  // button press signal
unsigned int toggle_count;  // allow for faster clock update
unsigned int blink, timer;  // control blinking functionality
unsigned int data, temp, temp_type, temp_count, temp_done;  // temperature display
unsigned int togglestate;  // auto toggle state machine
unsigned int toggletimer;  // 2 second auto toggle timer
unsigned int awake;  // sleep mode (FALSE = sleep, TRUE = awake)
unsigned int sleepcount;  // counts 15 seconds until sleep
unsigned int autowakecount; // counts 20 minutes until auto wake up

// temperature table
unsigned int temp_array[28][2] = {{0, 320}, {14, 345}, {28, 370}, {42, 395},
                                  {56, 420}, {69, 445}, {83, 470}, {97, 495},
                                  {111, 520}, {125, 545}, {139, 570}, {153, 595},
                                  {167, 620}, {181, 645}, {194, 670}, {208, 695},
                                  {222, 720}, {236, 745}, {250, 770}, {264, 795},
                                  {278, 820}, {292, 845}, {306, 870}, {319, 895},
                                  {333, 920}, {347, 945}, {361, 970}, {375, 995}};

// Main Function
void main(void)
{
    initialize();  // initialize ports, timerA and variables

    for (;;) // main state machine - run continuously
    {
        switch(mainstate) {
            case SHOW_TIME:  // display time
                if(buttonpress & (buttonpush == TOGGLE)) {
                    mainstate = SHOW_DATE;
```

```
            buttonpress = 0;
        }
        else if(buttonpress && (buttonpush == MODE)) {
            mainstate = SET_HOUR;
            buttonpress = 0;
        }
        break;
    case SHOW_DATE:  // display date
        if(buttonpress && (buttonpush == TOGGLE)) {
            mainstate = SHOW_YEAR;
            buttonpress = 0;
        }
        else if(buttonpress && (buttonpush == MODE)) {
            mainstate = SET_MONTH;
            buttonpress = 0;
        }
        break;
    case SHOW_YEAR:  // display year
        if(buttonpress && (buttonpush == TOGGLE)) {
            mainstate = SHOW_TEMP;
            buttonpress = 0;
        }
        else if(buttonpress && (buttonpush == MODE)) {
            mainstate = SET_YEAR;
            buttonpress = 0;
        }
        break;
    case SHOW_TEMP:  // display temperature
        if(buttonpress && (buttonpush == TOGGLE)) {
            mainstate = SHOW_TIME;
            buttonpress = 0;
        }
        else if(buttonpress && (buttonpush == MODE)) {
            mainstate = SHOW_TEMP;
            buttonpress = 0;
            temp_type ^= 0x01;
        }
        break;
    case SET_HOUR:  // set the hour
        if(buttonpress && (buttonpush == TOGGLE)) {
            mainstate = SET_HOUR;
            buttonpress = 0;
            hour++;
            if(hour == 12)
                ampm ^= 0x01;
            if(hour > 12)
                hour = 1;
```

```
        }
        else if(buttonpress && (buttonpush == MODE)) {
            mainstate = SET_MIN;
            buttonpress = 0;
        }
        break;
case SET_MIN:     // set the minute
    if(buttonpress && (buttonpush == TOGGLE)) {
        mainstate = SET_MIN;
        buttonpress = 0;
        min++;
        if(min == MIN_PER_HR)
            min = 0;
    }
    else if(buttonpress && (buttonpush == MODE)) {
        mainstate = SHOW_TIME;
        buttonpress = 0;
    }
    break;
case SET_MONTH:  // set the month
    if(buttonpress && (buttonpush == TOGGLE)) {
        mainstate = SET_MONTH;
        buttonpress = 0;
        month++;
        if(month > DEC)
            month = JAN;
    }
    else if(buttonpress && (buttonpush == MODE)) {
        mainstate = SET_DAY;
        buttonpress = 0;
    }
    break;
case SET_DAY:  // set the day
    if(buttonpress && (buttonpush == TOGGLE)) {
        mainstate = SET_DAY;
        buttonpress = 0;
        day++;
        // February calculation
        if((year % 4) == 0) {  // leap year
            if((month == FEB) && (day > 29))
                day = 1;
        }
        else {
            if((month == FEB) && (day > 28))
                day = 1;
        }
```

```
            // 30 day months
            if(((month == APR) || (month == JUN) || (month == SEP) || (month
== NOV)) && (day > 30))
                day = 1;
            // 31 day months
            if(day > 31)
                day = 1;
        }
        else if(buttonpress && (buttonpush == MODE)) {
            mainstate = SHOW_DATE;
            buttonpress = 0;
        }
        break;
    case SET_YEAR:  // set the year
        if(buttonpress && (buttonpush == TOGGLE)) {
            mainstate = SET_YEAR;
            buttonpress = 0;
            year++;
            if(year >= MAX_YEAR)
                year = MIN_YEAR;
        }
        else if(buttonpress && (buttonpush == MODE)) {
            mainstate = SHOW_YEAR;
            buttonpress = 0;
        }
        break;
    case AUTO_TOGGLE:  // auto toggle state
        if(buttonpress && (buttonpush == TOGGLE)) {
            mainstate = togglestate;
            if(mainstate == SET_HOUR)  // Celcius exception case
                mainstate = SHOW_TEMP;
            buttonpress = 0;
        }
        else if(buttonpress && (buttonpush == MODE)) {
            awake = 0;  // system sleeps with MODE push
            P1OUT = 0x00;  // turn off analog system
            P2OUT = 0x00;
            P3OUT = blank;
            CCR0 = 32767; // slow to 1 Hz interrupt speed
            buttonpress = 0;
        }
        break;
    }

    if(awake) fillbuffer();  // fill the 4-digit buffer
    LPM3;
```

```
    }
}


// PORT1 Push Button interrupt service routine - run on button press


#if __VER__ < 200
interrupt[PORT1_VECTOR] void PORT_1 (void)
#else
#pragma vector=PORT1_VECTOR
__interrupt void PORT_1(void)
#endif
{
    int i;

    // Clear Interrupt Flag
    P1IFG = 0x00;

    // Software Debounce Delay
    for(i=0;i<64;i++);

    // Save button press
    if(awake) {
        if((P1IN & 0xC0) == 0x40) { // P1.7 = TOGGLE
            buttonpush = TOGGLE;
            toggle_count = 0;
            sleepcount = 0;
            buttonpress = 1;
        }
        else if((P1IN & 0xC0) == 0x80) { // P1.6 = MODE
            buttonpush = MODE;
            toggle_count = 0;
            sleepcount = 0;
            buttonpress = 1;
        }
        else {
            buttonpush = NOPUSH;
            buttonpress = 0;
        }
    }
    else { // wake system up
        if((P1IN & 0xC0) != 0xC0) {
            toggle_count = 0;
            sleepcount = 0;
            awake = TRUE;  // system wakes up - button press
```

```
            P1OUT = 0x18;  // turn on analog system
            CCR0 = 63; // return to 512 Hz interrupt speed
        }
    }
}


// Timer A0 interrupt service routine - run at 512 Hz if awake, 1 Hz if asleep


#if __VER__ < 200
interrupt[TIMERA0_VECTOR] void Timer_A (void)
#else
#pragma vector=TIMERA0_VECTOR
__interrupt void Timer_A(void)
#endif
{
    // Clock Logic - run at 1 Hz
    // always runs
    if(awake) { // running at 512 Hz
        clock_count++;
        if(clock_count >= CLK_PER_SEC) {
            clock_count = 0;
            clocktick();
        }
    }
    else clocktick(); // running at 1 Hz

    // Temperature Controller - run at 1 Hz
    if(awake) {  // runs only when system is awake
        if(temp_done) {
            temp_count++;
            if(temp_count >= CLK_PER_TEMP) {
                temp_count = 0;
                temp_done = 0;
            }
        }
    }

    // 2 second Auto Toggler - run at 0.5 Hz
    if(awake) {  // runs only when system is awake
        if(mainstate == AUTO_TOGGLE) {
            toggletimer++;
            if(toggletimer >= 1024) {  // 2 second auto toggle
                togglestate++;
                togglestate %= 5;
```

```
                toggletimer = 0;
            }
            if(togglestate == SHOW_TEMP_F)
                temp_type = 1;
            if(togglestate == SHOW_TEMP_C)
                temp_type = 0;
        }
        else toggletimer = 0;
}


// System Sleep Controller
if(awake) {
    if((P1IN & 0xC0)== 0xC0) {
        sleepcount++;
        if(sleepcount >= 15360) { // 512 Hz * 30 seconds = 15360 cycles
            awake = FALSE;  // system sleeps after 15 seconds
            autowakecount = 0; //reset auto wake clock
            P1OUT = 0x00;  // turn off analog system
            P2OUT = 0x00;
            P3OUT = blank;
            CCR0 = 32767; // slow to 1 Hz interrupt speed
        }
    }
    else sleepcount = 0;
}


// Push Button Debounce - run at 32 Hz
if(awake) {
    if((P1IN & 0xC0) == 0x40) { // P1.7 = TOGGLE
        sleepcount = 0;
        if((mainstate == SET_HOUR) | (mainstate == SET_MIN) |
                (mainstate == SET_MONTH) | (mainstate == SET_DAY) |
                (mainstate == SET_YEAR)) {
            toggle_count++;
            if(toggle_count >= 80) { // update button held for 80/512 seconds
                toggle_count = 0;
                buttonpress = 1;  // send another button press signal
            }
        }
        if((mainstate == SHOW_TIME) | (mainstate == SHOW_DATE) |
                (mainstate == SHOW_YEAR) | (mainstate == SHOW_TEMP)) {
            toggle_count++;
            if(toggle_count >= 1024) {  // 2 second hold
                toggle_count = 0;
                togglestate = mainstate;
                mainstate = AUTO_TOGGLE;
```

```
                        if(togglestate == SHOW_TEMP) {
                            if(temp_type)
                                togglestate = SHOW_TEMP_F;
                            else togglestate = SHOW_TEMP_C;
                        }
                    }
                }
            }
        else if((P1IN & 0xC0) == 0x80) { // P1.6 = MODE
            sleepcount = 0;
        }
        else P1IFG = 0x00; // Button Released
    }

    // Display a digit - run at 512 Hz
    if(awake) { // runs only if system is awake
        display();
        LPM3_EXIT;
    }
    else {  // clear LEDs and shut down analog since system is sleeping
        P1OUT = 0x00;
        P2OUT = 0x00;
        P3OUT = blank;
    }
}


// System Clock - runs at 1 Hz
void clocktick(void)
{
    sec++;

    // update minutes
    if(sec >= SEC_PER_MIN) {
        sec = 0;
        min++;
    if ((min == 20) || (min == 40) || (min == 60)) {
        autowakecount = 0;
        toggle_count = 0;
        sleepcount = 0;
        awake = TRUE;  // system wakes up - auto wake
        P1OUT = 0x18;  // turn on analog system
        CCR0 = 63; // return to 512 Hz interrupt speed
        }
    }
    // update hours
```

```
        if(min >= MIN_PER_HR) {
            min = 0;
            hour++;
            if(hour == 12) {
                ampm ^= 0x01;
                if(!ampm) {
                    day++;
                }
            }
        }
        if(hour > 12) {
            hour = 1;
        }
        // update days
        // February update
        if((year % 4) == 0) {  // leap year
            if((month == FEB) && (day > 29)) {
                day = 1;
                month++;
            }
        }
        else {  // non leap year
            if((month == FEB) && (day > 28)) {
                day = 1;
                month++;
            }
        }
        // 30 day month update
        if(((month == APR) || (month == JUN) || (month == SEP) || (month == NOV)) &&
(day > 30)) {
            day = 1;
            month++;
        }
        // 31 day month update
        if(day > 31) {
            day = 1;
            month++;
        }
        // update year
        if(month > DEC) {
            month = 1;
            year++;
        }
        // reset year when max is reached
        if(year >= MAX_YEAR) {
            year = MIN_YEAR;
```

```
    }
}


// initialize ports, Timer A0 and variables - run once at start
void initialize(void)
{
    WDTCTL = WDTPW + WDTHOLD;  // Stop watchdog timer
    TACTL = TASSEL_1 + TACLR;  // ACLK, clear TAR
    CCTL0 = CCIE;              // CCR0 interrupt enabled
    CCR0 = 63;                 // TimerA interupts at 512 Hz

    digcount = 0;  // reset digit buffer
    dig1 = 0;
    dig2 = 0;
    dig3 = 0;
    dig4 = 0;

    clock_count = 0;  // reset clock vars
    sec = 0;
    min = 0;
    hour = 12;
    ampm = 0;

    month = JAN;    // reset date vars
    day = 1;
    year = MIN_YEAR;

    mainstate = SHOW_TIME;   // reset state machines
    togglestate = SHOW_TIME;
    buttonpush = NOPUSH;
    buttonpress = 0;
    toggle_count = 0;
    toggletimer = 0;
    awake = TRUE; // system is awake
    sleepcount = 0;
    autowakecount = 0;

    timer = 0;  // initialize blinking functionality
    blink = 0;

    temp = 0;   // set temp output
    data = 0;
    temp_type = 1;
    temp_count = 0;
    temp_done = 0;
```

```
      P1DIR = 0x3F; // Set I/O ports
      P2DIR = 0xFF;
      P3DIR = 0x7F;

      P1IES = 0xC0;
      P1IFG = 0x00;
      P1IE = 0xC0;
      P1OUT = 0x18;   // Turn A/D off

      TACTL |= ID_0;
      TACTL |= MC_1;   // Start Timer_a in upmode
      _EINT();         // Enable interrupts
}


// Fill 4-digit buffer - runs continuously
void fillbuffer(void)
{
    unsigned int counter;

    // determine clock digits
    if((mainstate == SHOW_TIME) || (mainstate == SET_HOUR) || (mainstate == SET_
MIN) ||
          ((mainstate == AUTO_TOGGLE) && (togglestate == SHOW_TIME))) {
        dig1 = hour / 10;
        dig2 = hour % 10;
        dig3 = min / 10;
        dig4 = min % 10;

        if(ampm)  // AM/PM lights
            P1OUT = 0x1F;
        else P1OUT = 0x1B;
    }
    // determine date digits
    else if ((mainstate == SHOW_DATE) || (mainstate == SET_MONTH) || (mainstate
== SET_DAY) ||
              ((mainstate == AUTO_TOGGLE) && (togglestate == SHOW_DATE))) {
        dig1 = month / 10;
        dig2 = month % 10;
        dig3 = day / 10;
        dig4 = day % 10;

        P1OUT = 0x1A;
    }
    // determine year digits
    else if ((mainstate == SHOW_YEAR) || (mainstate == SET_YEAR) ||
```

```
                        ((mainstate == AUTO_TOGGLE) && (togglestate == SHOW_YEAR))) {
    dig1 = year / 1000;
    dig2 = (year / 100) % 10;
    dig3 = (year / 10) % 10;
    dig4 = year % 10;

    P1OUT = 0x18;
}
// determine temp digits
else if ((mainstate == SHOW_TEMP) || ((mainstate == AUTO_TOGGLE) && (tog-
glestate >= SHOW_TEMP))) {
    if(!temp_done) { // update temp once per second
        data = 0;
        P1OUT &= 0x17;  // Turn A/D on
        for(counter = 10; counter > 0;)
        {
            data = data << 1;
            if((P3IN & 0x80) == 0x80) // P3.7 = A/D temp input
            {
                data |= 0x01;
            }
            counter--;
            P1OUT |= 0x20;  // send a clock tick to A/D converter
            P1OUT &= 0x17;
        }
        P1OUT &= 0x1F;   // Turn A/D off

        if(data <= 712)
            temp = temp_array[0][temp_type];
        else if (data <= 820) {
            temp = temp_array[(data-712)/4][temp_type];
            if(temp_type)
                temp += ((data-2)%4) * 6;
            else temp += ((data-2)%4) * 3;
        }
        else if (data != 1023) {
            temp = temp_array[27][temp_type];
        }

        temp_done = 1;   // temp recorded
    }

    // Fill Temperature Buffer
    dig1 = temp / 100;
    dig2 = (temp / 10) % 10;
    dig3 = temp % 10;
```

```
        if(temp_type)
            dig4 = 10;  // display Fahrenheit
        else dig4 = 11;  // display Celcius

        P1OUT = 0x1A;
    }
    else {  // system has gone into invalid state
        dig1 = 0;
        dig2 = 0;
        dig3 = 0;
        dig4 = 0;
        P1OUT = 0x18;
    }
}


// System Display Routine - runs at 512 Hz
void display(void)
{
    P2OUT = 0xFF;  // clear digit
    displaydigit(-1);

    // update digit counter
    digcount++;
    digcount %= 4;

    // update blink controller
    if((clock_count % 256) == 0)
        blink ^= 0x01;

    switch(digcount) {
        case 0:  // display digit 1
            if(blink || (mainstate == SHOW_TIME) || (mainstate == SET_MIN) ||
(mainstate == SHOW_DATE) || (mainstate == SET_DAY) ||
                (mainstate == SHOW_YEAR) || (mainstate == SHOW_TEMP)  || (main-
state == AUTO_TOGGLE)) {
                if(dig1 || (mainstate == SHOW_DATE) || (mainstate == SET_DAY) ||
(mainstate == SET_MONTH) || (mainstate == SHOW_YEAR) ||
                (mainstate == SET_YEAR) || ((mainstate == AUTO_TOGGLE) && (tog-
glestate != SHOW_TIME)))
                displaydigit(dig1);
            else displaydigit(-1);
            if(blink && ((mainstate == SHOW_TIME) || ((mainstate == AUTO_TOGGLE)
&& (togglestate == SHOW_TIME))))
                P2OUT = 0xDE;  // Display Digit 1 without clock colon
            else
                P2OUT = 0xCE;  // Display Digit 1 with clock colon
```

```
        }
        else P2OUT = 0xCF; // Digit off for blinking
        break;
    case 1:  // display digit 2
        if(blink || (mainstate == SHOW_TIME) || (mainstate == SET_MIN) ||
(mainstate == SHOW_DATE) || (mainstate == SET_DAY) ||
            (mainstate == SHOW_YEAR) || (mainstate == SHOW_TEMP) || (mainstate
== AUTO_TOGGLE)) {
            displaydigit(dig2);
            P2OUT = 0xFD;  // Display Digit 2
        }
        else P2OUT = 0xFF; // Digit off for blinking
        break;
    case 2:  // display digit 3
        if(blink || (mainstate == SHOW_TIME) || (mainstate == SET_HOUR) ||
(mainstate == SHOW_DATE) || (mainstate == SET_MONTH) ||
            (mainstate == SHOW_YEAR) || (mainstate == SHOW_TEMP) || (mainstate
== AUTO_TOGGLE)) {
            displaydigit(dig3);
            P2OUT = 0xFB;  // Display Digit 3
        }
        else P2OUT = 0xFF; // Digit off for blinking
        break;
    case 3:  // display digit 4
        if(blink || (mainstate == SHOW_TIME) || (mainstate == SET_HOUR) ||
(mainstate == SHOW_DATE) || (mainstate == SET_MONTH) ||
            (mainstate == SHOW_YEAR) || (mainstate == SHOW_TEMP) || (mainstate
== AUTO_TOGGLE)) {
            displaydigit(dig4);
            P2OUT = 0xF7;  // Display Digit 4
        }
        else P2OUT = 0xFF; // Digit off for blinking
        break;
    }
}


// Routine for converting integer number into display value
// Runs when called
void displaydigit(int number)
{
    switch(number) {
        case 0:   P3OUT = (P3OUT & 0x80) | zero;
                  break;
        case 1:   P3OUT = (P3OUT & 0x80) | one;
                  break;
```

```
    case 2:    P3OUT = (P3OUT & 0x80) | two;
               break;
    case 3:    P3OUT = (P3OUT & 0x80) | three;
               break;
    case 4:    P3OUT = (P3OUT & 0x80) | four;
               break;
    case 5:    P3OUT = (P3OUT & 0x80) | five;
               break;
    case 6:    P3OUT = (P3OUT & 0x80) | six;
               break;
    case 7:    P3OUT = (P3OUT & 0x80) | seven;
               break;
    case 8:    P3OUT = (P3OUT & 0x80) | eight;
               break;
    case 9:    P3OUT = (P3OUT & 0x80) | nine;
               break;
    case 10:   P3OUT = (P3OUT & 0x80) | fahrenheit;
               break;
    case 11:   P3OUT = (P3OUT & 0x80) | celcius;
               break;
    default:   P3OUT = (P3OUT & 0x80) | blank;
               break;
    }
}
```

A Refresher

The purpose of this appendix is to provide basic information for the reader who needs some help with the fundamental concepts contained in this book. It is a "refresher" on some of the concepts to make sure that the readers' level of understanding is improved to be able to absorb more of the discussion in the text.

The source of this material is in *Chapter 3* of *Basic Communications Electronics*, J. Hudson, J. Luecke, ©1999 Master Publishing, Inc., used by permission of and courtesy of Master Publishing, Inc.

Ohm's Law

The law of electricity most used in electronic circuit design is Ohm's Law. It is named in honor of Georg Simon Ohm, who formulated the relationship between voltage, current and resistance in the 19th Century. Ohm's Law states:

"The current in an electrical circuit is directly proportional to the voltage applied to the circuit, and inversely proportional to the resistance." In equation form, Ohm's Law is:

$$I = \frac{E}{R} \qquad \text{where:} \quad I \text{ is current in } \mathbf{amperes}$$

$$E = I \times R \qquad\qquad E \text{ is voltage in } \mathbf{volts}$$

$$R = \frac{E}{I} \qquad\qquad R \text{ is resistance in } \mathbf{ohms}$$

Simple Aid for Using Ohm's Law

A simple aid for remembering Ohm's law is shown in *Figure D-1a*. Just cover the letter in the circle that you want to find and read the equation formed by the remaining letters. When the current is unknown, but the voltage and resistance are known, the basic equation to be solved for I is found by using the aid of *Figure D-1b*. The result is:

$$I = \frac{E}{R}$$

Similarly, knowing the current and resistance, the voltage can be calculated by using the equation shown in *Figure D-1c*:

$$E = IR$$

Figure D-1d shows the aids when the current and resistance are in values other than amperes and ohms, respectively.

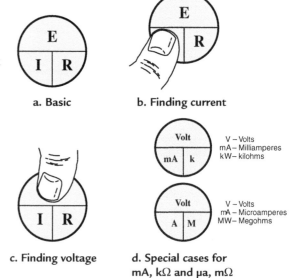

a. Basic

b. Finding current

c. Finding voltage

d. Special cases for mA, kΩ and μa, mΩ

V – Volts
mA– Milliamperes
kW– kilohms

V – Volts
mA – Microamperes
MW– Megohms

Figure D-1: Ohm's law circle

Source: Basic Electronics, G. McWhorter and A.J. Evans, ©1996, Master Publishing, Inc., Lincolnwood, IL.

Decibel—A Quantity to Describe Gain

Almost all technicians and engineers, as well as marketing personnel in communications systems, use the term "decibel." The decibel (abbreviated dB) is one-tenth of a bel. It is a standard unit for expressing the ratio between output power and input power or, in special cases, output voltage and input voltage. (It also is used to express differences in sound levels (power levels) in audio systems). The decibel is expressed as:

<u>Power</u>

$$dB = 10 \log_{10} \frac{P_{out}}{P_{in}}$$

<u>Voltage</u>

$$dB = 20 \log_{10} \frac{V_{out}}{V_{in}} \text{ when } R_{in} = R_{out}$$

To review its use, what power ratio is represented by 20 dB?

Remember: The logarithm of a number is the *exponent* to which the base of the logarithm must be raised in order to arrive at the number, for example, $10^x = Y$, therefore, $x = \log_{10} Y$. Also recall that $\therefore$ means "Therefore."

$$20 = 10 \log_{10} \frac{P_{out}}{P_{in}}$$

$$2 = \log_{10} \frac{P_{out}}{P_{in}}$$

$$\therefore 10^2 = \frac{P_{out}}{P_{in}}$$

$\therefore P_{OUT}$ is 100 times P_{in}

What voltage ratio is represented by 60 dB?

$$dB = 20 \log_{10} \frac{V_{out}}{V_{in}}$$

$$60 = 20 \log_{10} \frac{V_{out}}{V_{in}}$$

$$3 = \log_{10} \frac{V_{out}}{V_{in}}$$

$$\therefore 10^3 = \frac{V_{out}}{V_{in}}$$

$\therefore V_{out}$ is 1000 times V_{in}

There is a caution in using $dB = 20 \log_{10} V_{out}/V_{in}$. It is assumed that V_{out} and V_{in} are across the same value of resistance. If this condition is not met, $dB = 10 \log_{10} P_{out}/P_{in}$ must be used. The following reference table shows the equivalent power and voltage ratios to various decibels.

The advantage of using dB units is that they can be added and subtracted directly to obtain the final result. For example, if two amplifier stages each have 10 dB of power gain, the total power gain is 20 dB. 10 dB of power gain is a ratio of 10. The gains of two amplifiers cascaded multiply; therefore, $10 \times 10 = 100$ power gain for the two stages. 20 dB of power gain is a ratio of 100.

dB	P_{out}/P_{in}	V_{out}/V_{in}
3	2	1.4
6	4	2
10	10	3
20	100	10
30	1000	31.6
40	10,000	100
50	10^5	316.2
60	10^6	1000

Table D-1

Passive Devices

All resistors, inductors and capacitors have impedance (symbol Z) because they "impede" (resist) current in electronic circuits. Impedance is measured or expressed in ohms (symbol Ω). The schematic symbols for these passive devices are shown in *Figure D-2.*

1. Resistors impede (resist) current equally well in DC or AC circuits. Unless a resistor has inductance or capacitance, its impedance is $Z = R \underline{/0°}$, just the resistance with a zero phase angle. Because the phase angle is zero, it is left off and Z is just equal to R ($Z = R$).

2. Inductors are coils of wire that have inductance (symbol L), a capability to store energy in a magnetic field surrounding the coil when there is current through the coil. The stored energy opposes changes in the existing current through the coil. The opposition to the changing current is called inductive reactance (symbol X_L). The impedance of an inductor is made up of the coil's resistance and inductive reactance added as vectors at right angles to each other. Its value is $Z_L = \sqrt{R^2 + X_L^2}$ with a phase angle, θ, whose $\tan\theta = X_L / R$. When $R = 0$, $Z_L = X_L \underline{/90°}$. Inductors with zero or very small resistance are a short circuit (zero impedance) to DC current, but increase their inductive reactance as frequency increases according to the expression $X_L = 2\pi fL$, where f is frequency in Hz and L is inductance in henries. As frequency increases, the inductive reactance of an inductor increases. At 10 MHz (10,000,000 cycles), even a small inductance (0.1 millihenry) with little or no resistance has 6,280 ohms (Ω) of impedance.

3. Capacitors are made from two metal plates separated by an insulator that have capacitance (symbol C), a capability to store a charge in an electrostatic field. The stored energy opposes changes to the existing voltage across the capacitor. The opposition to the changing voltage is called capacitive reactance (symbol X_C). The impedance of a capacitor is made up of the capacitor's resistance and capacitive reactance added as vectors at right angles to one another. Its value is $Z_C = \sqrt{R^2 + X_C^2}$ with a phase angle (θ) whose $\tan\theta = -X_C / R$. The minus sign on X_C means that the right triangle leg is in the opposite direction from X_L. The R for capacitors is the DC resistance of the leads and the metal plates, which is very small. Therefore, $Z = X_C \underline{/-90°}$. Capacitors are open circuits (infinite impedance) to DC current, but decrease their capacitive reactance as frequency increases according to the expression $X_C = 1 /2\pi fC$, where f is frequency in Hz and C is capacitance in farads. As frequency increases, the capacitive reactance of a capacitor decreases. At 10 MHz, even a fairly large capacitor (0.1 microfarad) has only about 0.2 ohms (Ω) impedance.

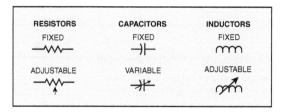

Figure D-2: Schematic symbols for resistors, capacitors, and inductors

Example 1. Impedances of R, L and C

A. What is the impedance of a 10,000-ohm resistor that has no inductance or capacitance, first to current in a DC circuit and second to current in a circuit powered by 60 VAC?

B. What is the impedance of a 1 millihenry inductor with zero resistance at 1000 Hz, 1 MHz, and 1000 MHz?

C. What is the impedance of a 1 microfarad capacitor with zero resistance at 1000 Hz, 1 MHz, and 1000 MHz?

Solution:

A. The impedance of a resistor is the same for a DC or an AC circuit—10,000 Ω. Since it has no inductance or capacitance, resistance is the only component and there is no phase angle.

$Z = 10,000\ \Omega\ /\ 0° = 10,000\Omega$

B. With R = 0, $Z = X_L\ \underline{/90°}$.

Use the equation $X_L = 2\pi fL$. Remember that π is a constant of 3.14.

1. For 1000 Hz: $X_L = 6.28 \times 1 \times 10^3 \times 1 \times 10^{-3} = 6.28$ ohms $\underline{/90°}$

2. For 1 MHz: $X_L = 6.28 \times 1 \times 10^6 \times 1 \times 10^{-3} = 6.28 \times 10^3 = 6280$ ohms $= 6.28$ kilohms $\underline{/90°}$

3. For 1000 MHz: $X_L = 6.28 \times 1 \times 10^9 \times 1 \times 10^{-3} = 6.28 \times 10^6$ ohms $= 6.28$ Megohms $\underline{/90°}$

C. With R = 0, $Z = X_C\ \underline{/-90°}$

Use the equation $X_C = \dfrac{1}{2\pi fC}$. Remember that π is a constant of 3.14.

1. For 1000 Hz: $X_C = \dfrac{1}{6.28 \times (1 \times 10^3) \times (1 \times 10^{-6})} =$

$\dfrac{1}{(6.28 \times 10^{-3})} = 0.159 \times 10^3 = 159$ ohms $\underline{/-90°}$

2. For 1 MHz: $X_C = \dfrac{1}{6.28 \times (1 \times 10^6) \times (1 \times 10^{-6})} =$

$\dfrac{1}{6.28} = 0.159$ ohms $\underline{/-90°}$

3. For 1000 MHz: $X_C = \dfrac{1}{6.28 \times (1 \times 10^6) \times (1 \times 10^{-6})} =$

$\dfrac{1}{(6.28 \times 10^3)} = 0.159 \times 10^{-3} = 0.159$ milliohms $\underline{/-90°}$

The Diode—A One-Way Valve for Current

A diode is a semiconductor chip (usually silicon) with a PN junction. The P material is the anode; the N material is the cathode. For a silicon diode, as shown in *Figure D-3*, conventional current will flow easily from the anode to the cathode when the voltage at the anode is 0.7V more positive than the cathode. If the anode is less than 0.7V more positive than the cathode, no current will flow. Therefore, the diode is a one-way valve for current. This characteristic is used extensively in electronic circuits, including rectifiers in power supplies, as well as in detection and mixing circuits. The 0.7V differential in a silicon junction is used extensively as a relatively constant voltage in amplifier circuits.

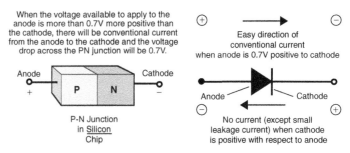

Figure D-3: A diode conducts current only in one direction

Active Devices

Electronic devices that provide gain are called active devices. The most important active device for electronic circuits is a transistor. There are two types of transistors—bipolar and field-effect transistors.

Bipolar Transistors

A bipolar transistor is a combination of two junctions of semiconductor material built into a semiconductor chip (usually silicon). There are two types of bipolar transistors—PNP and NPN—and their junction structures are shown in *Figure D-4a* and *D-4d*. For a transistor that is producing gain, the emitter-base junction is a forward-biased diode and the collector-base junction is a reverse-biased diode. The diode equivalents

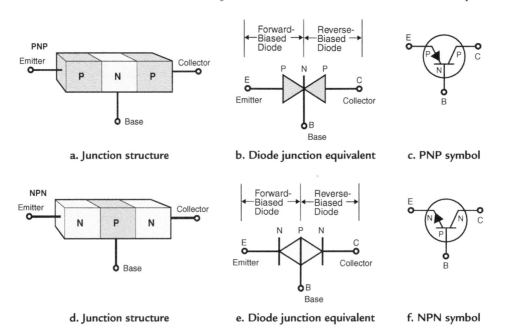

Figure D-4: Bipolar transistors—their construction and symbols

of a PNP and an NPN transistor are shown in *Figure D-4b* and *e*, respectively. The collector-base junction of a bipolar transistor is distinctly different from a reverse-biased diode. Normally, a reverse-biased diode conducts no current, except for a very small leakage current. The reverse-biased collector-base junction of a bipolar transistor conducts collector current that is *controlled* by the current into the base at the base-emitter junction. The ratio of the collector current to the base current is h_{FE}. Under normal operation, the collector current is greater than the base current; so there is a current gain and h_{FE} is a number greater than one—typically 50 to 200. However, there are special cases of operation or manufacture where h_{FE} is less than one.

NPN

The normal active device operation is shown in *Figure D-5*. An NPN silicon transistor P base is 0.7V more positive than its N emitter, and the N collector is several more volts more positive than the emitter, as shown in *Figure D-5a*. The emitter current, I_E, is the sum of the base current, I_B, and the collector current, I_C. The current gain of the transistor under any DC operating condition is h_{FE}, the ratio of I_C to I_B. h_{FE} current gains of 50 to 200 are common in modern day silicon transistors. h_{FE} is actually called "the common-emitter" current gain because the emitter is common in the circuit.

PNP

A silicon PNP transistor base is 0.7V *negative* with respect to its emitter in order to have the P emitter more positive than the N base. The P collector is several volts *negative* from the emitter to keep the collector-base junction reverse biased. As shown in *Figure D-5b*, the same current equations apply and the current gain, h_{FE}, is the same. The major difference is in the polarity of the voltages for operation. For the NPN common-emitter operation the base and collector voltages are *positive* with respect to the emitter; while for the PNP the voltages are *negative*.

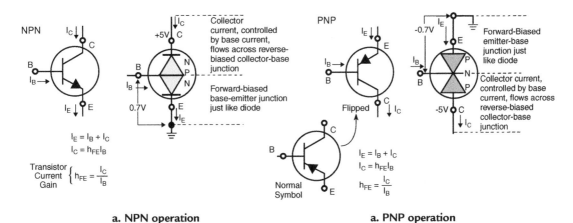

a. NPN operation   **a. PNP operation**

Figure D-5: Bipolar transistor operation

Field-Effect Transistors (FETs)

Unlike bipolar transistors, which depend on current into the base to control collector current, field-effect transistor current between *source* and *drain* is controlled by a *voltage* on a *gate*. Look at the basic structure of an N-channel MOSFET (metal-oxide semiconductor field-effect transistor), as shown in *Figure D-6a*. Heavily-doped N semiconductor material forms *source* and *drain* regions in a P semiconductor material substrate. The region between the *source* and *drain* is the *gate* region, where a thin layer of oxide insulates the P semiconductor substrate underneath from a metal plate that is deposited over the thin oxide. A thick oxide layer over the *source* and *drain* regions insulates metal connection pads from the substrate. Holes in this thick oxide layer allow the metal pads to contact the *source* and *drain*.

N-Channel Operation

Symbol D of the schematic symbol diagrams of *Figure D-6b* represents an N-channel enhancement-mode MOSFET. It indicates that a positive voltage is applied to the *drain* of an N-channel MOSFET with respect to the *source*. When no voltage is applied to the *gate* with respect to the *source*, no current flows from *drain* to *source*. However, applying a positive voltage to the *gate* with respect to the *source* produces a channel underneath the *gate* in the P-semiconductor substrate. This

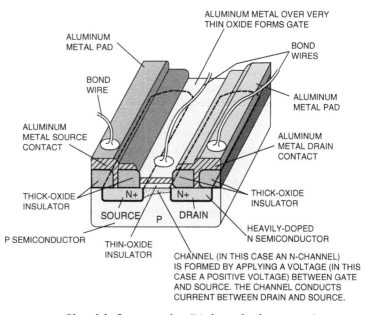

a. Pictorial of construction (N-channel enhancement)

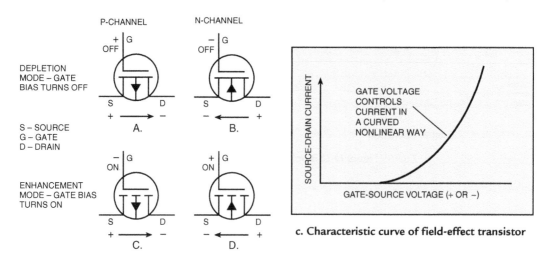

c. Characteristic curve of field-effect transistor

b. Schematic symbols of MOSFETs

Figure D-6: MOS (metal-oxide semiconductor) field-effect transistor

channel conducts current between *drain* and *source*. The characteristic curve that shows *drain*-to-*source* current plotted against *gate*-to-*source* control voltage is shown in *Figure D-6c*. Therefore, in field-effect transistors we have a voltage (between *gate* and *source*) controlling current between *drain* and *source*. The ratio of the *drain*-to-*source current* change to the *gate*-to-*source voltage* change that caused it is called the *transconductance* (abbreviated *gm*) of the field-effect transistor.

Four Common Types

Figure D-6b shows that there are four common types of MOSFETs: P-channel depletion and enhancement mode devices, and N-channel depletion and enhancement mode devices. P-channel devices have the semiconductor materials just reversed from the N-channel materials shown in *Figure D-6a*. The *drain* and *source* are P semiconductor material and the substrate is N semiconductor material. In an enhancement-mode MOSFET, current is produced and increased as an increasing voltage is applied between the *gate* and *source*. In a depletion-mode MOSFET, there already is current from *drain* to *source* when there is no *gate* to *source* voltage. Applying a *gate* voltage reduces (depletes) the current. Voltage polarities are reversed when using P-channel field-effect transistors from those used for N-channel. The *source* is positive with respect to the *drain* and the *gate* voltage is negative with respect to the *source*. Since the *gate* is insulated from the substrate, there is a very high impedance from *gate* to *source* for field-effect transistors. The *drain* to substrate and *source* to substrate junctions are just the same as any other semiconductor diode junction, and for proper operation they must be kept reverse biased.

About the Author

Jerry (Gerald) Luecke has almost 50 years experience in the design of semiconductor discrete-component and integrated circuits—32 of which were spent at Texas Instruments. At TI, he was an applications engineer, design engineering manager and development engineer for digital integrated circuits. He worked under Jack Kilby in the design of the first Minuteman integrated circuits and the initial TI integrated circuit families. He was instrumental in the initial development of the TI Series 54 T²L integrated circuit family, and in the design of the first ECL integrated circuits. He ended his career at TI in 1989 as manager of TI's University Program.

One of the founders of Master Publishing, Inc., he has spent the last 25 years writing, editing, and publishing books about the fundamental concepts of electricity and electronics, integrated circuits, and digital electronics. In addition, he is the author of several electronics reference books and an amateur radio operator, call sign KB5TZY. He earned a BSEE at the University of Iowa and an MSEE at Northwestern University. He is a member of Eta Kappa Nu, Tau Beta Pi, Sigma Xi and is a Life Member of IEEE.

Here are books he has authored:

1. *Semiconductor Memory Design and Application*, Gerald Luecke, Jack P. Mize, William N. Carr, ©1973 Texas Instruments, McGraw Hill Book Company.

   Published in Hungary ISBN: 963 10 3178 0. Published in Polish Informaatyka Series. Published as International Student Edition ISBN 0-07-038975-6

2. *Understanding Microprocessors*, Don L. Cannon, Gerald Luecke, ©1979, 1984, Texas Instruments Incorporated.

3. *Understanding Communication Systems*, Don L. Cannon, Gerald Luecke, ©1981, Texas Instruments Incorporated.

4. *Installing Your Own Telephones*, Gerald Luecke, James B. Allen, ©1986, 1987 Prentice-Hall, ©1987, 1989, 1992, 1997, Master Publishing, Inc.

5. *Basic Communications Electronics*, Jack Hudson, Jerry Luecke, ©1999 Master Publishing, Inc.

6. *Analog and Digital Circuits for Electronic Control System Applications: Using the TI MSP430 Microcontroller*, Jerry Luecke, ©2004 Butterworth-Heinemann, Div. of Reed Elsevier Inc.

Index